Designing Wireless Sensor Network Solutions for Tactical ISR

For a listing of recent titles in the
Artech House Intelligence and Information Operations Series,
turn to the back of this book.

Designing Wireless Sensor Network Solutions for Tactical ISR

Timothy D. Cole

ARTECH HOUSE
BOSTON | LONDON
artechhouse.com

Library of Congress Cataloging-in-Publication Data
A catalog record for this book is available from the U.S. Library of Congress

British Library Cataloguing in Publication Data
A catalog record for this book is available from the British Library.

ISBN-13: 978-1-63081-337-6

Cover design by John Gomes

685 Canton St.
Norwood, MA

10 9 8 7 6 5 4 3 2 1

To my father, Richard F. Cole, who instilled a sense of wonderment and passion for those things that mattered, and to my dear wife Brenda, who patiently supported my efforts throughout this entire undertaking

Contents

Preface

For many technical concepts, the associated vocabulary changes with the times. Such has been true as wireless sensor networks (WSNs) have merged with activities associated with the Internet of Things (IoT), and for tactical intelligence, surveillance, and reconnaissance (ISR) (T-ISR), the Internet of Battlefield Things (IoBT). This is not too surprising as both technologies are considered products of technical advances within several fields of engineering that coalesced at the beginning of the 21st century, as foretold by Mark Weiser [1] in his paper addressing ubiquitous computing (ubicomp). In the 1990s, when National Instruments released its mouse pad, exclaiming, "The software is the instrument," it became clear that computer code was becoming the fabric connecting physical sensing to instrument operation. To be precise, in using the "WSN" and "IoT" (IoBT) monikers, one may consider that WSN may or may not involve the internet and, alternatively, that IoT may or may not involve WSN. WSN technology forms a self-sufficient, ad hoc (multihop) network of sensor nodes that operate and report to an operations center with, or without, connecting to the Internet. Certainly when a WSN accesses the Internet, two major benefits result. First, the WSN-based system is provided with global connectivity via a well-established, mature infrastructure. Second, the WSN gains access to innumerable services and capabilities that offer sophisticated control, data processing, and visualization. A conceptual view is to see a WSN as a group of sensory inputs to IoT, and IoT as a processing capability for a WSN. As a result, although this text focuses on T-ISR applications of WSN, concepts, mathematical equations, technical details, and implementation are directly relatable to WSN-IoT systems associated with industrial, health, environmental, agricultural, and civilian (urban and transportation systems) applications as well.

WSN owes its beginnings to the seed funding provided by the Defense Advanced Projects Agency (DARPA). The Department of Defense (DoD) anticipated that WSN capabilities would solve a number of pressing issues, particularly sensor issues regarding T-ISR. As a result, DARPA was responsible for significant advances in the development of middleware functions, especially through the Networked Embedded Systems Technology (NEST) program. To gain acceptance by the DoD (i.e., warfighters), however, the technology had to demonstrate a level of maturity. During the early phases of research and testing, emphasis was on meeting low-cost and low-power specifications. One result of that was the quality of sensor node (mote) hardware suffered during this period [2]. Additionally, given that wireless network connectivity is already difficult in most applications, the placement of a radio frequency (RF) transceiver near, or at, ground level served to exacerbate link reliability [3]. During initial WSN development, a plethora of analytical and simulation results addressing WSN middleware and operation were published. Although these works were well-received, a comprehensive text (i.e., handbook) for WSN never appeared, and open literature continues to neglect significant work conducted during this seminal decade. Concurrent with the NEST effort, collaborations involving DARPA, the U.S. Special Operation Command (USSOCOM), and other government agencies conducted successful field tests and demonstrations. This book aims to compile all aspects of WSN technology into a single volume and to describe each subsystem associated with WSN in detail along with providing references that supply in-depth details. To achieve this goal, the book was planned with each subsystem defined and discussed alongside critical enabling technologies. Figure P.1 presents the book's topics by chapter.

The continued evolution of WSN research focuses on results from activities regarding software-defined networking and deep neural nets. In addition, an effort to imbue WSNs with logical complexity to operate autonomously and to comprehend and anticipate situation awareness is underway. The goal is for WSNs to become adroit at operating while learning how to respond to unexpected failures, interruptions, or intentional attacks by providing the capability to reconfigure resources, even if doing so results in degraded operational capability. The U.S. Army Combat Capabilities Development Command's Army Research Laboratory (ARL) has ongoing programs addressing such aspects of WSNs [4–6]. With continuing development and success with wearable IoT, interconnectivity with unmanned

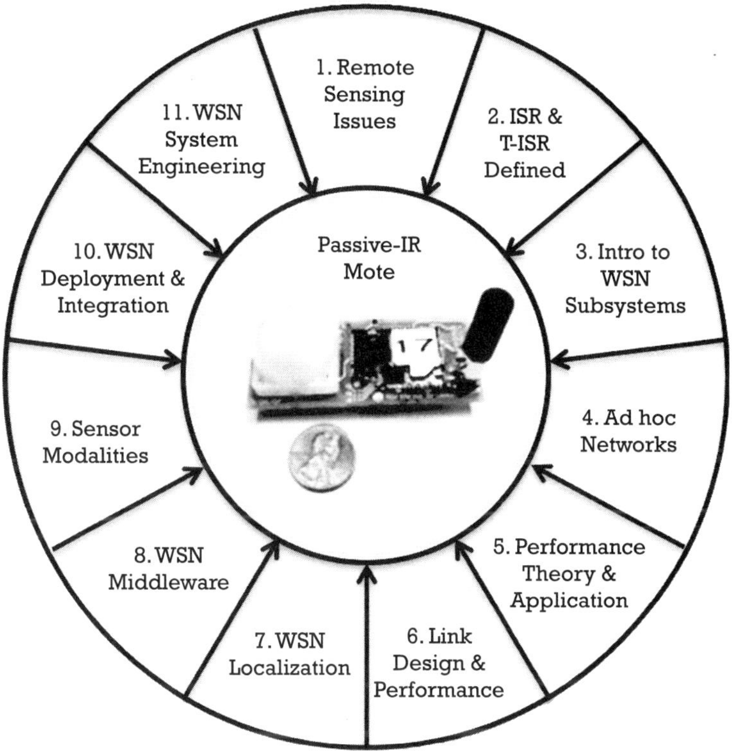

Figure P.1 Decomposing a WSN-based system into this book's chapters.

platforms, and the infusion of increased logic, the development of WSNs is emerging from its infancy.

I acknowledge those who provided background, capability, and guidance for me to successfully complete this project. I also recognize Drs. William Huggins and Jan Minkowski (both with The Johns Hopkins University) for their mentoring of me as a young system engineer and physicist, respectfully. I also thank Northrop Grumman IT (NGIT) for providing me with the opportunity to lead technical programs that required me to perform fundamental WSN research and testing alongside actual T-ISR end users. Not insignificant toward the creation of this text was the never-ending support provided to me by David Michelson of Artech House, who gave me quick answers to my endless questions and conducted weekly conference calls to ensure that all was on-track. Last—but certainly not least—I thank my wife Brenda for her unending encouragement, support, and for managing the home front when I was too busy to help.

References

[1] Weiser, M., "The Computer for the 21st Century," *Scientific American*, 1991.

[2] Ganeriwal, S., L. K. Balzano, and M. B. Srivastava, "Reputation-based Framework for High Integrity Sensor Networks," *ACM Transactions on Sensor Networks*, Vol. V, 2007.

[3] Polastre, J., R. Szewczyk, and D. Culler, "Telos: Enabling Ultra-Low Power Wireless Research," *IPSN 2005: Fourth International Symposium on Information Processing in Sensor Networks*, 2005, pp. 364–369.

[4] Kanowitz, S., "Army tests smart-city communications tool," *FCW*, 2019.

[5] Saccone, L., "Army Studies Smart Cities for New Communication Methods," *In Compliance*, 2019, https://incompliancemag.com/army-studies-smart-cities-for-new-communication-methods/, May 15, 2019.

[6] U.S. Army CCDC Army Research Laboratory Public Affairs, "Army Researchers Develop Innovative Sensor Inspired by Elephant," 2020. U.S. Army CCDC Army Research Laboratory Public Affairs, May 8, 2020, https://www.army.mil/article/235400/army_researchers_develop_innovative_sensor_inspired_by_elephants.

1

T-ISR Sensor Systems: Background and Overview

The design and development of sensor systems to the level of performance required by ISR missions can be accomplished through various paths. ISR missions span a gamut of objectives while presenting numerous and often conflicting requirements. ISR systems are typically designed to detect and track a variety of targets under harsh and variable operating environments. Successfully developing a sensor system to address ISR missions draws on a thorough understanding of mission objectives, target characteristics, and operational conditions. The variability of target characteristics and evolving mission objectives are to be expected and should be accommodated by an ISR system. Designers need to understand and fully consider each aspect, including the target, operating environment, and mission characteristics, from the conceptual level through the system design stages.

Further, designers need to scrutinize ISR system requirements and fully develop appropriate acceptance criteria to support effective and thorough evaluation during system integration and testing (I&T). Moreover, a WSN designer should consider, and plan in advance, operational evaluation criteria. Several ISR systems and operations developed without such attention to end-to-end system engineering have been designed, integrated, and tested only to falter during initial operations. System dependencies and the satisfactory allocation of available resources [size, weight power, and price (SWAP2)] are best accomplished through trade studies and awareness of available technologies. System design and user questions may include the following:

- Is an adequate data communications infrastructure (i.e., one with sufficient throughput, delay, and availability) in place to support the ISR system design?
- Are all interfaces identified, understood, and viable?
- Has a reasonable and responsive concept of operations (CONOPS) been fully developed?
- Have target characteristics and behaviors been completely defined?
- What data products and/or intelligence information are to be extracted?
- Are all SWAP2 constraints identified to a reasonable confidence level and appropriately considered by the design?

To frame the above cautions, let's consider two ISR system examples that failed to reach their objectives: (1) the use of U-2 imagery for improvised explosive device (IED) detection and (2) the employment of an E-8C Joint Surveillance Target Attack Radar System (JSTARS) to collect ISR data from within an urban environment [1]. During Operation Iraqi Freedom, imagery data from U-2 spy aircraft were requested by ISR analysts to aid in the detection and location of IEDs. The planned approach was to have analysts and machine vision view successive images of an area to detect disturbances of the soil or emplacement of objects along coalition pathways. However, scheduling of the U-2 was complex and, as a result, overpass flights were separated by days. Unfortunately, the enemy planned and executed IED attacks based on a much shorter period, and the ISR task failed to provide timely, actionable input to warfighters.

The second example was a 2010 study to assess the utility of a JSTARS ground-moving target indicator (GMTI) platform to support asymmetric engagements in urban areas. Unfortunately, although a reasonably good concept, JSTARS GMTI usage was deemed to offer dubious value, because no effective organizational framework existed to integrate GMTI information to war fighters. Analysts failed to recognize limitations associated with GMTI and its inability to distinguish moving targets within an urban environment. Why consider these examples regarding our T-ISR system design? These were well-considered approaches that suffered from incomplete understanding of the system in its operational environment. Such shortcomings underscore the need to (1) understand all mission objectives, (2) carefully consider what observations are to be made and how to do so, (3) determine the effectiveness that final data products can provide to warfighters and decision-makers, (4) assess the existence of timing and resource constraints, and (5) generate, format, and disseminate the right data to the right people at the right time.

Before we embark on designing an ISR system based on WSN technology, let us consider the characteristics and requirements of T-ISR systems. A useful approach is to begin with a review of preceding ISR system designs. An example of a large-scale, networked, remote-sensing system is the use by the British of a series of connected radar stations, Chain Home, at the outbreak of World War II. Although this particular system was deployed over eight decades ago, it reveals an issue that continues to plague ISR systems today, namely the issue of excessive data volume. With this, and more recent examples, we begin to recognize how we should proceed to arrive at an effective WSN-based system for T-ISR applications.

Accordingly, this chapter examines the development of a generic T-ISR system data flow and presents necessary interconnectivity among the various levels of an integrated sensor system. Finally, the chapter provides an overview that describes recent (and still emerging) technologies that fortuitously arrived in a timely fashion to enable viable WSN solutions to today's T-ISR mission objectives.

1.1 T-ISR Challenge: Sensor System Data Volume

Employing distributed sensing elements capable of long-range detection and identification of targets has a long history, first exemplified with the installation of radar stations near London in 1937. These stations became the vanguard of the distributed radar system code-named Chain Home and represented the first operational radar system [2]. By the Second World War, Chain Home, shown in Figure 1.1, had expanded into an effective ring of coastal early-warning radar

Figure 1.1 Chain Home radar towers employed during WWII.

stations to detect and track all aircraft. While Chain Home was successful in long-range detection, users incurred issues with obtaining useful (actionable) information. The unwieldy data volume created an overwhelming number of reports, many of which were contradictory, reducing overall effectiveness for British pilots. Commander of the RAF Fighter Command, Air Chief Marshal Hugh Dowding, partially solved this problem by using a hierarchical reporting structure. A vast network of telephone lines reporting to a central filter room was devised and set up in London. The central filter process then sorted this information and provided clear, concise, deconflicted formatted reports to pilots. The data reduction and well-needed clarification with the reporting worked, but the fix required significant manual data sorting and processing.

Problems with managing and analyzing large-volume sensor data continued with subsequent ISR systems. Consider, for example, radar sensors based on orbiting space platforms designed to conduct strategic reconnaissance. The first spaceborne radar, the 1964 United States National Reconnaissance Office (NRO) Quill program, employed side-looking radar (SLAR) [3]. Unfortunately, the data volume produced by the Quill radar overwhelmed the available RF downlink technology. To regain high-fidelity data, an onboard cathode ray tube (CRT) was used to transfer SLAR measurements onto photographic film. The spacecraft would eject film canisters, slowing them with by using parachutes and subsequently intercepting them using specially equipped aircraft.

The 1977 US Navy SEASAT-1 spacecraft payload was comprised of multiple RF instruments, including an L-band synthetic aperture radar (SAR) and Ku-band radar altimeter. Although SEASAT-1 operated for a mere 104 days due to a catastrophic power failure [4], only an estimated 15% of collected SAR imagery data has ever been processed. The Ku-band altimeter also produced a significant data volume (1,684 hours of preprocessed waveforms) during the short mission duration, yet what processing and evaluation that did occur would engage oceanographic and topographic analysts for over a decade. The next generation Ku-band altimeter, GEOSAT-1, launched in 1985, again presenting an issue with data volume. Data collected throughout the 567-day GEOSAT-1 mission required aggressive use of two onboard tape recorders along with 8–10-minute downlink windows to transfer the daily 450 MB of data [5]. The extraction and transport of data volume remains a critical issue; NASA's 15 September 2018 launch of the ICESat-2 spacecraft with a photon-counting laser altimeter [Advanced Topographic Laser Altimeter System (ATLAS)] is projected to produce a data volume exceeding petabytes during its 3.2-year mission. Not surprisingly, the design and implementation of significant signal and data preprocessing was required and implemented onboard the ICESat-2 spacecraft to aid in handling such a large data volume. This includes collaborative receiver algorithms and hardware designed to judiciously throttle the data stream, which uses a 220-Mbps X-band downlink [6, 7].

As sensor technology matures, providing higher temporal and spatial resolution, data volume associated with ISR continues to expand exponentially. With the additional requirements pushing for increased coverage and faster measurement frame rates, T-ISR system designers must address the critical characteristic of data volume, an issue further exacerbated by the processing advantage afforded through the use of data obtained from multiple T-ISR sensors. This necessitates sophisticated processing, such as target validation and classification, forward from central operations centers to the sensor nodes. Critical decision-making algorithms—determining which data is valuable—must now occur within the sensor nodes to avoid taxing limited network throughput. Additionally, aggregation should be employed, as data migrates through a network, to balance data payload versus network overhead and to identify and remove message redundancies.

1.2 T-ISR Network Sensor Predecessor: Unattended Ground Sensor

Current (and previous) WSN designs and operational concepts borrow heavily from remote sensor systems that address T-ISR missions. In particular, WSN designs have effectively applied lessons learned through the design and operation of unattended ground sensors (UGSs) throughout the Vietnam War era (1967–1972). Numerous UGS systems were developed and deployed to operate autonomously, thereby avoiding direct human attention or intervention using self-contained power sources (e.g., lead-acid batteries). These UGSs used wireless VHF RF connectivity to connect ISR command centers with various sensor deployments for data retrieval and commanding capability. The deployed UGSs and RF equipment were camouflaged and deployed via both airdrop and hand-emplacement activities along highly probable enemy resupply routes. During the Vietnam War, Operation Igloo White extensively employed UGSs [8], predominantly those that employed acoustic, seismic, magnetic, and RF-signal sensing modalities, and, on a more limited basis, biochemical capabilities. Figure 1.2 illustrates an aerially delivered acoustic UGS. A hand-emplaced system that met with success was the Remotely Monitored Battlefield Sensor System (REMBASS). REMBASS, capable of being

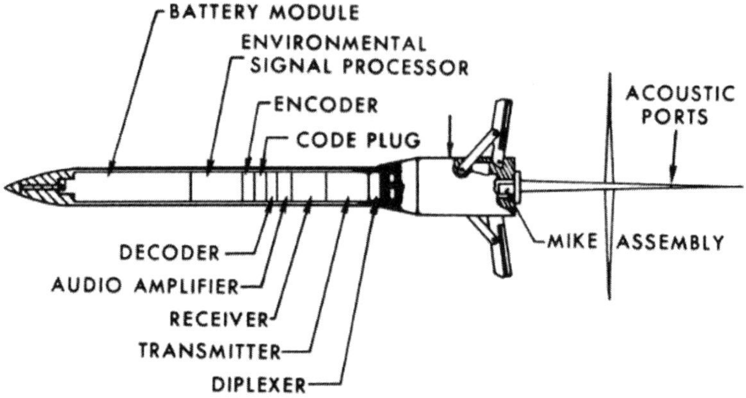

Figure 1.2 Airdropped unattended (UGS) acoustic sensor, 1970.

deployed and operated worldwide, could provide persistent (approximately 90-day) detection, classification, and tracking of targets, including personnel, wheeled vehicles, and tracked vehicles.

In 1999, the U.S. Army and U.S. Marine Corps began to deploy a follow-up to UGS, the AN/GSR-8 (V) REMBASS II [9]. REMBASS II was designed to gather data for enhanced situational awareness and to aid with force protection using a combination of seismic/acoustic, passive infrared, and magnetic transducers. Figure 1.3 displays a REMBASS II system, which

Figure 1.3 REMBASS II: programmer/monitor (CPU), with infrared, magnetic, and seismic sensors.

includes a mission controller (see Figure 1.3, upper right-hand laptop), a seismic/acoustic (SAS) sensor, an infrared sensor plug-in module (IPM), a magnetic sensor plug-in module (MPM), and a handheld receive-only monitor (AN/PSQ-16), with a 15-km (line of sight) repeater radio (RPTR) RT-1175C/GSQ [10]. REMBASS II sensors communicated target data messages using VHF communications, and with optional repeaters, extended the REMBASS II network communication range over 150 km. Other successful UGSs include Pathfinder [11], Scorpion2 [12], and Flexnet [13]. These systems employed similar sensor modalities and relied on Security Equipment Integration Working Group (SEIWG)-format VHF repeaters [14].

1.3 T-ISR System Data Processing Flow

To begin addressing the functionality that must be designed within a T-ISR system, it is necessary to understand the system data stream. A T-ISR sensor system data flow begins with sensor signals that are acquired and processed to extract key data. As data is collected and extracted from a sensor, the resultant signal is processed and digitized, and as appropriate, combined with metadata and ancillary data to form various ISR data products. Figure 1.4 summarizes the ISR processing flow from sensor measurements to intelligence product that would be distributed to the ISR system users. At the conclusion of the

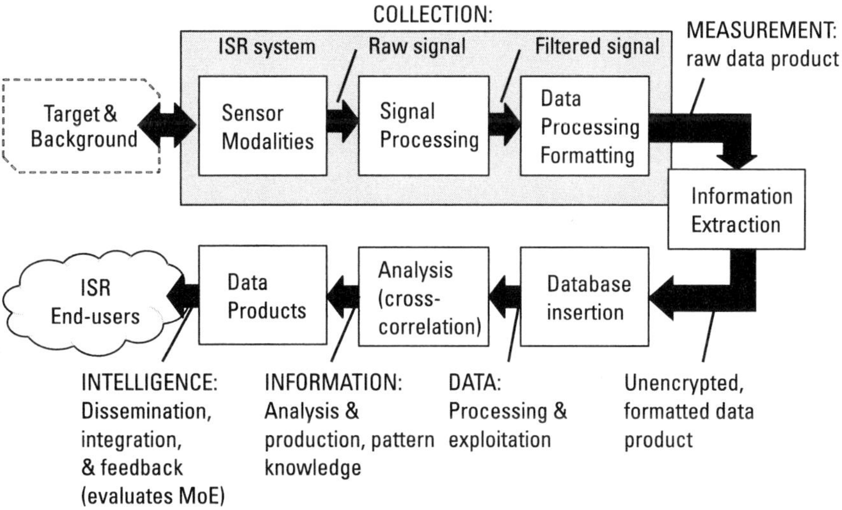

Figure 1.4 ISR system measurement processing flow.

collection process, data products are compressed to meet communication and storage constraints associated with the system and associated infrastructure used to transfer data products. Data products are designed to provide key measurements required of the ISR remote-sensing system. These products are compiled at the command center, correlated with supportive data and information, and analyzed for exploitation to produce actionable/intelligence products. The refined data is then formatted and distributed to the appropriate (and authorized) user community. Critically important to the ISR system designer is that this chain of events not only challenges the system-processing performance, but also stresses data communications, data transfer latencies and storage, and the operational capabilities of the sensor modalities.

1.4 ISR Overview: The Strategic, Operational, and Tactical ISR Levels

Defining levels of ISR coincides with defining strategic, operational, and tactical tiers [15–17]. In supporting these tiers, ISR sensor systems must encompass reliability, performance, and resilience. ISR system architectures are implemented to recover from communication interruptions through self-healing, adaptive dynamic routing of data and command communications, and use of built-in redundancies. These capabilities are realized by implementing overlapping functionality, fault-detection and correction, and fault-tolerant subsystems. This has been central to designs of early ISR systems and to T-ISR designs currently in use and planned for next-generation sensor systems [18, 19].

Strategic ISR missions are charged with the acquisition of data and intelligence that would support the definition of a strategy, large-scale direction, and/or overall allocation of resources. Due to the large coverage required, Strategic ISR systems require sophisticated sensors that operate worldwide with significant range capacity provided through the support of relatively large platforms [e.g., ships, aircraft/large unmanned aerial vehicles (UAVs), and tracked vehicles]. Strategic ISR assets access space, airborne, naval, and ground areas of operation. Excellent examples of strategic ISR assets include Pave Paws [20, 21], the USNS Howard O. Lorenzen (T-AGM-25) Missile Range Instrumentation Ship, which hosts two state-of-the-art Cobra King active electronically scanned arrays (AESAs) [22]; and the large-scale real-time networking exemplified by the AEGIS weapon system within the Cooperative Engagement Capability (CEC) [23, 24]. To provide end users with detailed and accurate data products (e.g., common operating picture (COP) displays) that monitor extensive areas of interest, these systems tightly integrated sophisticated sensor instrumentation with highly capable support platforms.

At the operational level, ISR data products are designed to address broad tasks that have as objectives the determination of whether the "right" things are being worked correctly, or as per Haugh [17], the Three ISR Rights: "Right Intelligence, Right Person, Right Time" (delivering the right ISR to the right person at the right time). At this level, all sensor and ISR capabilities are accessed by both strategic (global, theater) and tactical (regional or smaller) users. Analysis of ISR data at an operational tier is to provide guidance toward refinement of existing operations, to insert additional tasking and retasking controls, and to shape and define current/future missions. By accessing strategic and tactical ISR data products, the operational effectiveness of ISR operations can be continuously monitored, reviewed, and evaluated for use in estimating current issues. Operational ISR also helps to identify the need for new development, or upgrades, to existing ISR systems in response to new and pressing problems, such as a disruptive event, a threatening trend, or a gap in current ISR capabilities. Operational ISR analyzes measurement content to determine the significance of individual events and is used to discern patterns and trends that indicate the existence of imminent threats.

At the tactical level, ISR systems employ distributed networked sensors on a much smaller scale. T-ISR areas of interest (AOIs) are focused on a region, rather than larger theater areas. Ranges to targets and monitoring of T-ISR AOI are magnitudes smaller than that addressed by strategic systems (approximately 100 m^2 versus 100 km^2). Data extracted from tactical sensor systems emphasize time-critical operations and involve interrelated battlefield functions including: direction of force (firepower), determination of mobility (location and velocity) for both friendly and adversarial units, force self-protection (perimeter, incursion monitoring), overarching security, battle damage assessment (BDA), and direct support of rapid action (i.e., shock and awe). All data and processing results are typically combined through the use of a COP display monitor, with in-depth dependence on geographical information system (GIS) support. Subsequent chapters focus on this ISR tier; however, with the cross-linkage afforded by worldwide data linkage, the boundaries that define strategic-to-tactical sensor systems continue to blur.

As a preliminary introduction to WSN sensors, Figure 1.5 introduces the major components and characteristics of WSN sensor nodes that emphasize low cost, low power, and small size (volume). The WSN mote is designed around two critical subsystems, the RF transceiver (radio stack) and the microprocessor system (microcomputer with onboard memory, flash memory, and A/D converter). Figure 1.5 depicts a mote that supports multiple sensor devices, an external antenna, and input/output ports for testing and operation. In the upper right-hand corner of Figure 1.5 is a photograph of a mote based on the general design presented. The white-domed element is a Fresnel lens that operates with a passive infrared detector (PIR).

WSN-based systems differ from UGS in that WSN motes need to require significantly less in SWaP2. Constraints on SWaP2 results with diminished operating ranges for reliable RF connectivity and effective sensor operation for WSN motes. Despite the lack in RF range and sensor detection capability, WSN sensor nodes provide significant leverage through the deployment of a large number of networked cooperative nodes (>>100 nodes per deployment). WSN-based systems provide large area observation yet maintain both fine-grain spatial and temporal resolution. When exploited, WSNs' increased resolution performance results in an increase in detection capacity [detection probability (Pd)] despite the meager capability per sensor node. Approaches that have been successful strategically using large-scale sensor platforms has been,

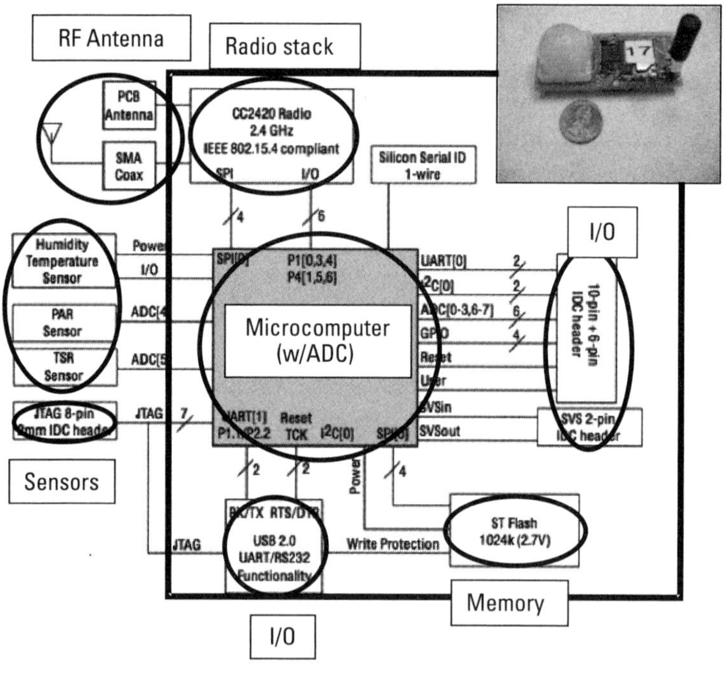

Figure 1.5 The concept of a WSN node (mote).

and continues to be, applied tactically through the use of WSN-based low-power, low-cost systems. Unfortunately, with the application of small autonomous processing sensors to T-ISR missions come unique issues, including:

- Large data volumes exist with fine-grained sensing, such as that provided using WSN. Getting data through a system and delivering data to appropriate mission operation center(s) requires a highly capable and coordinated network architecture.
- Transmission of critical T-ISR data products via low-power wireless channels are plagued with signal loss, congestion, or interference, presenting variable channel throughput and/or link continuity. This is emphasized using small WSN sensor nodes that are located close to the ground and have very limited transmission and receiver power available.
- Design of highly accurate and low-latency WSN code that meets performance and persistence requirements are hampered by the severely limited processing resources.
- T-ISR systems must maintain security, including low probability of detection (LPD), which is part of low probability of intercept (LPI).
- Reliable and robust node operation requires redundancy, multiple-process tasking, and minimal latency, with restricted power availability. With low cost being a central characteristic, the complexity and quality control of sensor node manufacturing become critical and stresses quality assurance.
- To achieve sufficient measurement performance at the system level using low-power (short range) nodes requires large numbers of (>100→10,000) sensor nodes, which in turn, requires sophisticated network management system (NMS) capabilities within very limited processing capability.

An additional benefit of WSN systems is derived from having multiple observations of a particular target. Simultaneous sampling implemented by spatially distributed sensors excels in detecting and discriminating target types by virtue of the physical extent of target and background stimulus. With numerous sensors dispersed over a large area, macroscopic noninteresting events, such as environmental factors (rain, wind, overhead aircraft), typically activate several sensors simultaneously. In contrast, sought-after targets only impact a subset of the distributed sensors. With extended physical area monitoring provided by a WSN-based system that uses several sensors simultaneously, multiple time-elapsed sequences of target measurements are available for processing, which is extremely useful for target classification and state-vector estimation.

Lessons learned from the use of UGSs in combat spurred much of the current generation T-ISR system design. Current engagements demand quick response, rapid adaptability, and persistent capabilities. Urban conflict, rules of engagement (RoE) limitations, and adaptation to terrain, water, forest, arctic venues, and tunnels require highly mobile deployment capabilities. Evolving warfare tactics also emphasize the need for highly flexible ISR sensors to successfully participate in asymmetrical conflict. In addition, providing a T-ISR system capability to lower military tiers (platoon through battalion levels) directly increases unit responsiveness and agility. The "permission" and "direction" chains of command associated with large-scale ISR assets require request and repositioning time and inject latencies that are not acceptable at the tactical level.

1.5 Confluence of Enabling Technologies for WSN

In the 1990s, simultaneous maturity of several key technologies enabled realization of extremely small-scaled sensor concepts. Although a few of these technologies had conceptual beginnings decades earlier, it was not until all of these critical technologies matured to a reliable level that feasible, low-cost, low-power WSN node designs became possible. The engineering and technology advances that spurred WSN capability include:

- *Packet-switched radio:* This concept and associated technology led to reliable data/voice networking and provided the key capability by which wireless nodes could form ad hoc network structures, which can be linked to larger packet-switched networks that operate globally (i.e., the internet and in particular, the DoD Information Networks (DoDIN)).
- *Microelectromechanical systems* (MEMS): This approach to manufacturing enabled the development of devices capable of extremely low-power and inexpensive subsystems to perform key functions associated with WSN nodes, such as sensors and accelerometers.
- *Advancement in the design and production of VLSI devices:* This led to low-power, low-cost designs for critical node subsystems, including pushing capability for sensors (e.g., CMOS imagers), fast-start and low-power RF transceiver chipsets, and high-density flash memory.

- *Advanced code development tools and real-time operating systems* (RTOSs): This enabled developmental support of embedded software for minimal processor and memory resources, specifically targeting middleware-based microcontrollers.
- *Alternative and improved portable power sources and power generation:* Power availability is troublesome for small sensor subsystems; however, power sources have experienced significant increases in energy density, allowing for battery-powered and adjunct forms of electrical supply and/or generation to satisfy long-term sensor node operation.

The emergence of these advances occurred almost simultaneously in the 1990s, which ignited the beginning of WSN research. The coincidence of the arrival for these technologies could be explained by various empirical laws for the underlying engineering:

- *Moore's law:* Integrated circuit (IC) transistor count doubles every two to three years with very-large-scale integration (VLSI), leading to the development of low-power, fast start-up RF transceiver chipsets, GPS receivers, and low-power memory [25].
- *Bell's law:* A new computing class arrives every decade, resulting in increasingly capable microcontrollers and associated subsystems [26].
- *Hendy's law:* Pixels per dollar doubles annually, which supports the implementation of small low-power and low-cost optical sensors for small systems, such as CMOS imager technology [27].

1.5.1 Packet-Switched Digital Networks

Research into survivable communication links through packet-switched networks started in 1962 to address Cold War fears concerning the loss of connectivity in the advent of all-out war. In 1962, a nuclear confrontation seemed imminent as the United States and the Union of Soviet Socialist Republics (USSR) were embroiled in what was referred to as the Cuban Missile Crisis. Each side was in the process of building nuclear ballistic missile systems, and each considered post-nuclear attack scenarios. At the time, the focus of Cold War-related military issues was survivability in the event of an attack. Questions that were considered included how, in the aftermath of a nuclear attack, could the United States' command and control network survive? Although it was believed that most of the links would be undamaged, it was assumed that centralized switching facilities would have been targeted and destroyed.

Paul Baran, a researcher at RAND, offered a solution: Design a robust communications network striving for redundancy using digital technology. Baran conceived a system that abandoned the need for centralized switches and that could operate even if numerous links and/or switching nodes were eliminated. Baran envisioned a network of unmanned nodes that would act as switches routing data node from one node to another to reach their final destinations, regardless of the timing of arriving data packets from User A to user A, as illustrated by Figure 1.6. Accordingly, packet-switched network design was adapted to accommodate network setbacks by sharing network resources. Links were provided the means to exhibit dynamic behavior, meaning that while source-to-destination network physical pathways were subject to intermittent link connectivity, or loss of a node, the links (which convey the information) would remain. Additionally, packet-network designs operate satisfactorily even with a reasonable occurrence of message errors, partially lost messages, and/or congested pathways [28].

Key to supporting T-ISR and in particular, successful transmission of a large data volume using WSN, is imbuing packet-switched networks with the ability to detect collisions and/or congestion and immediately reroute data packets to ensure complete and timely delivery. As a shared resource, the underlying network infrastructure is made available to a multitude of users

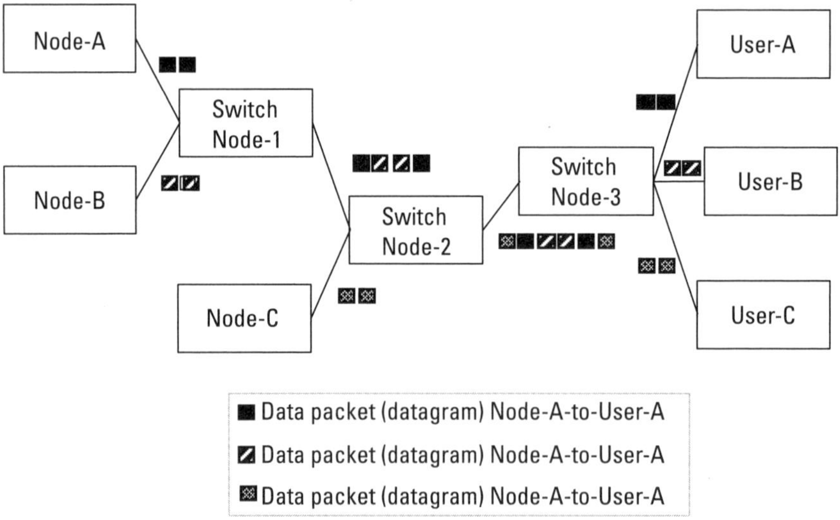

Figure 1.6 Packet-switched networking concept.

simultaneously. This technology was reliant on a packet-switched concept protocol, Transmission Control Protocol/Internet Protocol (TCP/IP), which was the basis on which the Advanced Research Projects Agency Network (ARPANET) and the descendant internet was built.

1.5.2 MEMS

MEMS device manufacturing made great strides throughout the 1980s and 1990s. Significant benefits of MEMS to WSN include (1) reduction of overall sensor and component volumes and (2) drastically reduced costs per sensor. As such, MEMS device manufacturing provided designers with access to miniaturized low-power, and low-cost yet highly capable sensors.

With MEMS, reduction of products to smaller volumes and power requirements were now available to small-scale sensor systems. MEMS-based elements include pressure sensors, magnetometers, thermal sensors, and chemical-biological-radiologic-nuclear (CBRN) detection devices [29, 30]. For systems use, MEMS provide cost-effective, miniaturized inertial sensors (accelerometers, gyroscopes), optical switches, audio transducers, dynamic RAM (DRAM) circuits, and miniaturized oscillators [31], which supports subsystem needs associated with a WSN sensor node. Additionally, MEMS is addressing power availability issues for nodes by opening a path to viable, microscale energy harvesting functions [32].

1.5.3 The Worldwide Grid and DoDIN

The enormous infrastructure known as the internet, and particularly the DoDIN, [formerly known as the Global Information Grid (GIG)], was directly derived from packet-switched networks and has had a critical impact on T-ISR systems. The DoDIN enables T-ISR systems having short communication range systems to be deployed and operated worldwide. Over recent decades, the DoD, having realized that the existence of a worldwide, reliable data communication architecture was critical to leveraging the United States' ability to aggressively pursue and maintain a strong defense, has focused on and significantly funded network-centric transformation programs. The availability of this worldwide infrastructure of digital data networking directly benefited from the advancement of terrestrial networks, the deployment of mobile networks using IP [e.g., the Joint Tactical Radio System (JTRS)], the development and deployment of Joint Tactical Information Distribution System (JTIDS) radios, and the development of teleports linking ground and space segments.

JTIDS, an L-band time division multiple access (TDMA) network radio system used by the United States military, implements Link-16, which is a survivable tactical data link (TDL) design that meets stringent requirements of providing a reliable situational awareness (SA) to various fast-moving forces. Throughout the late 1990s, the DoD combined programs and initiatives to form a secure network and a set of information capabilities fashioned after the internet. This deployed GIG-enabled rapid communications among warfighters, policymakers, and support personnel allowed for the realization of immediate strategic and tactical decision-making. Cyberspace capabilities were eventually added to the GIG, which formed GIG 2.0. In 2014, the GIG was absorbed by the Defense Information System Agency (DISA), and was refererred to as DoDIN [33, 34].

1.5.4 VLSI

Structured VLSI design is a modular methodology originated by Carver Mead and Lynn Conway with the objective of saving microchip area by minimizing interconnection areas through the repetitive arrangement of macro blocks that are joined using abutment wiring. Structured VLSI design was popular in the early 1980s but lost usage since placement and routing tools wasted chip area due to nonoptimized routing. However, this loss eventually was tolerated because of the progress associated with Moore's law.

VLSI brought forth field-programmable gate array (FPGA) structures, through which VLSI enabled the effective design and development of systems-on-chip (SoC) devices. Unlike traditional motherboard-based PC architecture, SoC devices separate components based on function and connect these components through central interfacing circuitry. SoC designs could integrate computer components onto a single IC as if all these functions were built into the motherboard. These low-power SoC devices provided critical solutions to processing, radio chipsets, and mixed-signal circuitry, resulting in higher speed and lower power flash A/D converters. Significant device fabrication has been achieved to increase microprocessor/microcontroller processing power, high-speed data buses and communications, realization of low-power RF-transceiver chipsets, and solid-state focal planes (imagers). In particular, VLSI design and production produced several components and devices used by T-ISR nodes (e.g., efficient delta-sigma A/D converters). With sensor nodes critically dependent on limited power (e.g., batteries), VLSI has also been applied to power management hardware implementation rules at each design level to realize low-power design and efficient protocols [35].

1.5.5 Embedded Real-Time Coding (Middleware)

Mirroring the progress of other engineering activities, the design and development of embedded systems software and engineering witnessed significant strides over the past few decades [36]. Embedded systems were successfully integrated into critical applications by high-use elements such as WSN. However, with the increased dependence on embedded systems, security measures became critical due to the invasive nature of embedded processing. Simultaneous with the insertion of embedded systems into key processing roles, the reliance of embedded systems on the internet has proportionally increased. Cloud connectivity tools continued to evolve to simplify the process of connecting embedded systems with cloud-based services by reducing underlying hardware complexities.

Of high interest to WSN designers is middleware. Middleware, the software layer that stands between the networked operating system and the application, provides well-known reusable solutions to frequently encountered problems including heterogeneity, interoperability, security, and dependability. As more and more embedded systems work with remote, and unattended, systems, operation with limited energy resources emerged as a high priority. Development and use of energy monitors and visualizations have been produced to support the development of low-power operation for WSN nodes. Additionally, real-time visualization tools have been implemented to provide in-depth capability to review embedded software execution and allow for tuning. Finally, building on recent research, deep learning and artificial intelligence are beginning to achieve success with complex applications central to WSN node processing, such as image processing. As these capabilities mature, additional processing should be expected to migrate upward toward the sensor node to address the never-ending issue of data volume.

1.5.6 Portable Power Source and Generation

Advances in power capacity and management, including improvements to battery and solar chemistry, have significantly extended the operating life of low-cost sensor nodes. Alkaline battery chemistry became increasingly available in the late 20th century, and in recent decades, the high-energy density of lithium ion batteries arrived. The key to sensor node power is that the power source has to be available upon deployment; recharging is not an option unless a viable (sufficient and amendable to key node design rules or low-cost, low SWaP) recharging capability is packaged with the sensor node.

Use of miniaturized solar arrays and thermal spike systems (i.e., thermal electricity generated by thermal gradients along a heat-generation spike inserted into the ground) have been investigated as a means to extend power beyond that available from a fixed (battery) source. These and similar techniques of providing power through chemical interactions with the atmosphere have been investigated. However, current technologies are still at a level of maturity that restricts their use in most T-ISR scenarios [37].

1.5.7 Technology Confluence: WSN Research and Development

The timeliness of these key technologies and the arrival of low-rate wireless networking significantly accelerated WSN capabilities to meet (and exceed) requirements associated with T-ISR mission objectives. Figure 1.7 indicates the serendipitous arrival of key technologies with the development of WSN systems. The thought of having complete sensor systems, such as motes, became

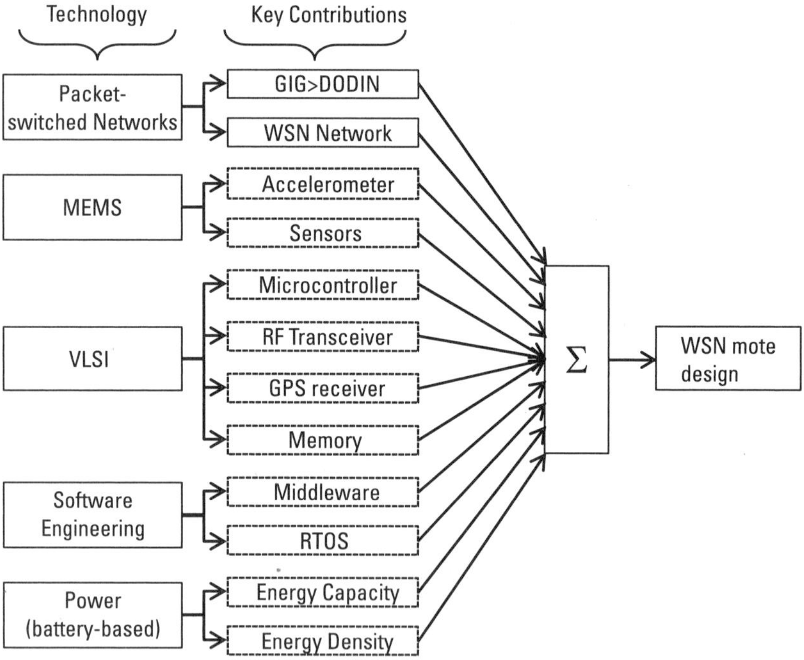

Figure 1.7 Critical technology maturity and key contributions to WSN.

a reality at the turn of the twenty-first century. However, improving all of these technologies to attain a level of technical maturity for use in the field as persistent, unattended, and reliable sensor components for a T-ISR system required another 10–20 years.

References

[1] Brown, J., "Strategy for Intelligence, Surveillance, and Reconnaissance," AFRL paper 2014-1, Air University Press, Air Force Research Institute, 2014.

[2] Neale, B., "CH-The First Operational Radar," The GFC Journal of Research, Vol. 3, No. 2, 1985.

[3] Butterworth, R., QUILL The First Imaging Radar Satellite, declassified 9 July 2012, revised 2004.

[4] Evans, D., et al., "Seasat—A 25-Year Legacy of Success," *Remote Sensing of Environment*, 2005, pp. 384–404.

[5] McConathy, D., and C. Kilgus, "The Navy GEOSAT Mission: An Overview," *Johns Hopkins APL Technical Digest*, 1987.

[6] "Ice, Cloud, and Land Elevation Satellite (ICESat-2) Project," NASA Algorithm Theoretical Basis Document (ATBD) for ATL02 (Level-1B) Data Product Processing.

[7] Accessed: 13 May 2020, https://directory.eoportal.org/web/eoportal/satellite-missions/i/icesat-2.

[8] Rosenau, W., "Special Operations Forces and Elusive Enemy Ground Targets: Lessons from Vietnam and the Persian Gulf War," *USAF, RAND Report*, 2001.

[9] Accessed: 13 May 2020, https://defense-update.com/20060107_rembass-ii-remotely-monitored-battlefield-sensor-system.html.

[10] Accessed: 13 May 2020 L3 Com REMBASS II, AN/GSV-8(V) Specifications, https://www2.l3t.com/cs-east/pdf/rembassii.pdf, 2004.

[11] Accessed: 13 May 2020ARA Pathfinder Specifications, https://www.ara.com/pathfinder/overview, 2018.

[12] Coster, M., and J. Chambers, "SCORPION II Persistent Surveillance System," *Proceedings of the SPIE*, 2010.

[13] Exensor, The Flexnet system, http://www.exensor.com.

[14] Coster, M., J. Chambers, and A. Brunck, "Updates to SCORPION Persistent Surveillance System with Universal Gateway," *Proceedings, Unmanned/Unattended Sensors and Sensor Networks V*, 2008.

[15] U. S. Joint Forces Command, Commander's Handbook for Persistent Surveillance, Version 1, Joint Warfighting Center Joint Doctrine Support Division, 2011.

[16] Roedler, G., and C. Jones (U.S. Army), Technical Measurement, A Collaborative Project of PSM, INCOSE, and Industry, Version 1.0 INCOSE-TP-2003-020-01, "Practical Software and Systems Measurement," INCOSE Measurement Working Group, 2005.

[17] Haugh, T., and D. Leonard, "Improving Outcomes Intelligence, Surveillance, and Reconnaissance Assessment," *Air & Space Power Journal*, 2017.

[18] Quattrociocchi, W., G. Caldarelli, and A. Scala, "Self-Healing Networks: Redundancy and Structure," *PLOS ONE*, 2014.

[19] Schaffner, M., "On the Automation of Intelligence, Sensing, and Reconnaissance Systems in Low UDR Operations," *26th Annual INCOSE International Symposium (IS 2016)*, 2016.

[20] Military Space Programs, Space Policy Project, AN/FPS-115 PAVE PAWS Radar, 2000.

[21] Korda, M., and H. Kristensen, "U.S. Ballistic Missile Defenses," *Bulletin of the Atomic Scientists*, 2019.

[22] Taylor, D., "USAF Missile Defense—From the Sea," *Air Force Magazine*, 2015.

[23] Johns Hopkins Applied Physics Laboratory (APL) Digest, 2012.

[24] "The Cooperative Engagement Capability," *Johns Hopkins Applied Physics Laboratory (APL) Digest*, 1995.

[25] Mollick, E., "Establishing Moore's Law," *IEEE Annals of the History of Computing*, 2006.

[26] Bell, G., "Bell's Law for the Birth and Death of Computer Classes: A Theory of the Computer's Evolution," *IEEE Solid-State Circuits Newsletter*, 2008.

[27] https://commons.wikimedia.org/wiki/Fi.le:Hendys_Law.jpg.

[28] Baran, P., "On Distributed Communications: I. Introduction to Distributed Communications Networks," The RAND Corp, Memorandum RM-3420-PR, United States AF Project Rand, 1964.

[29] "History of Microelectromechanical Systems (MEMS)," Southwest Center for Microsystems Education and The Regents of University of New Mexico, 2008–2010 Southwest Center for Microsystems Education (SCME), 2013.

[30] Accessed: 13 May 2020, https://www.lboro.ac.uk/microsites/mechman/research/ipm-ktn/pdf/Technology_review/an-introduction-to-mems.pdf.

[31] Warneke, B., K. Pister, "MEMS for Distributed Wireless Sensor Networks," IEEE Electronics, Circuits and Systems, 2002. 9th International Conference on, 2002.

[32] Stollan, N., "On-Chip Instrumentation: Design and Debug for Systems on Chip," New York: Springer, 2011.

[33] Chairman of the Joint Chiefs of Staff Instruction, Defense Information Systems Network (DISN) Responsibilities, Directive, 2015, https://www.jcs.mil/Portals/36/Documents/Library/Instructions/6211_02a.pdf?ver=2016-02-05-175050-653, p. A-2.

[34] Exhibit R-2, RDT&E Budget Item, Justification: PB 2015 Defense Information Systems Agency Date: March 2014, Appropriation/Budget Activity, Research, Development, Test & Evaluation, Defense-Wide / BA 7: Operational Systems Development R-1 Line #194, Program Element, 2014.

[35] Cook, B., S. Lanzisera, and K. Pister, "SoC issues for RF Smart Dust," *Proceedings of the IEEE*, 2006

[36] Romer, K., O. Kasten, and F. Mattern, "Middleware Challenges for Wireless Sensor Networks," *Mobile Computing and Communications Review*, 2002.

[37] Arman, H., and S. Kim, "Ultrawide Bandwidth Piezoelectric Energy Harvesting," *Applied Physics Letters*, 2011.

2

Designing a T-ISR System

Throughout history, as well as today, ISR capabilities are critical ingredients in the planning and execution of any successful campaign. Activities associated with ISR have provided critical sources of information for decision-making throughout our collective past. Obtaining relevant information enables leadership to determine if, how, when, and where to strike an opposing foe. Countless examples of ISR exist throughout history. In the Bible's *Book of Numbers* [1], Moses relied upon advanced surveillance to ascertain details of Canaan, which helped him determine what course of action to take, what the land was like, and whether the people who lived there were strong or weak, few or many.

During the American civil war, the absence of J. E. B. Stuart's cavalry to General Robert E. Lee in 1863 significantly decreased Lee's ability to know how best to disperse his troops upon arrival at Gettysburg in planning his offensive [2]. Without the cavalry, Lee was "blind" concerning the Union's strength and location, which meant loss of strategic battlefield positioning. Although the "ISR" absence may not have changed the outcome of this pivotal battle, it certainly didn't help the army of northern Virginia. Today, highly specialized ISR units, such as the 75th Ranger Regiment's Regimental Reconnaissance Company and 1st Special Forces Operational Detachment-Delta (1st SFOD-D), fulfill critical roles of providing surveillance and reconnaissance for the U.S. military. Having updates to critical information concerning the opposition provides valuable data to allow decision-makers time for critical adaptability, because preplanned directives are based on the best, but dated, intelligence available.

2.1 ISR Definitions

The objective of ISR sensor systems is the timely acquisition of data that, when processed, results in accurate, relevant, and coherent information for commanders (decision-makers) to direct current and next-step operations. The focus of ISR is to discern the whereabouts and characteristics of targets of interest. As a result, functions that implement data collection and processing required of ISR systems must address and perform object detection, object tracking, target classification, and extrapolation of a target trajectory to account for finite time duration between target acquisition and application of a response. ISR is a generalized term that indicates the activities of intelligence operations, reconnaissance, and surveillance, which are described within military doctrine, as the following.

Intelligence is, according to [3], "(1) the product resulting from the collection, processing, integration, analysis, evaluation, and interpretation of available information concerning foreign countries or areas; and (2) information and knowledge about an adversary obtained through observation, investigation, analysis, or understanding."

For use in intelligence, ISR system designs stress how to direct operations, over time, into steps leading to the successful achievement of mission objectives. Remote system objectives can range from identifying recurrent target patterns to indicating the presence (or absence) of targets in a protected area, within a level of uncertainty. Processing, associated with extracting intelligence, requires long-term collection of observations to form correlations to pair with other informational inputs and/or to reveal underlying patterns useful to intelligence analysts and activities. The core to intelligence, data processing, is conducted to determine the existence or structure of any exploitable pattern. The emphasis on using ISR for intelligence depends on sensor accuracy and measurement repeatability. Surveillance is the "systematic observation of aerospace, surface or subsurface areas, places, persons, or things, by visual, aural, electronic, photographic or other means" [3].

The objective of ISR systems for surveillance is twofold: (1) to provide actionable data, which emphasizes minimal latency associated with data delivery from observation to the decision-maker [e.g., mission operations center

(MOC)], and (2) to provide data for use in trending analyses to identify and/or characterize underlying patterns that might be of interest.

Reconnaissance is a mission "undertaken to obtain by visual observation or other detection methods, information about the activities and resources of an enemy (or potential enemy), or to secure data concerning meteorological, hydrographic, or geographic characteristics of a particular area" [3].

Surveillance is highly correlated with the monitored AOI and may be modified based on what is detected. Reconnaissance missions operate on a specific AOI and require mobility to obtain multiple points of observations. Reconnaissance measurements seek intelligence beyond targets of interest by providing data that characterizes topography, weather conditions, and dynamics of the AOI is core to surveillance missions.

Missions involving intelligence, surveillance, and/or reconnaissance employ various sensing and observation data functions conducted at high levels of sophistication. ISR systems are relentlessly challenged by the need to scale to address target dynamics and complexity, increased target numbers, and challenging performance goals resulting from adversarial measures to mislead or conceal. ISR missions can be as simple as perimeter monitoring and pipeline protection. Alternately, ISR missions may require operations employing a large array of sensor platforms and weapon systems to observe and track large numbers of complex targets in a significant engagement volume and occurring over an extended period of active period. The U.S. Navy's CEC [4] is an excellent example of an ISR design that addresses the latter mission.

2.2 T-ISR Objectives

The goal for T-ISR operations is to acquire data to provide tactical teams with a current awareness of adversary intentions and strength. To discuss tactical ISR systems adequately, we need a working description of what ISR systems intend to accomplish, particularly systems designed for T-ISR missions [5]. By defining the mission description and associated objectives, system-level requirements can be developed and flow down to individual subsystems. Categories of intelligence-collection efforts are based on methodologies and the type of data collection. Figure 2.1 presents a hierarchy that shows the interrelationships of each of the overarching intelligence concerns. There are differences in how various intelligence operations are defined and ordered hierarchically, but Figure 2.1 represents the prominent structure and helps to explain the individual ISR organizations and objectives.

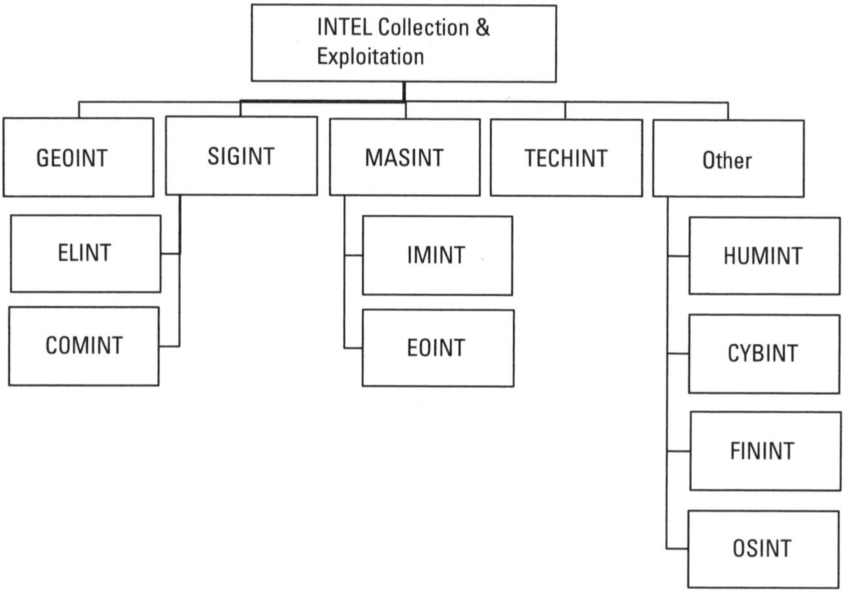

Figure 2.1 Hierarchy and relationships of INTEL organizations and operations.

The ISR categories are summarized as follows:

- *Geospatial intelligence* (GEOINT) is intelligence concerning human activity derived from exploitation and analysis of imagery and geospatial information that assesses the formation, or existence, of a threat. GEOINT visually describes and depicts physical features and geographically referenced activities on Earth [6].
- *Signal intelligence* (SIGINT) addresses information extracted from the interception of adversarial signals. SIGINT activities are divided into two disciplines, electronic intelligence (ELINT) and communications intelligence (COMINT), as conducted at U.S. Army base Ft. Huachuca in Arizona [7]. ELINT investigates the source of a signal, works to identify source platforms involved, and by extension, estimates the adversarial force at-hand. ELINT can be used to identify weapon systems (including current operational mode), reveal enemy encampments, and can be used to locate mobile enemy assets. COMINT addresses information content transmitted by RF (optical) communication carriers and works to decipher the contents of intercepted communication signals.

- *Measurement and signature intelligence* (MASINT) is technical intelligence gathering that serves to detect, track, identify, or describe distinctive characteristics (signatures) of fixed or dynamic targets. MASINT data is dependent on accuracy provide by sensor measurements and modalities, including radar, acoustics, seismic detection, imagers, nuclear sensors, and chemical and biological detection [8]. MASINT is defined as scientific and technical intelligence derived from an analysis of measurement data obtained with sensing instruments to isolate distinctive features associated with targets. This is the major area to which WSN contributes. It should be noted, however, that WSN can also be applied to other intelligence operations, such as SIGINT (i.e., sensors are replaced with RF receivers and data storage to acquire and store enemy transmission).
- *Technical intelligence* (TECHINT) is intelligence about weapons and equipment used by adversaries. A related term, scientific and technical intelligence, addresses information collected at the strategic (i.e., national) level. TECHINT includes identification, assessment, collection, exploitation, and evacuation of captured enemy materiel (CEM) in support of national and immediate technical intelligence requirements [9].
- *Human intelligence* (HUMINT) is intelligence gathered by means of interpersonal contact, as opposed to the remote sensing or technical data collection disciplines. HUMINT employs the human element to provide information not readily available through other means. While providing versatile and powerful information for situation awareness and decision-making, HUMINT can be time-consuming; it may take several months to initially set up effective contacts in a particular environment, and it is susceptible to deception [10].
- *Cyber intelligence* (CYBINT) acquires data gathered throughout cyberspace connectivity and systems (e.g., servers and laptops) [11]. Unlike other intelligence-gathering disciplines, CYBINT is not formally defined in any service-specific or joint doctrine. The approach to CYNINT is to access and pull significant information from the web-based messaging and data repositories.
- *Financial intelligence* (FININT) refers to gathering information about the financial affairs of entities of interest to understand the financial patterns and capabilities and to predict intentions [12]. FININT uncovers nefarious cash flow links and backtracks the flow of funds used to support adversarial activities. A notable program that directly addressed FININT is the Counter-Narco Tactical Program Office

[13] under the Department of Homeland Security (DHS). FININT strives to identify and audit financing associated with terrorist groups.
- *Open-source intelligence* (OSINT) is data collected from publicly available sources to be used in an intelligence context [14]. In the intelligence community, the term "open" refers to overt, publicly available sources (as opposed to covert or clandestine sources). OSINT has been around for hundreds of years (recall Moses and Lee's cavalry). With the advent of social media, instant communications, and rapid global information transfer, a great deal of actionable and predictive intelligence is obtained directly from public, unclassified sources.

T-ISR missions are tasked with determining if particular targets have entered an AOI and focus on maximizing extraction of data that can aid in the assessment of current battlefield situation. For missions associated with forward reconnaissance, key functions are to pursue detection and classification of targets, to determine the number of targets, to map target location (and direction), and to track particular targets of interest. For surveillance, the mission becomes one of persistent monitoring of an area or perimeter to provide base defense missions. Paramount for a successful T-ISR mission is accuracy in the identification of objects—and minimization of false alarms caused by noise, interference, or intentionally false targets (decoys). Figure 2.2 presents a generalized architectural diagram to display what is required of a T-ISR system. Specifically, Figure 2.2 indicates various logical activities and services that are to occur within a sensor node, and has been categorized into four major functionalities: node/system initialization and networking, node mode services, sensor data services, and sensor signal/data processing. Within each functional category are processes residing and operating onboard the sensor node.

For node/system initialization and networking functional group, logic modules provide initial conditions, network routing, path linkage, localization of the node (in a registered coordinate grid), and time synchronization. A preplanned operational service exists to provide a priori information, alongside the resident (nonvolatile) boot code, in the initialization process. This group is where routing schemes are initiated and performed through use of the various media access control protocols that are necessary to establish a sensor network. Once links are formed, additional sensor nodes and linkage are set up from each node to an access point (AP), which provides external control and data exfiltration. This enables an overarching network management process to exist that can monitor performance of the network and determine if corrective

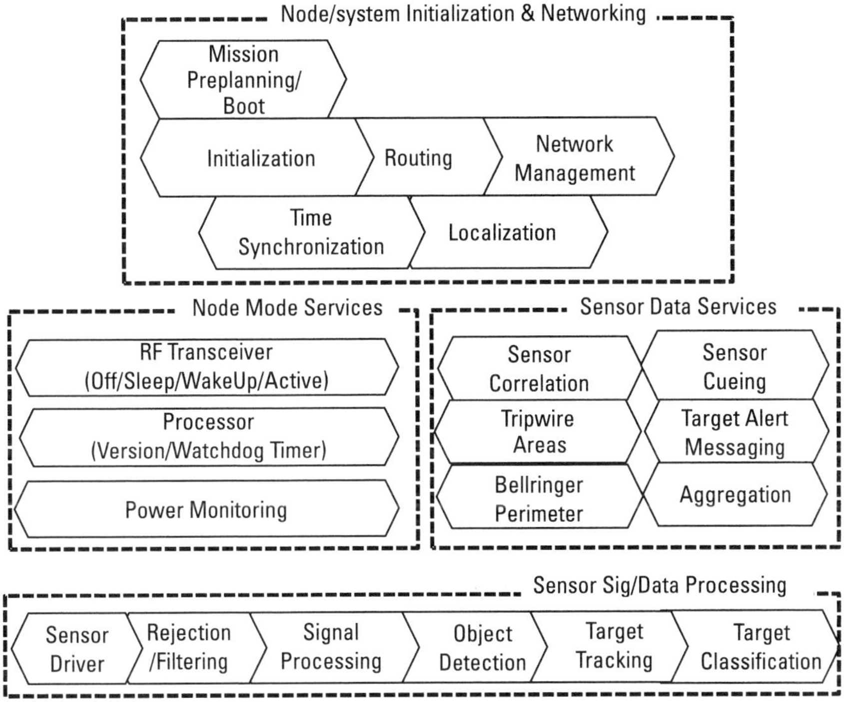

Figure 2.2 Simplified T-ISR system architecture.

action (e.g., self-healing) is necessary. As prerequisites to network formation, both time synchronization and node localization procedures are conducted to support the creation and optimization of the ad hoc network. As part of this functional group, security processes would be embedded in the network management module to protect the integrity of the network against various cyber attacks as well as to safeguard messaging packets to prevent unauthorized access to sensor data, sensor/network control, or housekeeping data.

In the node mode services group, logic modules provide management of the transceiver, processor, and continuous monitoring of onboard power consumption. The various modes of the transceiver are managed in this group with an emphasis on conserving power. The processor is managed to allow for priority events to usurp control as warranted, while watchdog timers ensure proper operation of the code if caught in a loop, or idle condition, to provide a means to exit, resetting the processor back to initial code. The power usage of a node is continuously monitored, and control of power distribution is adjusted periodically to promote long operational capability. As the main power draws within sensor nodes are associated with the RF transceiver and processor, interactions of modules within this group are tightly integrated.

The sensor signal/data processing contains those modules necessary to operate the sensor, process the signals, and digitize, extract, and preprocess data obtained from the signal. Modules within this group handle the key ISR functions of detection, classification, and tracking. The sensor input stage (detector) has been devised to reject spurious signals as much as possible through a combination of physical and electronic (signal) filtering. Physical filtering can be accomplished based upon the physics of how the signal is transformed by the detection circuitry. For example, for electro-optical cameras, narrowband optical filters are placed at the entrance aperture to reflect (reject) incoming optical signals except those associated with targets of interest (e.g., solar filter). For electronic filtering, matched filtering is conducted based on assumed noise characteristics of the observation and the expected target signal. These two signals are summed and presented to a threshold detection function determined using probability modeling. The objective is to reduce false alarms while supporting a high probability of detecting a target when present. Detection involves hypotheses testing: Is the target there, but not being detected (a missed target)? Is the target there and being detected correctly? Is the target absent while a false target is generated (false alarm)? Is the target absent and the system is correctly indicating so? These detection cases draw on probability and statistical theory and modeling (discussed in Chapter 5) and constitute the initial process in determining the presence (or absence) of an object. To determine if an object entering the AOI is a target of interest involves the next processing function, classification. The classification function is highly dependent on a priori information to discriminate one signal from another through a multitude of algorithms (e.g., spectral analyses, time correlation, and two-dimensional image correlation). Once a target of interest is indicated, tracking begins, and characteristics associated with the target are measured (e.g., location, direction and velocity). For networked sensors, this data has to be properly formatted and correlated by appending status indicating the level of confidence associated with the data product. In parallel to the target processing, multiple targets may have to be handled. This implies that the system should be designed to process parallel functions that realize classification and tracking for each unique detection. Issues such as closely spaced objects (CSOs) must be considered, as well as unique track identification and estimation (i.e., crossing tracks). Chapter 8 presents and describes the various algorithms employed to establish and maintain consistent tracking of targets.

The fourth category, sensor-data services, involves those modules devised to support various setups for sensor node behavior. Sensor correlation allows for adjacent sensor nodes to provide input that can be used to verify detection events, support target classification, and directly support target tracking. Tripwire and bellringer behaviors are provided to allow a sensor node to purposely remain inactive until a particular alert message is received from a tripwire sensor area, or from a bellringer node. Sensor cueing allows a sensor node to interface and alert sensor systems unrelated to that associated with the sensor node. The ability to integrate inhomogeneous sensor nodes with other ISR assets provides leverage based on legacy systems already emplaced or available. Target alert messaging is responsible for the generation of the sensor node data payload that gets forwarded to the transceiver for formatting, compression, encryption, and transmission. A final logic module is aggregation, which considers the network overhead associated with packet-switched networks that produce inefficiencies of data volume and associated energy use. The purpose behind aggregation logic is to perform efficient data packing by accumulating messages sent by other nodes to a node used as a relay through the network (dependent on the network hierarchy) and to combine these messages into a single message for the exfiltration.

2.3 ISR Reach: Worldwide Versus Localized

With novel but proven sensing system capabilities resulting from the confluence of maturing technologies, the practice of synchronously operating multiple sensor systems at a worldwide level has become commonplace. However, accessing widespread ISR systems has been around for decades. In 1949, the initial concept and design of large-scale ISR systems existed [i.e., the Sound Surveillance System (SOSUS), a passive acoustic system developed by the U.S. Navy to track Soviet submarines] [15]. Another large-scale (and more recent) ISR architecture with a completely different set of system hardware was implemented to detect and identify the launch of intercontinental ballistic missiles. The U.S. ballistic missile defense (BMD) system employed sensors including over-the-horizon E/F–band radar (e.g., AN/SPQ-11 Cobra Judy) as well as the continuously operating satellite system known as the Defense Support Program (DSP). The initial DSP launch in November 1970, which employed infrared camera technology onboard geosynchronous satellites, is part of the worldwide early warning system. DSP continues to operate, although it is now tasked with collection of data on volcanoes and forest fires. For ISR, the DSP satellites are being replaced with the Space-Based Infrared System (SBIRS) and Space Tracking and Surveillance System (STSS), formerly SBIRS-low. As of January 2018, a total of 10 satellites carrying eight SBIRS or two STSS payloads had been launched [16].

T-ISR occurs at a much smaller, but perhaps more penetrating level. As any ISR system, T-ISR addresses a plethora of mission objectives but provides a unique capability: actionable responses. Actionable ISR systems emphasize fast-reaction responses to identified targets of interest and observed enemy behavior. These systems are used for generating alert conditions and/or for timely provision of critical information to decision-makers.

T-ISR systems have been repeatedly proven in combat situations associated with fast-response teams, such as quick reaction force (QRF) [17]. Certainly, field commanders and in situ decision-makers can, and do, avail themselves of large-scale ISR contextual information provided by assets operating at a strategic level such as satellite systems (e.g., Keyhole), long-range and highly capable UAVs, and specialized aircraft (e.g., Global Hawk, AWACS). However, localized T-ISR information is tunable to the commanding level to operate at a much lower (and immediate) level than that associated with strategic ISR assets. At the lower command and control levels, it is possible to react against an immediate situation and initiate appropriate reaction.

2.4 Leveraging Target Characterization: Signature Extraction

Target characterization has been, and continues to be, actively studied through field-testing, data collection, and simulation. Target characterization, and in particular, classification analysis, addresses attributes derived from sensor measurements that provide unique correlation(s) between targets and associated signatures. Target characteristics drive sensor-performance equations, especially, detection, target identification (discrimination), and tracking capability. As one result, identifying target attributes and characteristics are initial tasks that T-ISR system designs aim to address, which is accomplished through calibration. Target signatures depend on direct observations of characteristics associated with physical shape, size, projected area cross-section, materials employed, target velocity, sensor-to-target angles, and environmental considerations.

Characteristics may include: electro-optical or RF-based wavelength-dependent properties (e.g., reflectivity, emissivity), target-released energy, and temporal physical (e.g., area cross-section) properties. Successful use of target signature(s) rely upon the effective operation of an ISR system and more importantly, how precise and accurate underlying ISR measurements and processing were in producing successful discriminates. This is why sensor repeatability, stability, and performance are evaluated through intense calibration.

Effective signature databases strive to collect identifying and useable characteristics while avoiding details that ISR systems are insensitive to, or that do not provide value to downstream ISR product generation. An excellent example of signature database efficiency is the use of laser vibrometry sensors (LVSs) against maritime targets. LVS systems respond to microscopic target surface vibrations through interaction of a transmitted, continuous-wave laser beam that interrogates the target surface. The beam, once backscattered from the target, is frequency-modulated (FM) by minuscule (0.5–1 micron amplitude) vibration waves that propagate across the entire target surface. These vibrational waves have been proven to be significantly unique to any particular target and are readily used as a discriminant for identification (akin to sonar signal processing). With this particular ISR sensor type, target profiles (cross-section), and other macroscopic details do not impact the effectiveness of the LVS to achieve target identification. Because of this insensitivity to physical characteristics, LVS can operate regardless of relative heading orientation between the target and sensor, unlike systems that operate off of projected area.

In support of the classification and identification process, specialized ISR-oriented signature-extraction programs and measurement efforts exist that focus on target attributes, collectively referred to as intelligence collection and exploitation. Key to this effort is seeking robust signatures, those that are repeatable readily observed and that present unique characteristics measurable by T-ISR sensors. Designing signature databases depends heavily on the successful identification of measurement parameters that correlate with unique targets. If a comprehensive, time-invariant database can be formed to provide separable parameter values that is statistically separate similar targets, the possibility of viable discrimination exists. An example uses data associated with IMINT collection. The key performance parameter (KPP) for electro-optical (EO) sensors is optical resolution. If points that constitute a two-dimensional image of a target can be collected over various look angles (angles between sensor and target), a profile can be generated to conduct object classification and identification. Not surprisingly, this depends on the fidelity of the image data and the underlying quality of the database, which is described using resolution. For a circular camera aperture, a definition of optimal resolution assumes the Rayleigh resolution criterion [18]. This criterion states that the maximum achievable resolution for an optical system is dependent upon the wavelength of light (λ) and the diameter of the optical system entrance pupil

(D). Expressed in appropriate units of length, Rayleigh's optical resolution is based on angular separation of two points separated by an angle (θ), where the maximum intensity diffraction pattern of one point overlaps the first minimum intensity point of the second pattern. This occurs at

$$\sin(\theta) = 1.22\frac{\lambda}{D} \tag{2.1}$$

Resolution of signals provides a vast richness to signature databases. This is not surprising as well-established classification systems, such as sonar, have access to extensive signature repositories. How well discrimination of targets can be attained is based on the level of resolution in processing, such as that achievable with low-frequency analysis representation [19]. As with sonar, temporal resolution is also critical with SIGINT. Intelligence, in this context, is information that provides an organization or individual with support for making decisions and possibly gaining a strategic advantage. A key performance characteristic associated with SIGINT is similar to that for sonar: having sufficient frequency resolution to distinguish frequency content presented within a received signal. Using T as the sampling window (duration), f_s as the sampling frequency, and N as the number of equally spaced signal samples, frequency resolution (δf) is representable as

$$\delta f = \frac{1}{T} = \frac{f_S}{N} \tag{2.2}$$

2.5 Target Identification Against Operational Backgrounds

To determine if a target is present or absent requires sensors to be able to distinguish a target from other objects, especially those presented within a stressing environment and/or against various backgrounds that present characteristics that may mistakingly be attributed to the target. These interference sources decrease the signal-to-noise ratio (SNR) and can result in hiding a target and/or confounding sensing systems through the generation of false alarms. As presented in discussions of imagery (Section 2.1) and frequency analyses (Section 2.2), overall measurement resolution is critical. The selection of sensor, signal-processing approach, and overall architectural design require a comprehensive understanding of spatial (or spectral) characteristics associated with targets of interest, ambient operating environments, and anticipated backgrounds.

A well-designed sensor response and signal processing suppresses (or rejects) background and clutter data within a scene under observation. Judicious sensor and sensor signal processing design strives to reject unwanted noise and background energy. Downstream signal and data processing further enhances selectivity, which reduces data volume and WSN message traffic. Finally, key measurements and ancillary data (e.g., proximity weather sensors) can be applied during data processing (e.g., at the operations center) to further refine sensor data products. For cameras, radar, seismic and acoustic sensors, there exists structured background processing using spectral-based algorithms (temporal and/or spatial frequency selectivity and filtering) [20, 21].

For active sensors, sensors that emit controlled (i.e., known) probing waveforms and process backscattered reflected from targets to extract signals (e.g., radars, sonars, laser radars), there exist techniques beyond the well-used short-time Fourier transform (STFT) algorithms to increase SNR performance. For example, the application of Cohen-class time-frequency distributions (TFDs) [22] has successfully extracted signals in SNR scenarios where FFT-based processing falters due to lack of SNR. Referring back to LVS, the application of TFD-based algorithms significantly improved capability of both signal demodulation and signal spectral analysis [23].

2.6 T-ISR System Data Product Formation

Subsequent to signal and sensor data processing, the final data products can be as simple as a detection alert capability, or as complex as sensor-to-sensor correlation displays (i.e., ISR-centric COP), requiring a well-defined logical combination of sensed data, database information, and precise scene registration sensor to sensor. Design of product formation must balance latency issues (time-to-report) against data precision and accuracy. Processing ISR information into data products must also consider communication bandwidth for collected data, network management functions, and incoming mission operation commands. Given the ever-increasing capability of sensor resolution and capability (especially for high-resolution cameras), care must be taken to avoid undue data volume. For this reason, along with design rules to assert system autonomy, tiered processing is employed in many T-ISR systems, where overall processing is distributed physically from the sensor node to the relay.

Tiered processing supports increased sensor autonomy, allowing sensors to acclimate rapidly to changing conditions or target characteristics without intervention from the MOC. First-order processing is relegated to sensor nodes, beginning with multimodality comparisons (sensor-to-sensor signals and event timing). Node-based processing directly influences overall system performance by reserving bandwidth over the network through elimination of noncritical data (e.g., false alarms). With decreased data volume, network energy is conserved and congestion minimized. Propagation of the data, from sensor to relay, provides data convergence, revealing opportunities to consider how T-ISR functions might benefit from additional processing. Finally, at an exfiltration relay, which typically possesses more processing capability than sensor nodes, additional data formatting and compression may be considered.

2.7 T-ISR Data Product Dissemination

With the generation of a well-defined, timely, and accurate data product (e.g., alert message), the next function to be completed is data dissemination. ISR data products are the valued commodities that must be supplied to T-ISR mission operations center(s) where larger-scale processing and analyses occur. Many missions require cross-correlation and verification using overlapping (albeit individual) observations. Additionally, an ISR system must consider how to securely disseminate data to certified and approved end-users.

Leveraging, via joint sensor, network, and display/control interoperability, is an established goal for supporting T-ISR missions. Accordingly, standards, such as MIL-STD-2525D (Common Warfighting Symbology), exist to ensure inoperability [24] with trusted workstations. This consideration in the design allows for the combination of service layers to provide immediate and yet separately verified data sources—a formidable capability when evaluating T-ISR system performance and design trade-offs.

2.8 T-ISR System Engineering

With well-defined mission objectives and requirements, a requirement verification matrix (RVM) should be generated for use in evaluation, verification, and validation of the system, which begins at subsystem/unit levels and culminates with end-to-end operational evaluation (OpEval). With a comprehensive and accurate RVM, requirements are linked to the verification process, which assure testability, well-planned test approach development, and coordinated test preparations using the well-established system "V" diagram [25]. Requirements derived at each design level are clearly defined and, through the RVM test matrix, are correlated to unique tests and parameter measurements.

The systems engineering process begins with the identification of a need to either modify an existing system or develop a new system to address an emerging threat or event. Initial studies and tests are conducted to determine the existence of feasible solutions. Issues and parallel solutions are identified, and trade studies are conducted to rank (and select) the best solution path. Upon developing a viable design concept(s), a conept of operations (CONOPS) is developed to identify critical stakeholders, to present top-level program requirements and objectives, to provide criteria to determine successful testing (and validation) of the system, to underscore operational considerations for the new/modified system, and to establish responsibilities among critical stakeholders [26, 27].

The V pattern of Figure 2.3 (left side of the V) emphasizes the flowdown of functions and associated requirements to each underlying design and development level. The decomposition of the system begins with defining interfaces and functions necessary to implement the capability defined by the CONOPS process. The decomposition, definition, and assignment of functions continue with development of system requirements (abbreviated as reqdts), allocation of system requirements to subsystems, and derivation of subsystem requirements to constituent units, which finally converge with specific hardware (H/W) and software (S/W) implementations.

As hardware components, software modules, and subassemblies are designed and fabricated, testing begins at each level (right side of the V).

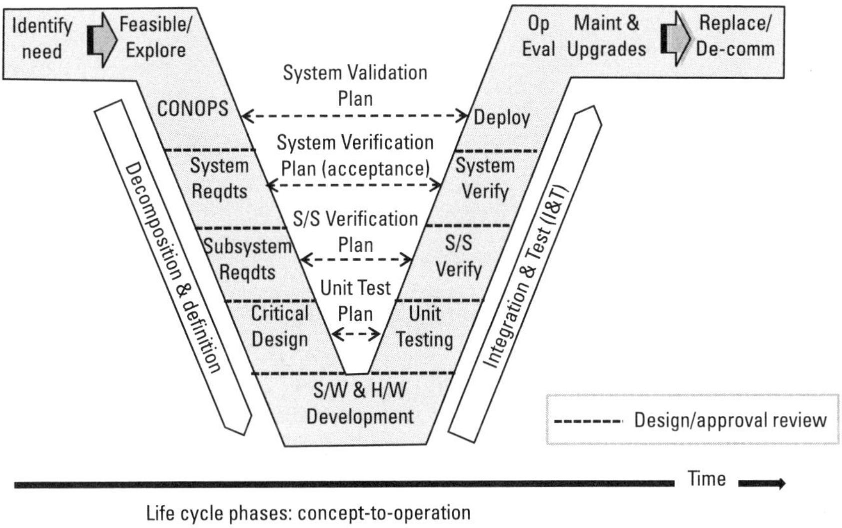

Figure 2.3 Correlation between system requirements and verification/validation.

With successful testing at an assembly level, the next tier of I&T is continued until all constituent elements are qualified. With problems and/or test failures, design and development stages may be revisited to mitigate the problem source, with reverification testing performed. (There are occasions where a technical or implementation problem resides at the CONOPS stage, and top-level requirement(s) must be modified. This is very costly and is the reason that significant time is expended on the early mission definition stage.) The correlation of requirements and approach to test planning and acceptance is accomplished through the test plans that link requirements to specific test parameters, which is represented by the two-ended arrow lines in Figure 2.3. Each requirement is certified via an approved approach (test, analysis, demonstration, combination) as satisfactorily meeting (exceeding) the established requirement threshold(s) to ascend to the next I&T stage.

The system test, when deemed successful, leads to deployment, and initiates the critical OpEval phase. OpEval determines if the designed system answers all top-level mission objectives (need) and ensures seamless and expected integration with all interfacing systems. Verification indicates successful design and development based on derived requirements, whereas OpEval indicates whether the system performs as desired. Subsequent to OpEval, system operations start along with maintenance (and upgrades, if appropriate) throughout the system lifetime. The dashed line on the V diagram indicates gate reviews (preliminary design review, critical design review, software requirements review, and similar events) that are typically required to provide a scheduled evaluation of technical capability and overall system progress. These events necessitate specific documentation that discusses interfaces, test plans, and procedures, along with design details for subject matter expert (SME) review. These gates exist to ensure that the design or implementations do not diverge from answering the requirements developed earlier in the project. The reviews also help provide guidance on how to handle interfaces and subsequent testing.

2.9 Monitoring Development and Testing Progress

Evaluating T-ISR system elements appropriate for a particular mission requires the translation of mission-specific objectives into measureable technical parameters. These parameters are used at various stages of integration and testing to indicate satisfactory progress toward attaining the desired technical capabllity at various stages of development and are denoted as technical performance measures (TPMs). A select subset of TPMs, or key performance parameters (KPPS), is critical for the successful operation of the system. TPMs are used to monitor performance throughout design, development, and test phases at the program manager level to gauge progress and to control program resources (funds, schedule, and personnel). KPPs are required to be verified to declare the system operational. KPP values receive high visibility and are considered at all program levels, including the funding source, to determine development go/no-go at each review milestone (e.g., reviews). TPMs and KPPs represent a means to continuously monitor progress to determine if development is going as planned, or if risks are increasing and warrant review or consideration of design contingencies.

2.10 Downstream Use of ISR Data

ISR systems operate by implementing multiple stages of information extraction and analyses. These stages serve to fully develop data products possessing accuracy, precision, and timing (latency) deemed useful to analysts and decision-makers in assessing whether the events for which the ISR system had been tasked to monitor are taking place. As sensor measurements are collected, downstream processing accumulates data and forms time series to enable time correlation and filtering on target signals. These time series enable processing to mitigate background interference, sensor artifacts, aid in target identification and discrimination, and reduce residual noise. Finally, using T-ISR data within offline analyses, aspects of signals collected from enemy targets are evaluated and exploited if possible. This is frequently accomplished with SIGINT- and COMINT-centric missions.

A prime example of the exploitation of target signatures is drawn from a combat engagement involving the NATO Operation Allied Force. In 2005, Colonel Zoltán Dani (250th Air Defense Missile Brigade of the Army of Yugoslavia) detected and shot down a F-117 stealth fighter using a Yugoslav version of the Soviet Isayev S-125 Neva missile system. According to Dani, "...we used a little innovation to update our 1960s-vintage SAMs to detect the Nighthawk" [28]. Dani declined to discuss specifics, keeping modifications to the SA-3 guidance system secret, but suggested that the modifications involved long wavelengths, which allowed them to detect the aircraft when the bomb bay doors opened. Dani realized that although the long wavelength system could not be used by itself to be effective, when combined as a handover sensor from a SAM RF system, an approach availed itself to provide a reliable target track and allow the shorter wavelength radar to execute the proportional steering that successfully intercepted the F-117. Colonel Dani knew that as radar frequencies decrease below 900 MHz, radar cross section exponentially increases.

References

[1] "Numbers Chapter 13," verses 17–20, *New International Version (NIV) Bible*, Biblica, Inc., 2011.

[2] Wert, J., *Cavalryman of the Lost Cause: A Biography of J. E. B. Stuart*, Simon & Schuster, 2008.

[3] Joint Publication 1-02, Department of Defense Dictionary of Military and Associated Terms, 2016.

[4] "The Cooperative Engagement Capability," *Johns Hopkins APL Technical Digest*, Vol. 16, No. 4, 1995.

[5] GAO, Intelligence, Surveillance, and Reconnaissance DOD Can Better Assess and Integrate ISR Capabilities and Oversee Development of Future ISR Requirements, Report GAO-08-374, 2008.

[6] "Geospatial intelligence (GEOINT) Basic Doctrine," National System for Geospatial Intelligence Publication 1.0, 2018.

[7] Henderson, L., "Operational and Technical Sigint—2020 Foresight," The Industrial College of the Armed Forces, 1993.

[8] Lynn, C., "Making the Most of MASINT and Advanced Geospatial Intelligence," MASTER OF MILITARY STUDIES, USMC Command and Staff College, 2012.

[9] "TECHINT Multi-service Tactics, Techniques, and Procedures for Technical Intelligence Operations," U.S. Army Field Manual FM 2-22.401, CD-ROM, Army Publishing, June 9, 2006.

[10] Pigeon, L., C. J. Beamish, and M. Zybala, "HUMINT Communication Information Systems for Complex Warfare," Defence Research and Development Canada, Valcartier, Quebec, 2020.

[11] Alsmadi, I., "Cyber Intelligence Analysis: Cyber Security Intelligence and Analytics," in *The NICE Cyber Security Framework*, Springer, 2018.

[12] Walton, A., "Financial Intelligence: Uses and Teaching Methods (Innovative Approaches from Subject Matter Experts)," *Journal of Strategic Security* 6, No. 3 Suppl, 2013.

[13] Levesque, N., "Fighting Narcoterrorism: A Counter Narcotic Approach to Homeland Security," Master in Management for Public Safety and Homeland Security Professionals Master's Projects, Pace University, 2012.

[14] Williams, H., and H. Blum, "Defining Second Generation Open Source Intelligence (OSINT) for the Defense Enterprise," *RAND Report*, 2018.

[15] Holler, R., "The Evolution of the Sonobuoy from World War II to the Cold War," *U.S. Navy Journal of Underwater Acoustics*, 2013.

[16] Office of the Secretary of Defense, "Status of the Space Based Infrared System Program," Report to the Defense and Intelligence Committees of the Congress of the United States, 2005.

[17] Chychota, M., and E. L. Kennedy Jr. "Who You Gonna Call? Deciphering the Difference Between Reserve, Quick Reaction, Striking and Tactical Combat Forces," *Infantry Online*, https://www.benning.army.mil/infantry/magazine/issues/2014/Jul-Sep/Chychota.html.

[18] Smith, D., *Field Guide to Physical Optics*, Bellingham, WA: SPIE, 2013.

[19] Martino, J., J. P. Haton, and A. Laporte, "Lofargram Line Tracking by Multistage Decision Process," *IEEE International Conference on Acoustics, Speech, and Signal Processing*, 2013.

[20] Rauch, H., W. I. Futterman, and D. B. Kemmer "Background Suppression and Tracking with a Staring Mosaic Sensor," *Optical Engineering* 20(1), 1981.

[21] Galambos, R., and L. Sujbert, "Active Noise Control in the Concept of IoT," *Proceedings of the 2015 16th International Carpathian Control Conference (ICCC)*, 2015.

[22] Boashash, B., *Time-Frequency Signal Analysis Methods and Applications*, Wiley, 1992.

[23] Cole, T., and A. El-Dinary, "Estimation of Target Vibration Spectra from Laser Radar Backscatter Using Time-Frequency Distributions," *SPIE Proceedings Applied Laser Radar Technology*, v. 1936, 1993.

[24] "Joint Military Symbology," Department of Defense Interface Standard, 2014.

[25] Engel, A., *Verification, Validation, and Testing of Engineered Systems*, Hoboken, NJ: Wiley, 2010.

[26] "JCIDS Process Concept of Operations (CONOPS)," *AcqNotes*, acqnotes.com, 2017.

[27] "Illustrating the Concept of Operations (CONOPs) Continuum and Its Relationship to the Acquisition Lifecycle," *Proceedings of the 7th Annual Acquisition Research Symposium*, 2010.

[28] Gregory, R., *Clean Bombs and Dirty Wars: Air Power in Kosovo and Libya*, Potomac Books, 2015, pp. 65–67.

3

The WSN as a T-ISR System

This chapter formally defines WSN and describes the associated sensor node. There are many variations to sensor node design; here we focus on the core aspects of WSNs as applied to T-ISR missions. As presented in Chapter 1, several technologies associated with low-power, low-cost sensor nodes (motes) simultaneously arrived at effective levels of maturity, enabling development of viable WSN systems. WSN systems have demonstrated significant success in key performance areas including multimodal sensing, real-time signal and data processing, autonomous network management, adaptive power management, and self-localization. All of this is realized without operator intervention and with the use of secure communications. In light of these technical achievements and successful demonstrations of WSN-enabled systems, application of WSNs to tactical missions expanded exponentially.

A network of motes, a motefield, consists of spatially distributed sensor nodes that employ multihop, packet-switched networking. All functions including initialization, operation, and ongoing network management are conducted without operator intervention. This includes fault detection and remediation at the node and network levels. Real-time sensor support and cooperative processing (through node-to-node data sharing and aggregation) are realized, with data extracted upon request based upon MOC-issued commands and/or through a priori instructions.

The continuing focus on WSNs as a segment of the Internet of Battle Things (IoBT) stems from the ability of WSN-enabled systems, with appropriate design tailoring, to autonomously monitor structures such as buildings, terrain, tunnels, and bridges within a single networked field. It is akin to having a sensor that extends across various directions yet remains linked together. This characteristic makes WSN adaptive to variable environments and terrain, as shown in Figure 3.1. Using Figure 1.3, we can observe that WSN sensors can be deployed and operated in treacherous locations and can integrate groups of sensor nodes, as a sensor fabric that operate as an unified sensor system, operated and controlled by minimal (or no) staff. As important to the versatility of extending sensing capability into areas of interest is the ability of WSN to scale. Once a WSN system is deployed, it can easily facilitate augmentation at a later time through additional mote deployments, or for temporary system capability, through the use of mobile sensors (e.g., mobile-platform sensors).

Initially, developments within the WSN field presented high expectations, setting a goal for the size (volume) of sensor nodes to that of dust (a few

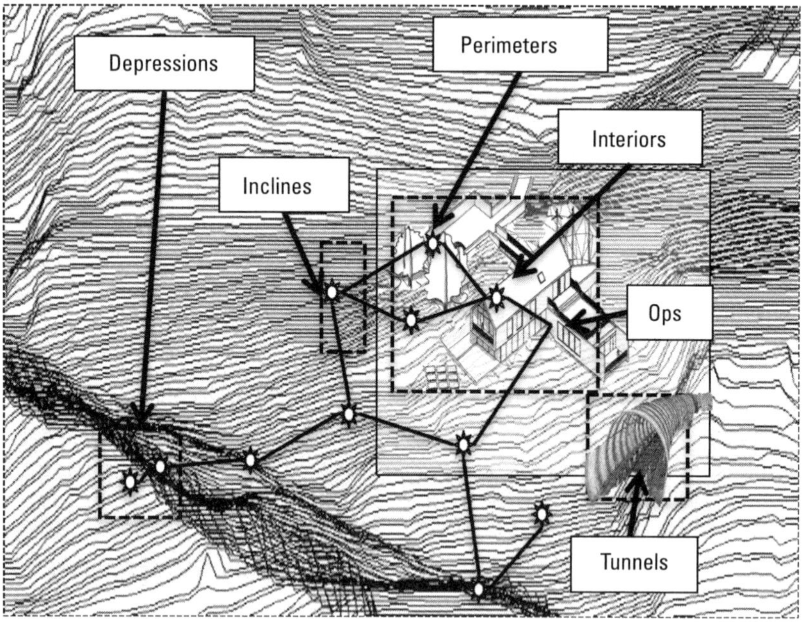

Figure 3.1 Inherent WSN motefield capabilities. Each circular icon represents locations (emplacement) of sensor nodes with network linkage shown by connective lines.

cubic millimeters) [1]; however, pragmatic motes design for T-ISR resulted in sensor size of at least three to four orders of magnitude larger to accommodate appropriate sensors, transceivers, antennas, and power sources (e.g., batteries). There has been progress toward the production of motes approaching dust-sized dimensions, but only for diagnostic applications within biomedical engineering, where node-to-node distances and operating durations are significantly less than those required for T-ISR use. These early technical breakthroughs and demonstrations do indicate an eventual ability to reach extremely small volumes, but the current capability has yet to achieve such a manufacturing capability to provide a complete low-cost sensor node solution.

What challenges sensor node design for T-ISR is the myriad of functions the underlying subsystems must provide in terms of performance and capabilities. Discussion of the sensor node functions required for T-ISR applications occurs in Sections 3.1 through 3.5.

3.1 WSN Node

Successful application of WSN to T-ISR missions is contingent on sensor node implementations capable of meeting measurement requirements that are robust, persistent, and compatible with autonomous operation within AOIs. T-ISR sensors are not thought of as sensor systems that are deployed and subsequently retrieved. This does not indicate that all WSN-based systems are considered disposable but for particularly challenging (and hazardous) applications, a WSN system should be considered expendable.

WSN attributes are well-matched to T-ISR tasking, which can require the monitoring of inaccessible terrain (e.g., Figure 3.1). In addition, the capability of a WSN-based system far exceeds that associated with the relatively simple sensor nodes. When deployed in large numbers, a WSN node field (motefield) presents system users with an exceptionally capable sensor. This is because WSN motefields have multiple sensors observing the same AOI and targets, but from different perspectives. The dimensional extent of a motefield allows for signal correlations and data processing based on a combination of local and field-wide stimulus (i.e., motefields can scale to square-kilometer-sized areas). The SWaP characteristic associated with motes, along with the self-organizing ability, is supportive of large mote deployments via mechanisms beyond hand-emplacement (e.g., aerial dispersal system). This allows WSN-based systems to provide a quick response solution to T-ISR system setup and operation.

3.2 WSN Node (Mote) Functions

There is an overarching restriction in the design of a WSN mote because of a severe competition for resources: the size (volume), weight, and power (SWaP). With large numbers (100→10,000 motes per motefield), price (or SWaP2) becomes an issue as well. All resources required by an individual mote must be self-contained, as no man-in-the-loop (MITL) capability is required with the exception of intermittent network commands being issued to implement on-demand data extraction or commanding/reprogramming of a motefield. T-ISR mote sensors must meet minimal TPM thresholds for range, resolution, and measurement accuracy as required by mission objectives and characteristics. In response, numerous mote designs for T-ISR consist of two or more sensors to provide improved detection probability and classification of targets.

Upon deployment, individual motes are expected to form ad hoc networks, and once networked, they are designed to operate as a singular sensor capability. Built-in network protocol and management capabilities continue to identify and include new motes upon introduction to the AOI, which promotes scalability. Upon detection of faults, network management logic (which is implemented via middleware) rapidly identifies faults and responds with self-healing of the network through falts resulting of the network topology and automated updating to associated routing tables. To support T-ISR missions, motes must be designed to operate for long periods, allow for dynamic and agile network configuration, and implement subsystems illustrated by Figure 3.2, which is typical WSN of sensor systems: wireless communications, signal/data processing via microcontroller (including A/D, D/A conversion), accommodation of sensors, and input/output (I/O) access for testing, updates, and external hardware connectivity.

3.3 WSN Mote Subsystems and Examples

With the addition of each subsystem associated with a WSN sensor node, critical constraints, SWaP2, are depleted. Each subsystem on a WSN sensor node exists because it provides a necessary capability. For many of the earlier motes, there were excessive subsystems added because WSN applications varied widely. However, as applications become more specific as to which functions are and are not required, mote design was tuned to support a particular application. A problem with T-ISR applications is the variability associated with performance and capability requirements. Most (if not all) of the early uses of WSN in T-ISR applications employed generic core subsystems, as presented

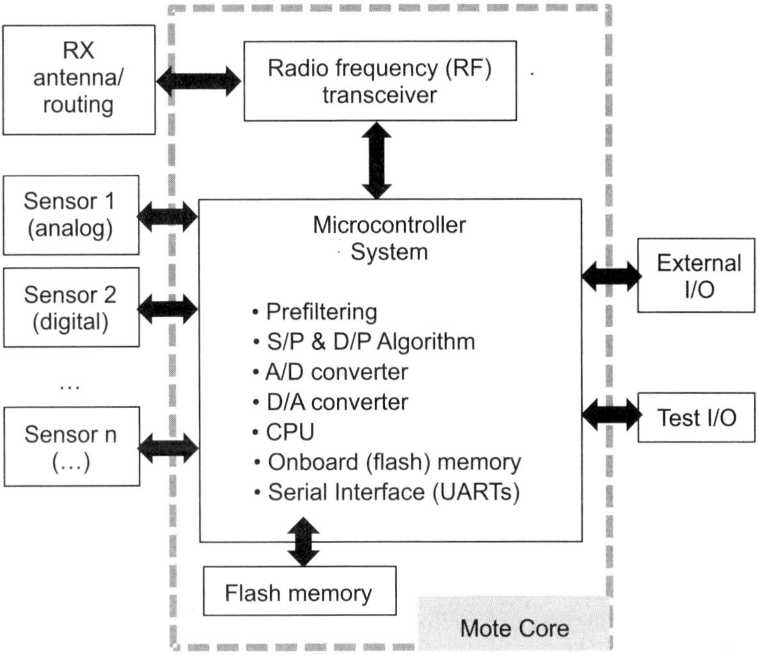

Figure 3.2 WSN subsystems diagram.

by Figure 3.2. However, throughout the past decade, WSN systems increased their dependence on internet connectivity to link sensor systems with MOC, becoming subsumed under the IoT, and now, the IoBT. With large deployments, such as industrial and medical systems, wireless sensor nodes are finely tuned to the application, shedding unused circuitry and I/O, thereby reducing the need for SWaP2.

WSN nodes provide a data acquisition system (DAS) that provides signal conditioning and digitalization through use of processor and analog-to-digital converter (ADC) hardware. Signal processing (data extraction and compression), storage, and formatting of measurements are conducted through microcontroller logic. In operation, a motefield maintains wireless connectivity to all network nodes and one or more pathways capable of communicating with a MOC. Robust wireless capability is required to allow for distributed positioning of motes without regard to a strict deployment pattern. Combining the above functionality required of a mote, with existing and defined subsystems, resulted in a mote core as depicted in Figure 3.3, and typically implemented as a SoC.

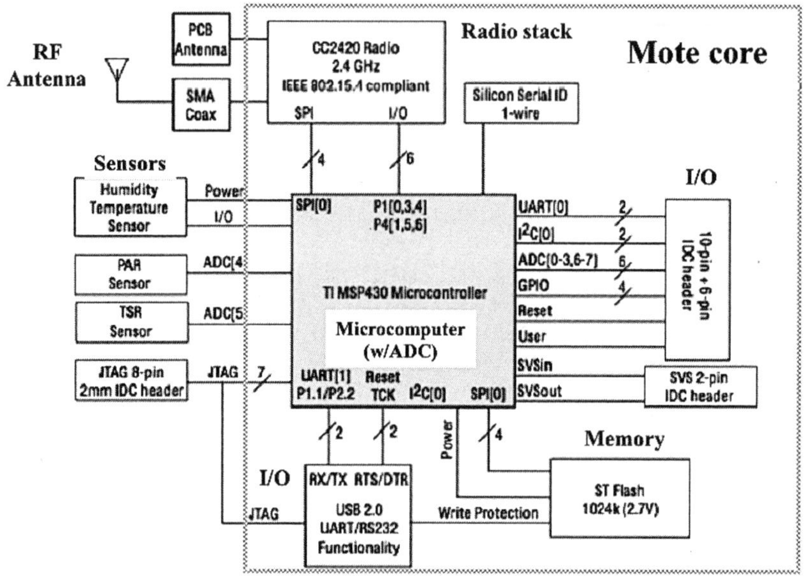

Figure 3.3 System diagram of a T-ISR mote.

Numerous research and test programs using motes have been conducted, and in support of these tests, numerous mote designs have been developed and fabricated. The Tmote Sky mote [2], presented in Figure 3.4, is an implementation of typical mote subsystems (Figure 3.3). The Tmote Sky is a low-power mote implementation, supportive of a small-sized, fast (802.15.4-compatible) RF radio startup operation (<6 microseconds) operating at 2.4 GHz, and a microcontroller (16-bit TI MSP430, with flash memory) well known by mote designers. The Tmote Sky provides integrated support functions, such as timers; supply voltage regulation; comparator, test and operational I/O ports; a DMA controller; and environmental sensors (humidity, temperature, and photodetectors) [3, 4].

In addressing the application of WSN to T-ISR missions, no particular mote design comes to the forefront as much as the Extreme Scale Mote (XSM) developed under DARPA's NEST program [5]. Figure 3.5 presents the XSM2 mote, with Figure 3.5(a) showing the extended RF antenna and the internal circuit bard supporting the subsystems, and Figure 3.5(b) showing the various sensors: passive infrared (PIR) detectors, acoustic microphone, and chip magnetometer. The objective for the XSM design was to provide a platform for middleware development in the pursuit of realizing persistent detection, target classification, false alarm discrimination, and tracking of

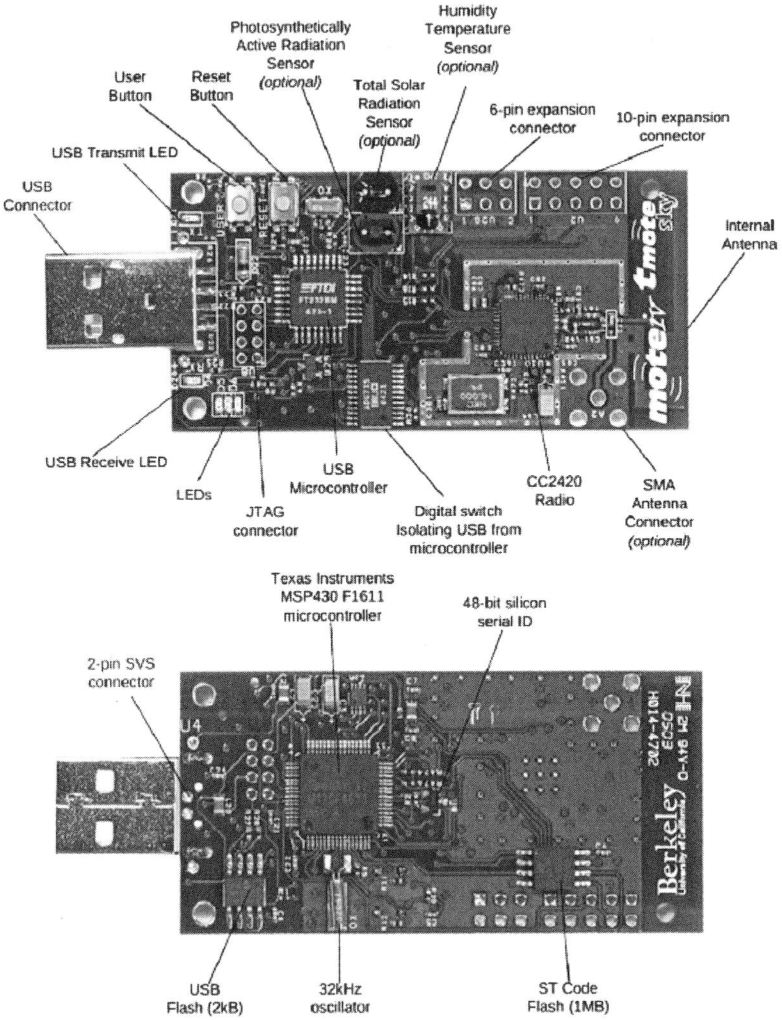

Figure 3.4 Tmote Sky. (Reprinted by permission of moteIV Corp. (J. Polastre, R. Szewczyk.)

targets of interest [6]. The XSM is backward-compatible with the University of California-Berkeley (UC-B) and Crossbow MICA2 mote [5]. The design is based on Atmel Advanced RISC (AVR) Architecture that contains an eight-bit microcontroller (ATmega128L) operating at a clock speed of 7.383 MHz. The microcontroller is integrated with a Chipcon CC1000 RF chipset that can transmit at 315, 433, or 868/916 MHz; the XSM2 was designed to operate at 433 MHz.

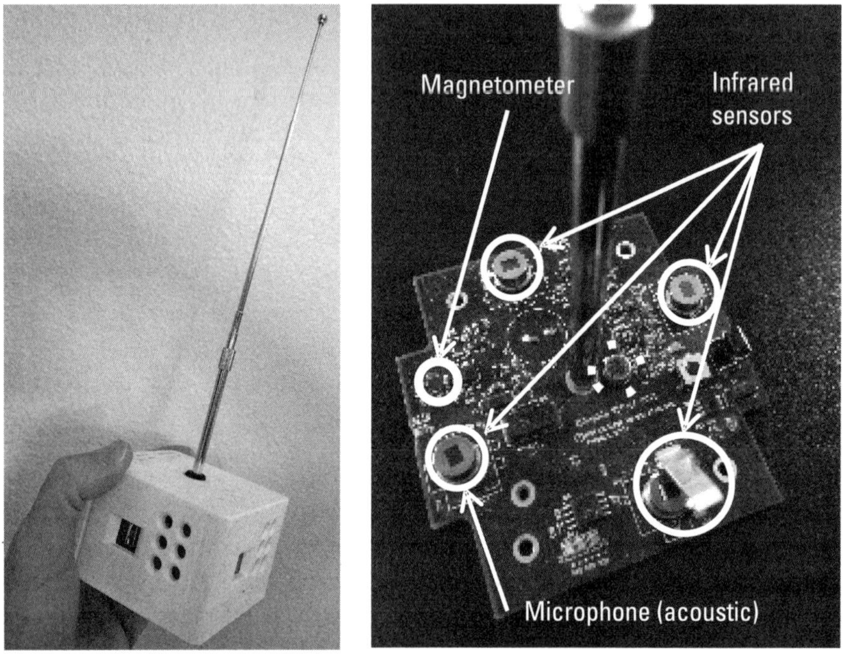

(a)
XSM2 mote
(3.5"x3.5"x2.5)"

(b)
XSM2 mote board

Figure 3.5 XSM2:(a) presents the DARPA NEST mote design, the XSM2, and (b) presents the internal layout of an XSM mote showing the magnetometer, microphone, and four passive infrared sensors.

3.3.1 WSN Microcontroller

The microcontroller system (SoC device) implements a microprocessor, A/D (and D/A) converters, RF transceiver, simple sensors (thermal, light detector), power management circuitry, numerous I/O ports, and onboard (flash) memory. T-ISR motes need to be able to conduct localization to establish relative locations within a network and may include a GPS receiver chipset. Cooperation among mote subsystems is managed using the microcontroller running a RTOS. With appropriate middleware (see Chapter 8), the microprocessor schedules event timers controlling mote operation, media access control (MAC) (see Chapter 4), and overall dynamics of network structure. Requested data is formatted, and via a RF transceiver, is transmitted through the network to end users through use of a relay to a long-haul communication network (e.g., DoDIN; see Chapter 1). Numerous microcontroller devices are employed by WSN mote designers; however, of 149 listed motes (as of July 2020), 42 mote cores use the TI MSP430 RISC microcontroller, whereas 46 employ the Atmel ATmega128 (L) AVR microcontroller [7]. Together these two SoC devices account for approximately 59% of all WSN-related microcontroller devices, with the third most-used device being the Micorchip PIC (10 motes).

Recall that microcontrollers (microprocessors) use either the Harvard or von Neumann architectures; for microcontrollers, Harvard is prevalent. Harvard architecture is where program memory and data memory are separated, whereas the von Neumann model has program and data stored within the same memory module. The XSM2 ATmega128 is a modified Harvard architecture based on a low-power CMOS 8-bit microcontroller using the AVR-enhanced RISC architecture (as with previous mote designs such as the MICAz and MICA2 mote). AVR was one of the first microcontroller families to use on-chip flash memory for program storage, instead of one-time programmable ROM, EPROM, or EEPROM. (The original AVR was developed at a local ASIC house in Trondheim, Norway, called Nordic VLSI, now Nordic Semiconductor, the developer of ultra-low power communication SoC devices such as the nRF52 Series) [8]. The ATmega128 can execute powerful instructions in a single clock cycle and can approach 1 MIPS per megahertz, allowing designers to optimize power consumption with respect to processing speed [9]. The Atmel AVR core combines a rich instruction set with 32 general-purpose working registers. All 32 registers are directly connected to the arithmetic logic unit (ALU), allowing two independent registers to be accessed in one single instruction executed in one clock cycle. The resulting architecture is more code-efficient while achieving throughputs up to 10 times faster than conventional CISC microcontrollers.

A power-down mode on the ATmega128 saves register contents but freezes the oscillator, disabling all other chip functions until the next interrupt or hardware reset. In power-save mode, asynchronous timer continues to run, allowing the user to maintain a timer base while the rest of the device is sleeping. An ADC noise-reduction mode exists that halts the CPU and all I/O modules except an asynchronous timer and ADC to minimize switching noise during ADC conversions. In standby mode, the ATmega128L crystal/resonator oscillator is running while the rest of the device is sleeping, allowing very fast start-up combined with low power consumption. In extended standby mode, both the main oscillator and the asynchronous timer continue to run.

The MSP430 architecture implements a von Neumann approach, which prevents instruction fetch and data operations from occurring simultaneously as they share a common bus. Simpler than a Harvard architecture machine, the von Neumann design has a dedicated set of address and data buses for reading data from and writing data to memory, as well as another set of address and data buses for instruction fetching. This dedicated address space is responsible for the oft-mentioned von Neumann bottleneck, which limits performance of this design. Texas Instruments (TI) developed the MSP430 in 1993 and is currently in its sixth generation. This architecture includes a processing unit containing an arithmetic logic unit and processor registers; a control unit with an instruction register and program counter, an on-device memory to store both data and instructions, and external mass storage; and input and output ports. Supportive of software-engineered programming techniques, the MSP430 implements calculated branching, table processing, full register access, and direct memory-to-memory transfers. All instructions have access to all addressing modes, with certain instructions being single-cycle register operations (RISC-like behavior).

3.3.1.1 Processor Performance, Latency

Processor speed (clock frequency, f) and instruction speed (millions of instructions per second, MIPS) are metrics that help determine the processing speed capability of microcontrollers. Alone, these measures do not provide an accurate measure of how fast actual code is executed, because the speed of executed code is significantly influenced by specific instruction set architectures (ISA). Single machine instructions may require more than a single clock cycle, especially for recursive looping code structures. Metrics attempting to gauge performance have been pursued [10], with one of the more accepted approaches being cycles (clocks) per instruction (CPI) which is determined using widely accepted benchmark codes [11]. If we measure execution time (ΔT) for the application program, performance (P) is defined as

$$P = \Delta T^{-1} \tag{3.1}$$

However, an issue exists: We have yet to select a microcontroller and code applications, which for T-ISR can vary significantly from application to application. To predict processor-application combined performance, we use CPI, destined as the product of the processor speed (f) and total execution time (ΔT) divided by instruction count (IC) associated with code used in the evaluation,

$$\mathrm{CPI} = \frac{\Delta T * f}{\mathrm{IC}} \tag{3.2}$$

As application code is yet to be generated during initial designing, IC is approximated by use of benchmark programs that attempt to simulate, at a high-fidelity level, code to be applied in implementing an application. To avoid bias, values for CPI are estimated by averaging performance over various instruction types that occur at specific frequencies of incurrence in the planned application code. The effective CPI_e, is the expected value (weighted average) of each instruction type CPI_i based on the instruction-associated occurrence probability (P_i) of the ith type of instruction. CPI_e becomes

$$\text{CPI}_e = \sum_i (\text{CPI}_i) * (P_i) \tag{3.3}$$

In (3.3), the individual CPI_i component of each instruction is evaluated using (3.2). Probability (P_i) in (3.3) is denoted as the normalized frequency of use of an instruction type in the planned code. For completeness, once an application program is produced, we can evaluate (3.2) for performance comparison, as

$$\Delta T = \text{IC}_a * \text{CPI}_e * \frac{1}{f} \tag{3.4}$$

The term introduced in (3.4), IC_a, is the actual instruction count associated with the specific benchmark used (assumed to be reflective of the application). Not having a value of IC_a does not totally impede comparing various microcontrollers, as comparisons can use ratios of CPI_e to determine the relative capability of each microcontroller candidate. Not surprising, there are other performance measures that can impact final selection, such as memory bandwidth (bytes per second), power dissipation, and associated peripheral support (i.e., ADC speed, bit size, and I/O support).

In addition to using processor speed values (CPIs) as a metric of speed, architectural design enhancements through implementation of pipelined instruction architecture directly impact overall speed capability. With a pipeline design, instruction execution overlaps in time. Having a pipeline divided into stages, and connected with one another to form a pipe-like structure, operates on data without being forced into a serial process by processing two or more levels simultaneously. Each pipeline stage hosts an input register to hold data followed by combinational logic to provide input to the following stage. Pipelining accumulates instructions from a processor to allow for the storage and execution of instructions in an orderly process and through instruction timing overlap, improves overall instruction throughput. Speed improvement in functional execution is also possible through processing parallelism through architectures based on multiple processing cores [12]. Amdahl's law expresses theoretical improvement of execution time of a task employing multiple processing streams. Having p defined as the proportion of code that benefits from parallelism, and s as the resultant speed-up factor, the theoretical improvement in latency, $S_L(s)$ is evaluated as [13]

$$S_L(s) = \frac{1}{(1+p) + \frac{p}{s}} \tag{3.5}$$

However, there remains a limit to the improvement due to task segment(s) that cannot benefit from parallel processing.

3.3.1.2 Thin Real-Time Operating Systems

The design goal for WSN real-time operative systems (RTOSs) is not high throughput, but rather a guarantee of predictable performance, which requires minimal interrupt latency and minimal thread switching latency. RTOSs are designed for real-time applications that process data upon arrival without buffer delays. The logic structure associated with real-time systems must be well-defined, with fixed time constraints requiring processing to be completed within a time constraint else the system fails. The implementation of RTOSs can follow event-driven or time-sharing approaches. Event-driven RTOSs switch between tasks based on task priorities, while time-sharing systems switch tasks based on clock interrupts. Most RTOS implementations use preemptive scheduling algorithms with the key characteristic of having consistent execution time to accept and complete an application task [14]. The merit of a RTOS for WSN sensor nodes is how quickly the OS can respond versus task completion capability within a given period. However, there is a maximum execution time required for a T-ISR microcontroller to ensure that sensor inputs and external interrupts are addressed.

WSN RTOS implementations do diverge from traditional RTOS design. For WSN, the RTOS design must stress power management, robust communication networking, and event-driven sensor data acquisition. RTOS candidates must provide an architecture that results in a small kernel size (small memory footprint). Additionally, the RTOS architecture for WSN must be flexible, with only application-required services loaded onto the system and extensions to the kernel as required. RTOS-based systems for T-ISR need to respond to real-time events quickly and efficiently, which emphasizes the need for an advanced interrupt system over latency performance. If interrupts require excessive clock cycles to respond due to CPU registers that need to be saved prior to calling the interrupt routine, real-time response suffer. Significant documentation exists that provides in-depth discussions and keen examples of RTOS implementations used by WSN systems [14, 15]. There are also in-depth reviews and comparisons of RTOS for WSN applications that have employed various benchmarks in their evaluation [16, 17]. Figure 3.6 presents a diagram of a WSN-oriented RTOS that illustrates OS functions and hierarchy among applications, board support package (BSP), and mote hardware. (BSP provides linkage for platform-independent RTOS through standard interface functions to initialize and control mote hardware.)

3.3.1.3 WSN Middleware for T-ISR

Middleware is a set of programs that reside above the operating systems and network protocols to provide a foundation for application development and operation. The objective is to have middleware aggregate details of low-level hardware that supports application development and management. Resource management at the middleware level is easier and more flexible than that at the RTOS level and application levels. For example, adding security features is more appropriate at the middleware level as it supports standardizes integration and provides a run time environment supporting and coordinating multiple applications [18]. Chapter 8 delves into specific middleware modules required by WSN-based systems for T-ISR, but for this discussion addressing

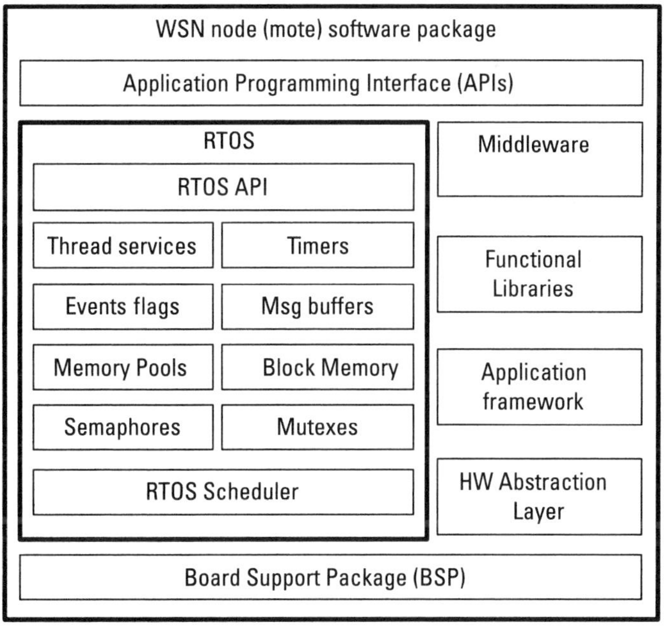

Figure 3.6 WSN node (mote) RTOS block diagram.

mote subsystems, middleware represents a collection of standardized system services in support of the development of applications. Through adaptive algorithms, WSN middleware can dynamically manage network resources and support efficient use of system and RTOS resources. This is a major benefit to WSN. The optimization of network management systems (NMS) capabilities is paramount to satisfying quality-of-service (QoS) requirements, which is conducted through negotiation between applications and low-level network protocols. WSN middleware correlates application-specific features and network protocols as directed by current network status and required QoS [19].

Mote-based middleware is also depended on to support mote deployment, setup, and maintenance of a network structure, including mote initialization and scaling of an existing network fabric. NMS operates continuously with detection of link failure and/or congestion identified and remedied quickly through the packet-switched network. WSN middleware must also handle: sensor operations data product generation and storage in real-time, power management, localization of the motes, and supporting architecture (e.g., relay, nearest neighbors, mote group gateway) [20]. Finally, reprogramming and middleware updates via over-the-air (OTA) code installation is addressed by middleware code. Figure 3.7 illustrates the middleware layers on a T-ISR mote, as well as interfaces required for physical devices, coordination with the RTOS, and interaction with a network.

3.3.2 Mote-Based Data Acquisition

Motes may host multiple sensor devices directly attached to the microcontroller SoC or through a sensor subsystem that connects directly to the microcontroller system. Most WSN sensors are analog transducers that respond to physical stimuli, producing voltage or current signals commensurate to sensory input strength. Typical functions associated with sensor signal processing implemented by the onboard processor include: signal prefiltering, discrete-time sampling, and signal-level digitalization (amplitude quantization). Regardless of the particular setup, signal conditioning is designed to occur early at detector output and includes a combination of amplification, filtering, and impedance matching. Subsequent to signal conditioning, a time-based sampling function occurs, followed by digitalization using an ADC function.

3.3.2.1 Sampling Theory

Sensors are considered transducers. An output signal (voltage or current) is produced based on the intensity of input energy. For imagers and other two-dimensional sensors, the sensor output (voltage, current) can be thought of an

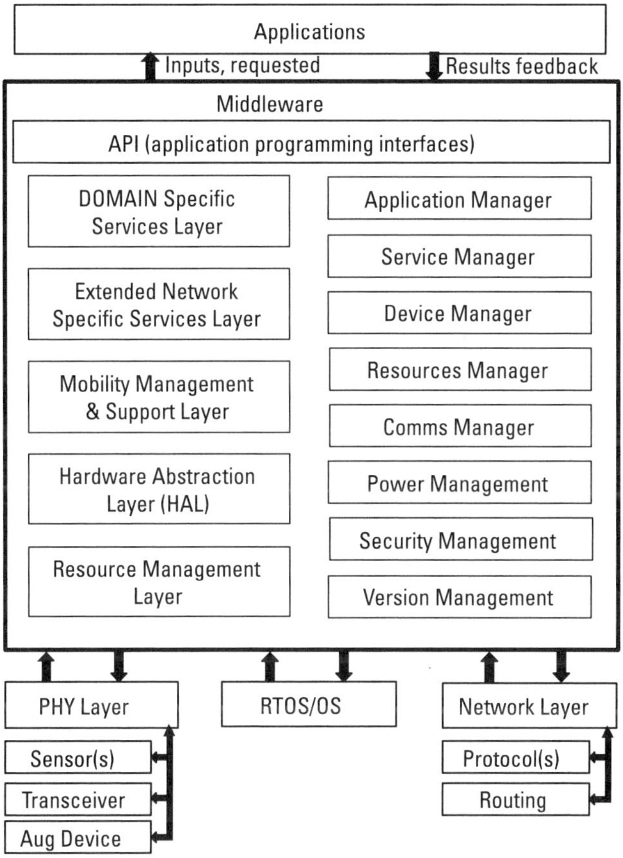

Figure 3.7 T-ISR mote middleware functions.

array of signal values where processing must preserve the 2-D spatial ordering. Motes sample and digitize responses produced by onboard sensors that are time-sampled as continuous data flow from sensor-to-end-user is restrained due to finite bandwidth of the communication network, as well as in the long-haul network. Equation (3.6) provides an algorithmic approach to uniformly sampling a signal $g(t)$ via a temporally sample function $s(t)$ represented as a series of Dirac impulse functions, $\delta(t)$, with a period T_s between samples to produce discrete-time values of the signal $g(t)$, as $g(k)$. Figure 3.8 presents a diagram of a sampled signal, $g(k)$ derived from $g(t)$. In Figure 3.8(a), a continuous signal, $g(t)$, exists and is equidistantly sampled at a period of T_s by the sampling function, $s(t)$. The result is Figure 3.8(b), which shows the digitized function, $g(k)$ at each sampling interval. The sampling function uses

$$g(k) = g(t)_{nT_s} = g(t)s(t) = g(t)\sum_{k=-\infty}^{\infty}\delta(kt - kT_s) \tag{3.6}$$

Sampling time-varying signals, as indicated by (3.6), is addressed using the well-known Nyquist-Shannon sampling theorem [21, 22], which indicates that the minimum sampling frequency of a signal that will not distort underlying information should be greater than (or equal to) double the frequency of the highest signal frequency component. If frequency components exist higher than this maximum value, the sampling process becomes compromised since the high-frequency components are reintroduced at incorrect frequencies due to aliasing (i.e., false assignment of signal energy at incorrect frequencies).

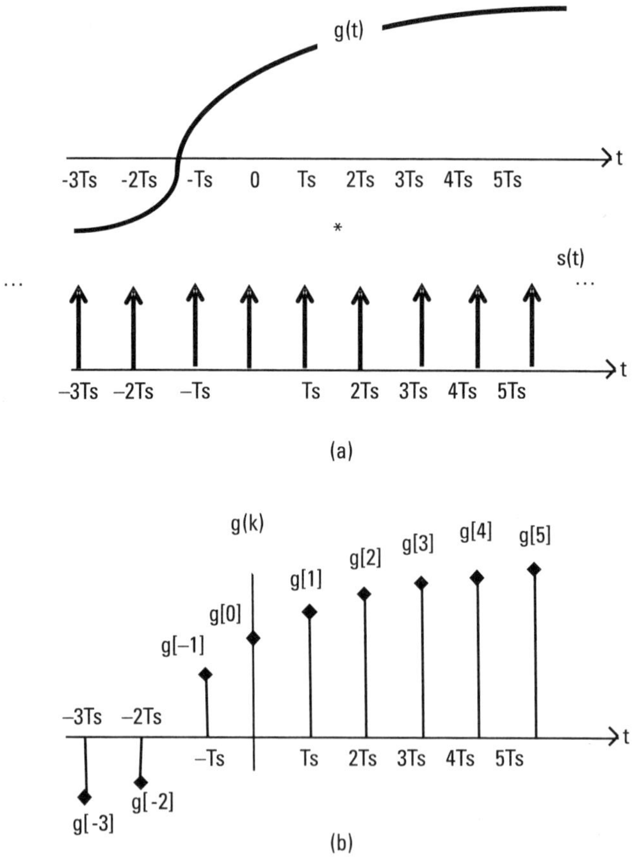

Figure 3.8 Sampling function, *s*(*t*), operating on signal *g*(*t*). (a) The sampling of a continuous signal, *g*(*t*), and (b) resultant sample series, *g*(*i*).

Equation (3.7) shows the path of translating the temporal sampling to the frequency (spectral) representation using the Fourier transformation, $F\{\}$. Recall that the temporal multiplication of functions results in the convolution of the individual Fourier transforms, as

$$g_s(t) = s(t)g(t) \rightarrow \text{Fourier} \rightarrow F\{g_s(t)\} = F\{s(t)\} * F\{g(t)\} \tag{3.7}$$

Using (3.6) and (3.7) results in the frequency representation of the sampled signal is determined as,

$$G_s(w) = F_S(w) * \sum_{k=-\infty}^{\infty} \delta(\omega - k\omega_s) \tag{3.8}$$

The sampled signal representation, (3.8), consists of the combination of the expected spectral content of the signal [Figure 3.9(a)], as well as multiple copies of signal spectra occurring at integer multiples of the sample frequency (ω_s), shown in [Figure 3.9(b)]. Figure 3.9 illustrates the impact of undersampling by not sampling at (or above) twice the highest signal frequency [Figure 3.9(c)]. As presented by (3.8) and Figure 3.9, the frequency distribution of the sampled function is scaled and repeated at integer values of the sampling frequency.

Reconstruction of the original signal is realized through the application of an appropriately scaled frequency-windowing function, which is the rectangle function that blocks all frequencies except those within the Nyquist limit. The rectangle function is defined as rect(x) with values of

$$\text{rect}(x) = \begin{cases} 0 \text{ for } |x| > \frac{1}{2} \\ \frac{1}{2} \text{ for } |x| = \frac{1}{2} \\ 1 \text{ for } |x| < \frac{1}{2} \end{cases} \tag{3.9}$$

Applying the Fourier transformation to the rect(x) function having a total width of W, results in the normalized sine cardinal function,

$$F\{\text{rect}(W)\} = W\,\text{sinc}(x) = W\frac{\sin(\pi x)}{\pi x} \tag{3.10}$$

In reconstructing the signal, multiple processing steps are indicated. As seen in Figure 3.9 [and through (3.8)], the frequency content of the sampled

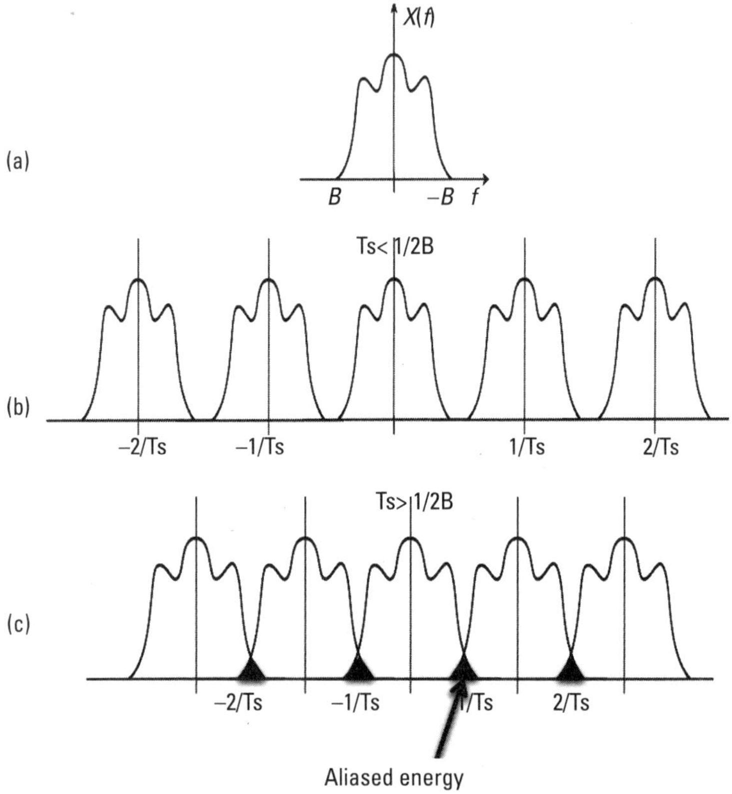

Figure 3.9 Nyquist criterion to prevent aliasing caused by undersampling.

signal, $g(t)$, is such that the spectrum is repeated with the spectral center located at integer intervals of ω_s. To reconstruct the signal using the temporal samples obtained with (3.6) requires filtering the signal to limit its frequency width to, $W = 2\omega_s$. When transformed (inverse Fourier, $F^{-1}\{\}$), the rect(ω) function is represented again by the cardinal sine function, sinc(t),

$$F^{-1}\left\{\frac{1}{\sqrt{2\pi a^2}}\mathrm{rect}\left(\frac{\omega}{2\pi a}\right)\right\} = \mathrm{sinc}(at) = \frac{\sin(at)}{at} \tag{3.11}$$

We now have enough to persue the idealized reconstruction of the signal, $g(t)$ using the samples obtained by $g(k)$ in (3.6). Implementing frequency limiting to the sampled frequency signal, (3.9), and performing the inverse transformation results in a series of sinc(t) functions, as

$$g_r(t) = \sum_{k=-\infty}^{\infty} g(k)\text{sinc}\left(\frac{t-kT_s}{T_s}\right) \tag{3.12}$$

The above represents the common foundation of data sampling using quantized Shannon representation. Figure 3.10 illustrates reconstruction of a signal from sampled data using (3.12).

Recently, significant research has gone into nonuniform sampling, adaptive sampling, and in particular, compressed sampling (CS). CS assumes two characteristics of the signal to be sampled: sparsity and incoherence. Sparsity means that the information rate of a signal may be reliably obtained by sampling at a rate less than the signal's bandwidth, or that the signal contains a smaller number of degrees of freedom than its finite length dictates. Incoherence is inferred in that the signal has an extremely dense representation when using the proper basis. CS approaches are most beneficial for sparse signals whereby frequency content is grouped at various spectral regions. Assuming that the transformation used on the signal has, at most, K nonzero entries, it has been indicated that sampling such a K-sparse signals] requires (order) O(K log(W/K)) samples per second to stably reconstruct the signal, which is exponentially lower than the Nyquist rate of *W* Hz [23, 24]. Of significance is that CS is not reliant on deletion or compression of information and is not required to operate at the Nyquist rate. Effective sampling can occur below the Nyquist rate, which saves ADC operating power.

3.3.2.2 ADC

Implementation of the ADC function onboard a sensor node is handled through separate ADC circuitry, or as a partitioned function of the microcontroller

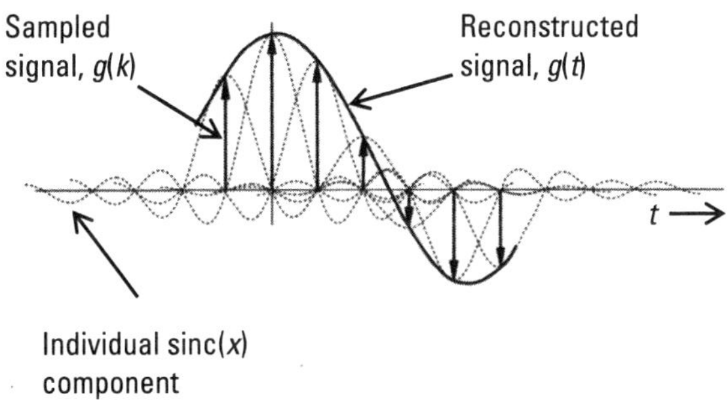

Figure 3.10 Signal reconstruction using sinc(*x*) series.

SoC. With either implementation, there are a variety of ADC algorithms and approaches available, most based on Nyquist rate. Notable ADC approaches include the following:

- Successive-approximation (SAR);
- Sigma-delta (Σ–Δ);
- Pipeline;
- Slope converters;
- Flash;
- Folding-interpolating.

The SAR converter is the ADC architecture of choice for many multiplexed data acquisition systems and sensor applications. SAR converters are relatively easy to use, present no pipeline delay, and are available with high accuracy and resolution (up to 24 bits) with sampling rates over 5 Msamples/sec (Msps) [25]. Recently, a single-channel hybrid pipelined-SAR achieved 300 Msps [26]. Although SAR converters are slow in comparison with other ADC designs, they are also lowest in power usage. SAR architecture includes a digital-to-analog (DAC) comparator and control logic (successive approximation register); however, the overall form factor remains small compared to other ADC architectures. A concern is that to maintain high accuracy, the associated DAC must meet that accuracy level as well.

Although sigma-delta ADC converters require fast oversampling clocks, which render low output rates, these converters are preferred for high-level resolution. Sigma-delta converters operate by executing one-bit conversions at high sample rates and subsequently averaging results to provide high-resolution output. The inherent advantage of sigma-delta converters is that special trimming or calibration is not needed, even with 16 bits of resolution. Alternatively, an issue with sigma-delta conversion is the need for increased sampling rates to produce an acceptable output value, which requires components to operate at higher rates, making this architecture power-hungry.

Pipeline ADC uses a parallel structure of multiple stages, each simultaneously operating on one or a few bits of successive sample. The parallel architecture increases throughput; however, the multiple stages cause an increase in both power consumption and conversion latency. The pipeline architecture allows for reduced accuracy required from each flash-comparator stage by providing digital error-correction logic. This architecture does require significant silicon and if accuracies beyond 12 bits are required, trimming and/or calibration is indicated.

The simplest slope converter is a single-slope integrating ADC that compares input signals against a well-known reference voltage level. The time duration expended to trigger a comparator is proportional to the unknown input voltage. This ADC requires low power consumption but is highly dependent on tolerances of the integrator components. To overcome this sensitivity, a dual-slope integrating architecture is used to integrate an input voltage, then deintegrates using a known reference voltage for a variable amount of time. In comparison to SAR ADC, single-slope ADC presents a narrower bandwidth range and cannot equal the SAR conversion speeds; slope converters are limited to approximately 100 samples/sec. However, a benefit associated with slope converters is common mode rejection. With slope converters, undesired frequencies can be filtered out. In addition, since integration is simply averaging, slope converters have relatively good noise performance when spurious noise occurs at the input. Finally, the integrator ramp of a slope converter is set by the value of the integrating resistor and can be used to match input signal range to the ADC [27].

The flash (direct-conversion) converter approach implements a linear voltage ladder with a comparator at each ladder step to compare input voltage to successive reference voltages. Reference ladders are constructed of resistors or capacitive voltage divisions at each step, providing input to a comparator and a subsequent digital encoder that converts inputs into a binary value. To increase precision requires significantly more comparators compared to other converter structures. For k-bit conversion, flash converters require 2^k-1 comparators, which consumes power. Also, the cost of having a large number of comparators makes flash converters impractical (costly) for precisions much greater than 8 bits unless high-frequency conversion is required.

Folding-interpolating (F-I) ADC architectures address power consumption and large layout issues, especially when compared to flash converters. F-I converters, using single-step conversion, approach or meet conversion rates associated with flash converters. For F-I conversion, output from each of two adjacent preamplifiers is connected to a middle comparator in a way that compares the input signal with a reference voltage midway between reference voltage steps. Although the number of latched comparators in a F-I ADC is equivalent to flash ADC (assuming that both having the same resolution), interpolation reduces preamplifiers and resistors required by the reference ladder by 50% (or less, depending on the interpolation factor). Overall, this saves total area and reduces power consumption. Also, due to the interpolation process, F-I converters demonstrate high linearity due to averaging and distribution of errors [28]. Figure 3.11 summarizes the SAR, sigma-delta, and flash converter architectures. In addition to the ADC architectures summarized, WSN researchers have investigated ADC approaches for the specific purpose of operating with mote-sized sensors. These approaches include ADC

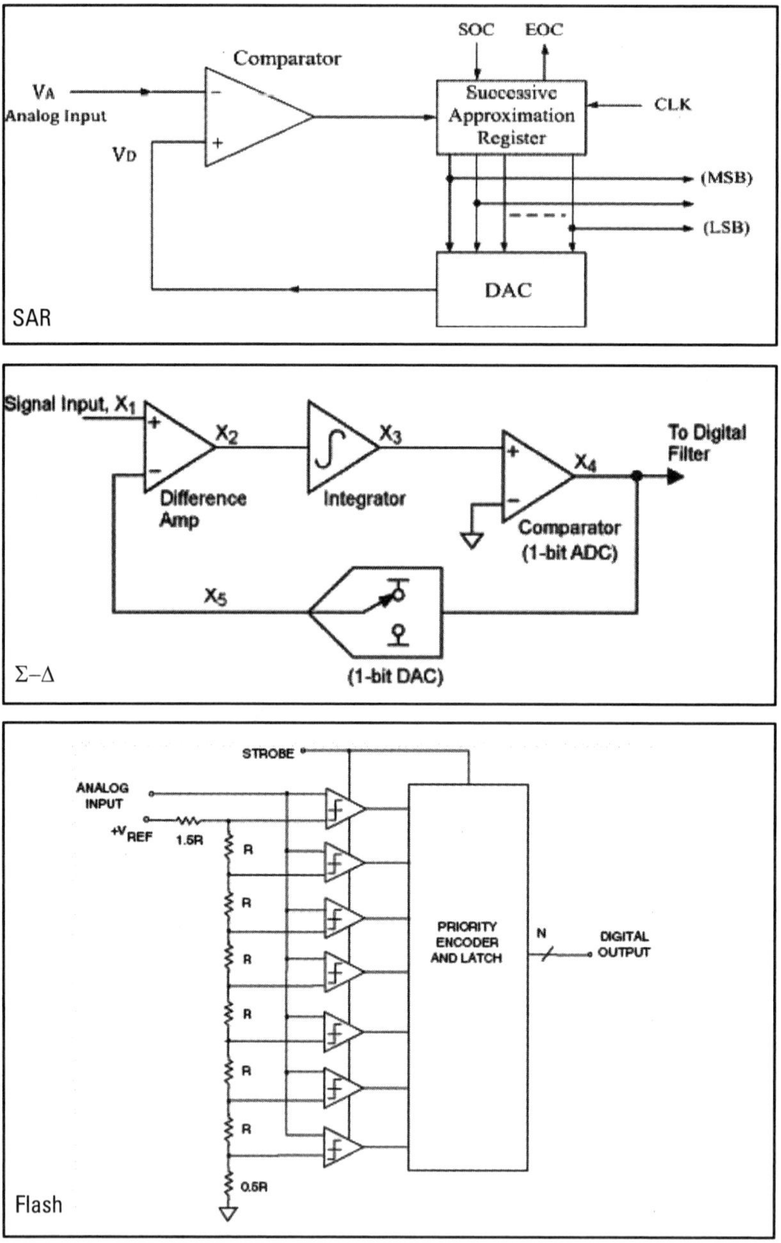

Figure 3.11 Prominent ADC architectures (SAR, Σ–Δ, and flash) employing Nyquist sampling.

structures as: entropy-coding [29] reconfigurable [30], time-interleaved/parallel, residue-type, and oversampled converters.

3.3.3 RF Transceivers

With the large number of motes associated with WSN-based persistent systems comes the requirement for extremely low-power (LP), low-cost, short-range wireless transceivers. Low-cost embedded LP transceiver design has sparked a new generation of WSN mote designs effective for T-ISR use. In devising the design of the transceiver, it is necessary to consider energy for each of its operational states: transmit, receive, idle, sleep, and transition. Transmission is the heavy user of energy, as both the power amplifier and transmitter electronics are energized. Reception has the receiver electronics operating, and typically presents the second-most energy use. Idle is when the transceiver is operating, but neither transmission nor reception is occurring and only a subset of transceiver components is powered. With sleep state, most if not all of the transceiver subsystems are unpowered. The final state, transition, represents powering up (or down) of transceiver elements as the operation moves from one state to another (e.g., sleep-to-receive). Transitions can require a variety of component energized durations and present varying energy use depending which direction the transition is oriented (e.g., off-to-transmit and transmit-to-off).

To ascertain if a proposed communication link would meet requirements, various models are employed, typically starting with the more simplistic Friis free-space transmission model [31]. The Friis equation assumes a host of conditions and is relatively accurate for the simple case presented by ideal surroundings, such as those for satellite communications. Otherwise, multipath issues arise from objects in the transmission path as well as attenuation due to atmospheric absorption, requiring more complexity than the Friis approach allows.

Given a power level measured at transmitter antenna terminals (P_t), the power available at receiver antenna terminals (P_r) depends upon areas associated with the reflective receiving antenna (A_r), the transmitting antenna (A_t), the distance between transmitter and receiver (d), and the wavelength of the RF transmission (λ). The Friis equation is summarily,

$$P_r = P_t\left(\frac{A_r A_t}{\lambda^2 d^2}\right) \tag{3.13}$$

Equation (3.13) assumes that powers and length measures are in identical units and that the distance (d) is such that plane wave propagation can be assumed, which requires distance to be greater than $2a^2/\lambda$, where a is the largest linear dimension of either of the antennas involved. Additionally, (3.13) assumes omnidirectional transmission of the RF power with energy loss over distance varying as $1/d^2$. This will be shown not to be the case regarding WSN propagation when in proximity to the ground (see Chapter 6). Considering antenna gains for transmitter and receiver (Gt and Gr, respectively), losses for transmitter and receiver (Lt and Lr, respectively), and introducing miscellaneous losses to account for fading, polarization mismatch, and parasitic losses as Lm, an equation for the RF link budget can be provided, Applying the logarithmic equivalent of both sides of (3.13) results with

$$\mathrm{Pr} = Pt + Gt + Lt + Gr - Lr - Lm - 20\log\left[\frac{4\pi d}{\lambda}\right] \tag{3.14}$$

Equation (3.14) is considered a basic link budget equation, where all terms are provided in decibel units, as defined within Section 3.3.3.1. Using decibels allows for simple addition/subtraction of terms to determine the link margin (that portion of the SNR exceeding unity).

3.3.3.1 Review: RF Decibel (Logarithmic) Units

One of the more significant failures that occur with WSN systems is unreliable communication links. As seen in (3.14), a discussion of RF system invariably makes use of decibel terminology to indicate link margin, antenna gain, RF power, antenna gain, and voltage reference. Let's quickly review decibel units and the associated definitions. RF power, measured in units of decibel-milliwatts (dBm), where P2 is power measured in watts (W) and P1 = 1 milliwatt (mW), is computed as

$$10\log\left[\frac{P_2}{P_1}\right] \tag{3.15}$$

Antenna gain relative to an isotropic (4π) radiator, compares antenna gain value (G_a) to the isotropic value (G_i) in decibels (dBi), and is defined as

$$10\log\left[\frac{G_a}{G_i}\right] \tag{3.16}$$

Voltage reference (dBv), has a measured voltage value, V2, referenced to V1 = 1; however, regarding RF measurements, V1 is usually set to 1 microvolt (uV),

$$20\log\left[\frac{V2}{V1}\right] \tag{3.17}$$

As an example of using decibel units, the maximum effective radiated power (ERP) transmission permitted in the European Union (EU) for particular short-range devices (SRDs) such as motes is 25–100 mW. Converting 25 mW to dBm results in 13.98 dBm. ERP is the total power radiated by an actual antenna relative to a half-wave dipole rather than a theoretical isotropic antenna. A half-wave dipole has a gain of 2.15 dB compared to an isotropic antenna (3.16). A related antenna term, effective isotropic radiated power (EIRP), is the total power radiated by a hypothetical isotropic antenna if confined to a single direction. It gives the signal strength in the direction of the antenna's strongest beam. The relation between EIRP and ERP, in decibel units, is provided by

$$\text{EIRP } [\text{dB}] = \text{ERP } [\text{dB}] + 2.15 \text{ dB} \tag{3.18}$$

or if using watts directly, (3.18) becomes,

$$\text{EIRP } [\text{W}] = 1.64 \times \text{ERP } [\text{dB}] \tag{3.19}$$

On the component (subsystem) level, to determine EIRP for a system given transmitter power (P_t) and antenna gain (G_a), we have

$$\text{EIRP} = P_t \text{ [W]} \times G_a \tag{3.20}$$

Typical WSN receiver sensitivities are between −85 and −110 dBm. Equation (3.20) can also be written as an electric field, E, at a given distance, d, in meters between two points. The EIRP using units of volts/meter is determined by

$$\text{EIRP} = \frac{(E \times d)^2}{30} \tag{3.21}$$

As a final note concerning decibel units, when working with power spectral density (PSD), decibel units employed are dBm/Hz or dBm/MHz.

3.3.3.2 Review: The Industrial, Scientific and Medical Radio Bands

The Industrial, Scientific and Medical (ISM) radio bands are frequency bands reserved internationally for short-range use of RF for data research and medical purposes. ISM bands were designated to provide a means of RF communication without a need for formal licensing. These unlicensed RF bands were established at the International Telecommunications Conference of the International Telecommunication Union (ITU) held in Atlantic City in 1947, with current U.S. standards being updated and defined in 1985 (FCC Rules, Part 15.247). Unlicensed ISM band use includes applications associated with medical devices, wireless computer networks, Wi-Fi, WLAN, Bluetooth devices, ZigBee, Z-wave, WirelessHD, WiGig, radio frequency identification (RFID), near-field communications (NFC), and of course, WSN nodes. ISM users are not required to hold a radio operator's license thereby permitting rapid and novel growth of RF-based technologies to afford data transfer and access. On the negative side, ISM devices are not granted regulatory protection against interference from other users of ISM bands. Researchers can have rapid access to the ISM bands, but results may vary given local interface by others.

WSN node designs have been developed using the ISM bands. This does not indicate that such would be true of deployed T-ISR systems, but for our consideration, we will limit our focus to motes operating within ISM bands. UCB and other universities made significant use of ISM bands in support of WSN research for DARPA (NEST program). Table 3.1 presents the prominent ISM bands and the applications and power levels associated with these bands. The frequency and power level values are subject to the associated country regarding frequency allocation and power levels. Various applications (devices and systems) that have matured using the ISM bands are also listed. For example, the U.S. Food and Drug Administration (FDA) has extensive coordination with the FCC and regulations associated with RF-based (ISM) medical devices [32]. For T-ISR, the ISM bands employed include the 433-MHz, 908-MHz, 2.4-GHz, and 5-GHz bands. However, system design tends toward operation using the lower ISM bands (e.g., 433-MHz) due to propagation losses when operating outdoors in presence of vegetation (see Chapter 6) and low data rates required by individual motes.

3.3.4 Mote-Based Sensor Modalities

A plethora of WSN-based sensor modalities have been developed and used in various WSN demonstrations, deployments, and operations. WSN nodes have derived scaled-down versions of sensor approaches used by UGS configurations,

Table 3.1
ISM Bands: Frequencies, Applications, ERP Permissible

ISM Bands	Application(s)	EIRP	BW (MHz)
6.765–6.795-MHz	Medical, RFID	16.4 mW	30 kHz
13.553–13.567-MHz	HF RFID, NFC	100 mW	14 kHz
26.957–27.283-MHz	Citizen's band (CB), monitors	820 mW	326 kHz
40.66–40.70-MHz	Perimeter protection, control systems	100 mW (500 mW)	40 kHz
433.05–434.79-MHz	RFID,MICA mote, XSM (TXSM) mote, Low-power hand-held radio (FRS)	50 mW (100 mW)	1.84 MHz
863–870-MHz	UHF RFID, Zwave	3280 mW	3 MHz
902–928-MHz	MICA2 mote, Measurement systems (amateur), Zwave	1W	26 MHz
915-MHz	ZigBee	4W	26 MHz
2.4000–2.4835-GHz			
	Bluetooth, Wi-Fi (802.11b/g)	100, 500 mW	100 MHz
	MicaZ, TelosB, SunSPOT motes (802.15.4)		
	ZigBee, RFID, NFC	1W	100 MHz
5.650–5.925-GHz			
	Wi-Fi-802.11a/n/ac	4W	275 MHz
	U-NII 5 GHz bands		275 MHz
	Wi-Fi-802.11a/n	1W	275 MHz
5.15–5.25-GHz	WLAN	250 mW	100 MHz
5.25–5.35-GHz	WLAN, wireless access system (WAS)	200 mW	100 MHz
5.725–5.825-GHz	WLAN, wireless access system (WAS)	1W	100 MHz
57–64-GHz band	WirelessHD, WiGig (802.11 ad)	10W	7 GHz
122.020–123.000-GHz	SiGe transceivers (sensor/comms), SoC, AmSat	500 (1000) mW	1 GHz
244.0000–246.000-GHz	AmateurSat	500 (1000) mW	2 GHz

such as: seismic, acoustic, magnetic, passive infrared, and/or optical (camera) instrumentation. Additionally, WSN research and demonstrations have evaluated and employed ultra-wideband (UWB) radars, micro laser rangefinders, and chemical, biological, radiological, nuclear, and explosive (CBRNE) sensors. Chapter 9 discusses the various sensor modalities suited for WSN, complete with a physical description and derivation of performance equations for each modality.

3.4 Adapting WSN Functionality to Address T-ISR Missions

To meet T-ISR mission minimal requirements, motes must provide for and support numerous operational modes. These modes (phases) are implemented using middleware modules, which are discussed in detail in Chapter 8. The operational phases are presented in this section to provide a preview of what T-ISR motes must be capable of in performing the array of autonomous functions. These operational modes are associated with tasks categorized as: setup, network management, signal processing, data/status communications, and power management. Additionally, T-ISR motes must be capable of seamless integration with existing standardized interfaces and compatible with appropriate legacy systems and adhere to physical constraints to meet LPD thresholds. Figure 3.12 provides an overview of mote phases with top-level mote activities associated with each phase. As illustrated by Figure 3.12, the setup phase focuses on individual motes, from predeployment checks to individual mote turn-on and operation. Beginning immediately after mote initialization, operational focus shifts to forming and maintaining the entire WSN mote field, mote and motefield sensing operations, extraction of motefield data products, mote and network persistence, interconnectivity, and secure operations.

Figure 3.12 provides the timing relationship among WSN operational modes, showing the phase behavior of the individual mote as it evolves from initial start-up through to operational capability within a newly established network. The top-level operational modes associated with mote and WSN operational phases shown by Figure 3.12, include the following (grouped by the phase functions): setup, NMS, Sensor Processing, Data/Status Communications, and Power Management.

- Setup:
  - Automated premission readiness verification;
  - Deployment effectiveness and ease;

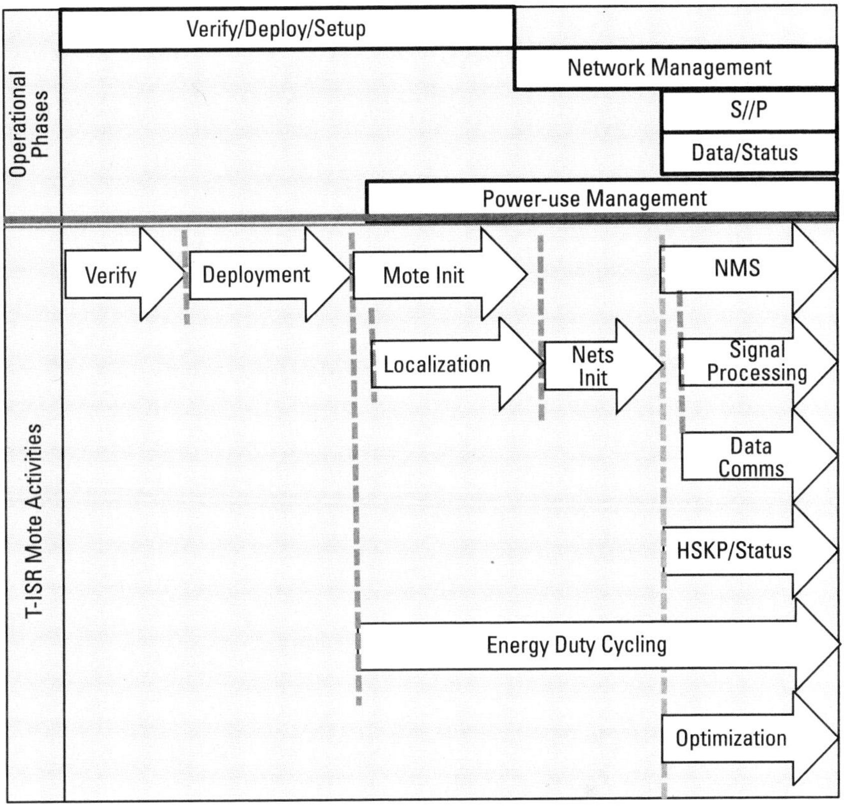

Figure 3.12 Alignment of WSN system operational phases with mote activities.

- Self-initialization;
- Localization.

- Network management system (NMS):
  - Self-organization;
  - Identification of nearest neighbor;
  - Network initialization
  - Routing optimization;
  - Network maintenance;
  - Self-healing.
- Sensor processing:
  - Data acquisition and conversions;
  - Real-time sensor signal processing.

- Data/status communications:
  - Data communications (exfiltration);
  - Secure communications and self-protection;
  - Mote housekeeping data acquisition and reporting.
- Power management:
  - Energy duty cycling;
  - Operation optimization.
- Standardization and legacy:
  - Interoperability with exfiltration relay capabilities;
  - Standardized message formats;
  - Seamless interface with communication infrastructure and mission operations centers.
- Physical attributes:
  1. Minimal size (volume) and weight;
  2. Environmental durability;
  3. Concealment.

Network management and power management phases work collaboratively to establish the WSN while striving to minimize power use through mote-level and motefield energy duty cycling. With the established network, operational sensor processing and data/status communications begin and continue until the WSN mote field is decommissioned or ceases to operate due to failure (or battery depletion). Standardization and legacy activities consider how WSN data is accepted externally from its mote field. WSN data must be delivered and successfully interpreted by a mission operations center (MOC) to make use of the hard-won data sets. Additionally, motes should meet remote sensing standards, such as IEEE1491-99, Standard for Harmonization of Internet of Things (IoT) Devices and Systems [33]. Physical attributes address physical appearance, constrain dimensions (for detectability and for weight), ease of deployment, and provide specifications that affect logistics.

3.4.1 Predeployment Considerations

Prior to any WSN system deployment, all subsystems, motes, relays, and communication systems are verified as being fully functional. As with any large-volume production line, not every mote can be tested [34]. Instead, as with large production runs, a statistically significant sample size (based on random selection) is used to estimate underlying failure rates of mote subsystems. Given the characterization of mote reliability based on low-cost fabrication, reliability analysis methods (Markov chains, Monte Carlo simulation, reliability block diagramming) can be applied to ascertain overall WSN system reliability. The network design expects a percentage of nodes to fail, ergo the self-healing capability exists and is required at the start of operations. Early in WSN predeployment, mote-level testing was shown of value as several mote subsystem failures have been observed. This was not surprising as the core objectives of WSN are to employ low-cost motes to operate in harsh environments. Using real-world WSN deployments, an investigation into sensor faults revealed an impact on the net performance and provided a background and means to characterize these failures [35]. Detailed analyses categorized fault types (e.g., permanent, intermittent, temporary/transient, and potential) [36] and subsystem failures [sensor, microcontroller system, radio transceiver, and power (battery)] [37].

In operating and using a WSN system, faults and/or degraded operation can be monitored as it pertains to mote middleware, sensor data flows and speed, and network management error(s) to determine the existence of faults, to provide isolation of possible cause(s), and to employ techniques to resolve incapacitated WSN components [38–40]. The reliability aspects of WSN nodes due to mote hardware failures, software errors, or RF communication link issues have been studied through simulations and analyses as well actual field demonstrations. In situ testing with early mote devices (2000–2010) was required due to hardware and software problems—recall that these devices were being built with cost constraints while answering a very challenging application of low-power, persistent sensor operation within only a few years of development.

3.4.2 Network Management System

Network management is highly reliant on the relative localization of a mote to all its nearest neighbors and to a path that allows a mote to connect to an exfiltration relay. With T-ISR applications, it is assumed that the selected WSN system is a multihop-capable system to allow for reliable system operation and deployment of large mote numbers allowing for long-term operation (fault tolerance). Significant to designing a successful WSN T-ISR system is inclusion of self-organizing behavior, which improves reliability, scalability, and availability of WSN systems through mote coordination and collaboration. Instead of requiring a centralized control over the network, self-healing networks operate via peer-to-peer processes wherein motes distinguish and connect with their nearest neighbors. This avoids the necessity for having robust connectivity to a centralized network management function. Linkage established at the local, nearest neighbor level propagates throughout the network, and widespread connectivity becomes an emergent behavior.

Scalability is addressed when the number of nodes in the system is to increase. Additional motes may be deployed due to attrition, to augment a current system with updated mote capability, and/or to expand AOI monitoring. For T-ISR in particular, the AOI typically involves large dimensions, which implies that large numbers of motes are required. (Also of note, these areas may or may not be contiguous.)

3.4.3 Sensor Signal Processing

As previously described, WSN motes typically employ multiple sensor modalities (see also Chapter 9) to support the reduction of false alarms and improve target classification capacity. As one result, signal-processing considerations must fit within WSN resources while satisfying basic objectives such as battery lifetime, latency, measurement accuracy, and sensor calibration. As the preponderance of simpler (less complex) WSN sensors are analog, analog filtering, gain control, and A/D processing, all must be implemented within the mote core.

Use of magnetometers, compact magnetoresistive devices, begins as analog signals (usually derived using a Wheatstone bridge arrangement) and requires instrumentation amplifier circuitry (employing a voltage divider function). With analog signal processing, voltage thresholds are instrumental in obtaining reliable analog levels. For WSN, threshold adjustments can be implemented via middleware to enable user-controlled amplifier bias. Optical sensors, both passive and especially active, require a large variety of signal processing depending on the components used. For example, PIR sensors require power control, power supply filtering, infrared optical detectors and attendant circuitry, active bandpass filters, a summing operational amplifier, and window comparators.

3.4.4 Data/Status Communications

Characteristics desired by a WSN include a high QoS, significant fault tolerance, network scalability, low power consumption, sufficient security, programmability, ease of maintenance, and low total costs. QoS (e.g., latency), operability, detection and remediation from network faults require sufficient instrumentation of system health monitoring. Knowing parameters such as the RF signal strength (RSSI), battery voltage, and active/inactive sensors can be combined to render a formidable capability to assess the health and status of a mote and of the overall WSN field as well.

3.4.5 Power Management

The selected communication protocols used in a WSN implementation directly influence power usage. However, the complete power budget requires an understanding of power distribution, operational profiles, and the underlying circuitry and approach implemented by mote hardware. Power management begins with mote initialization and formation of a WSN and is critical during the early setup stages to enforce duty cycling, sleep modes, and built-in functionality (mote hardware, firmware and middleware). A difficulty with Figure 3.12 is that operational phases require multiple interfaces with other activities to conduct the assigned tasking, such as power management requiring LP_MAC protocols as well as access to monitor (and control) sensor and transceiver operations

3.4.6 Standardization and Legacy

Standardization and compatibility are important factors in interconnecting a WSN motefield via a relay (exfiltration point). Injecting WSN-processed data into the GIG (DoDIN) requires adherence to standard message formatting and protocols, as well as adherence to interfaces associated with legacy communication and display systems. Legacy issues arise when considering the following:

Chapter 11 discusses T-ISR WSN test system architecture and provides examples of T-ISR WSN test systems and demonstrations.

- Performing bellringer activation for sophisticated sensors (e.g., UGS): How best to connect to an existing UGS capability (e.g., LP-SEIWG formatting)?
- Injection and display of processed WSN data at the mission control center for verified target detections, tracks, and classifications (e.g., symbology needs to adhere to MILSTD-2525D).
- Seamless integration of the WSN physical packaging with deployment mechanisms and/or platforms (dispensers, aircraft pods).
- Adherence to overarching security concerns (such as chip identification numbers, manufacturing data, and use of authentication protocols).

3.4.7 Physical Attributes

Recall that early in WSN development, motes were conceptualized as having dimensions on the order of "dust." The initial excitement to convey the WSN concept as a military application went far beyond what was then realizable. The use of a large dipole antenna, shiny surfaces, acoustic annunciators, and flashing lights (LEDs) detract from the usability (and acceptance) by military end users in the field (where maintaining a very low detection profile can mean life or death). Additionally, designing a mote to require hand emplacement and button or switch operations are considered nonstarters for T-ISR missions per Special Force personnel recruited to support the various field test and demonstrations (see Chapter 11). Those involved with T-ISR asset delivery and emplacement simply do not wish to have a sensor system that requires altering existing gear (e.g., removal of gloves, moving significant weights from mission start to the AOI, and time-consuming man-in-the-loop initialization and verification of correct operation). When considering T-ISR, much of what is taken to the field occurs by individual personnel and is carried in backpacks. Thus, personnel question the importance of every item they carry and must prioritize accordingly.

3.5 Cooperative (Tiered) Architecture

For T-ISR, WSN systems are expected to rely on relays to connect to worldwide data communications. This differs from internal relays within WSN networks that provide a higher level of processing as the data flows toward the top level of a hierarchical architecture [41]. The low-power, low-cost aspect of WSN nodes allows for dense sensor coverage but at a cost of communication range. To effectively connect with worldwide systems, such as DoDIN, WSN systems are reliant on exfiltration relays for operations controlled by remote mission operation center(s). WSN systems have frequently been tasked to work with sophisticated sensors (e.g., UGS), which depend on such relay capability. Exfiltration relays may serve a dual purpose by providing communication connectivity to capable sensors (e.g., imagers) and WSN motefield(s) simultaneously to end users. Through this connectivity, with appropriate control logic injected into the system, motes could be tasked to control operation of the energy-hungry sophisticated sensors without intervention from the MOC. In this way, a tiered sensor configuration would rely on the motefield tier as a bellringer or tripwire function. With this capability, operation of the more sophisticated (and energy-consuming) sensors can realize extended operational durations. Figure 3.13 presents a tiered T-ISR implementation that employs a WSN system to collaborate with a variety of UGS devices.

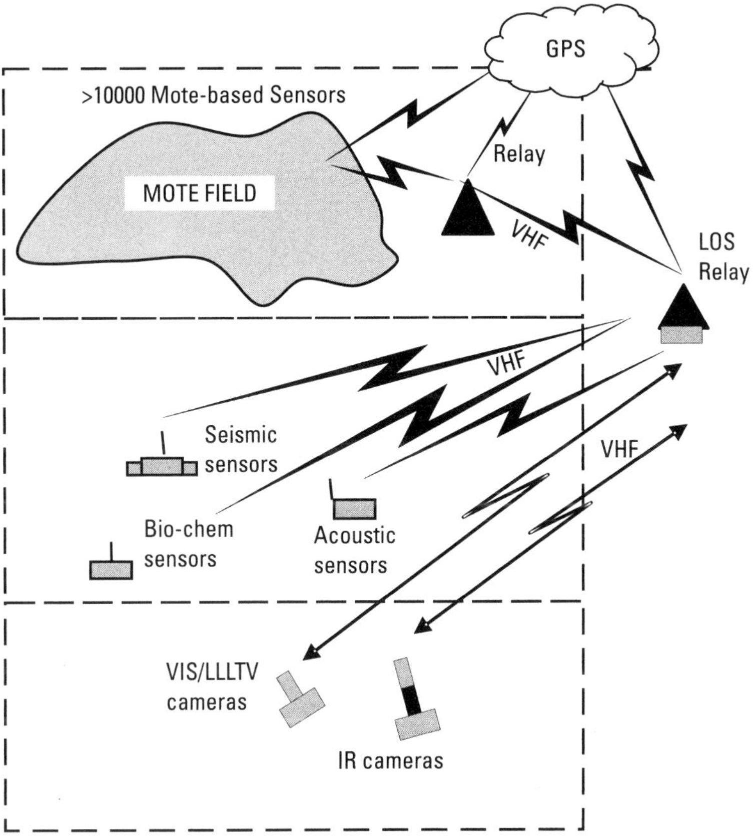

Figure 3.13 Tiered T-ISR system composed of multiple sensor tiers: WSN mote field, simple UGS, and complex UGS (imagers).

References

[1] Römer, K., "Tracking Real-World Phenomena with Smart Dust," *Wireless Sensor Networks*, EWSN 2004. Lecture Notes in Computer Science, Vol. 2920, Springer, 2004.

[2] https://usermanual.wiki/Sentilla/TMOTESKY/html, used by permission, Joe Polastre, Rob Szewczyk (UCB), email, June 29, 2019.

[3] TMOTESKY Tmote Sky User Manual tmote-sky-datasheet-102, Sentilla, 2005.

[4] Gajjar, S., et al., "Comparative Analysis of Wireless Sensor Network Motes," in *2014 International Conference on Signal Processing and Integrated Networks (SPIN),* Noida, 2014, pp. 426–431.

[5] Polastre, J., R. Szewczyk, and D. Culler, "Telos: Enabling Ultra-Low Power Wireless Research," *Fourth International Symposium on Information Processing in Sensor Networks*, 2005.

[6] Dutta, P., et al., "Design of a Wireless Sensor Network Platform for Detecting Rare, Random, and Ephemeral Events," *ACM/IEEE Information Processing in Sensor Networks*, 2005.

[7] List of wireless sensor nodes, https://en.wikipedia.org/wiki/List_of_wireless_sensor_nodes.

[8] Nain, N., and S. Vipparthi," Internet of Things and Connected Technologies," *4th International Conference on Internet of Things and Connected Technologies (ICIoTCT)*, 2019, 2020.

[9] "8-bit Atmel Microcontroller with 128KBytes In-System Programmable Flash," Atmel Corp, 2011.

[10] Buturuga, A., R. Constantinescu, and D. Stoichescu, "A Practical Approach to Microcontroller Performance Evaluation," *Proc. SPIE 10977, Advanced Topics in Optoelectronics, Microelectronics, and Nanotechnologies IX*, 2018.

[11] Jeffers, J., J. Reinders, and A. Sodani, Chapter 14 in *Intel Xeon Phi Processor High Performance Programming* (second edition), Knights Landing Edition, Elsevier, 2016.

[12] Farber, R., *CUDA Application Design and Development*, Elsevier, 2011.

[13] Amdahl, G., "Validity of the Single Processor Approach to Achieving Large Scale Computing Capability," *AFIPS Spring Joint Computer Conference*, 1967.

[14] Reddy, V., et al., "Operating Systems for Wireless Sensor Networks: A Survey," *Int. J. Sensor Network*, 2009.

[15] Farooq, M., and T. Kunz, "Operating Systems for Wireless Sensor Networks: A Survey," *Sensors*, 2011.

[16] Tan, S., B. Anh, and T. Nguyen, "Survey and Performance Evaluation of Real-Time Operating Systems (RTOS) for Small Microcontrollers," *IEEE Micro*, 2009.

[17] Hadim, S., and N. Mohamed, "Middleware: Middleware Challenges and Approaches for Wireless Sensor Networks," *Proceedings IEEE Computer Society*, 2006.

[18] Ajana, M., et al., "Middleware Architecture in WSN," in *Wireless Sensor and Mobile Ad-Hoc Networks: Vehicular and Space Applications*, Springer, 2015.

[19] Sohraby K., D. Minoli, and T. Znati, "Chapter 8: Middleware for Wireless Sensor Networks," in *Wireless Sensor Networks Technology, Protocols, and Applications*, John Wiley & Sons, 2007.

[20] Romer, K., O. Kasten, and M. Friedemann, "Middleware Challenges for Wireless Sensor Networks," *Mobile Computing and Communications Review*, 2002.

[21] Chen, Y., Y. Eldar, and A. Goldsmith, "Shannon Meets Nyquist: Capacity of Sampled Gaussian Channels," *IEEE Transactions on Information Theory*, 2013.

[22] Stern, H., "Bandpass Sampling—An Opportunity to Stress The Importance of In-Depth Understanding," *American Journal of Engineering Education*, 2010.

[23] Schroeder, D., "Adaptive Low-Power Analog/Digital Converters for Wireless Sensor Networks," *IEEE: Third International Workshop on Intelligent Solutions in Embedded Systems*, 2005.

[24] Davenport, M., et al., "Introduction to Compressed Sensing," *Compressed Sensing: Theory and Applications*, Cambridge University Press, 2012.

[25] https://www.ni.com/en-us/innovations/white-papers/10/benefits-of-delta-sigma-analog-to-digital-conversion.html.

[26] Wu, C., and J. Yuan, "A 12-Bit, 300-MS/s Single-Channel Pipelined-SAR ADC with an Open-Loop," *IEEE Journal of Solid-State Circuits*, 2019.

[27] https://www.maximintegrated.com/en/design/technical-documents/tutorials/1/1041.html.

[28] Zahrai, S., and M. Onabajo, "Review of Analog-To-Digital Conversion Characteristics and Design Considerations for the Creation of Power-Efficient Hybrid Data Converters," *Journal of Low Power Electronics and Applications—Open Access Journal*, 2018.

[29] Marcelloni, F., and M. Vecchio, "A Simple Algorithm for Data Compression in Wireless Sensor Networks," *IEEE Communications Letters*, 2008.

[30] Razak, Z., A. Erdogan, and T. Arslan, "An Adaptive Algorithm for Reconfigurable Analog-to-Digital Converters," *2010 NASA/ESA Conference on Adaptive Hardware and Systems*, 2010.

[31] Shaw, J., "Radiometry and the Friis Transmission Equation," *Am. J. Phys.*, 2013.

[32] U.S. Food and Drug Administration, "Wireless Medical Devices," 2018.

[33] IEEE P1451-99—Standard for Harmonization of Internet of Things (IoT) Devices and Systems, 2016.

[34] Jiang, P., "A New Method for Node Fault Detection in Wireless Sensor Networks," *Sensors*, 2009.

[35] Sharma, A., L. Golubchik, and R. Govindan, "Sensor Faults: Detection Methods and Prevalence in Real-World Datasets," *ACM Trans. Sensor Network*, 2010.

[36] Khan, M., "Fault Management in Wireless Sensor Network," *GESJ: Computer Science and Telecommunications*, 2013.

[37] Deif, D., and Y. Gadallah, "A Comprehensive Wireless Sensor Network Reliability Metric for Critical Internet of Things Applications," *EURASIP Journal on Wireless Communications and Networking*, 2017.

[38] Rehena, Z., et al., "Detection of Node Failure in Wireless Sensor Networks," *IEEE Conference: Applications and Innovations in Mobile Computing (AIMoC)*, 2014.

[39] Shaikh, R., and A. Sayed, "Sensor Node Failure Detector in Wireless Sensor Network: A Survey," *International Journal of Emerging Technology in Computer Science & Electronics (IJETCSE)*, 2015.

[40] Mahapatro, A., and P. Khilar, "Detection of Node Failure in Wireless Image Sensor Networks," *ISRN Sensor Networks*, 2012.

[41] Lamont, L.,et al., "Tiered Wireless Sensor Network Architecture for Military Surveillance Applications," *SENSORCOMM 2011 : The Fifth International Conference on Sensor Technology*, 2011.

Selected Bibliography

Arora, A., et al., "A Line in the Sand: A Wireless Sensor Network for Target Detection, Classification, and Tracking," *Computer Networks Journal*, Vol. 46, No. 5, 2004, pp. 605–634.

Arora, A., et al., "ExScal: Elements of an Extreme Scale Wireless Sensor Network, Embedded and Real-Time Computing Systems and Applications (RTCSA)," in *IEEE International Conference on (Formerly Real-Time Computing Systems and Applications, International Workshop on)*, September 2005.

Avilés-López, E., and J. Antonio García-Macías, "TinySOA: A Service-Oriented Architecture for Wireless Sensor Networks," *Service Oriented Computing and Applications*, Vol. 3, No. 2, pp. 99–108.

Blanckenstein, J., J. Klaue, and H. Karl, "A Survey of Low-Power Transceivers and their Applications, *IEEE Circuits and Systems Magazine*, Vol. 15, No. 3, 2015, pp. 6–17.

Chen, Y., C. -N. Chuah, and Q. Zhao, "Sensor Placement for Maximizing Lifetime per Unit Cost in Wireless Sensor Networks," *IEEE MILCOM 2005–2005 IEEE Military* Communications Conference, DOI: 10.1109/MILCOM.2005.1605825.

Chong, C.- Y., and S. P. Kumar, "Sensor Networks: Evolution, Opportunities, and Challenges," *Proceedings of the IEEE*, Vol. 91, No. 8, August 2003.

Defense Information Systems Network (DISN) Connection Process Guide (CPG), Version 5.1, Defense Information Systems Agency (DISN), Risk Management Executive (RME), Risk Adjudication and Connection Division (RE4), September 2016.

FACT FILE, A Compendium of DARPA Programs, Defense Advanced Research Projects Agency, Revision 1, August 2003.

FACT FILE A Compendium of DARPA Programs, Defense Advanced Research Projects Agency, August 2003, p. 39.

Distributed Sensor Nets, Information Processing Techniques Office, Defense Advanced Research Projects Agency (DARPA), Proceedings of a Workshop, Carnegie-Mellon

University. Pittsburgh, PA, Session 1: System Organization, Rpt AD-A143 691, December 1978.

Gay, D., et al., "The nesC Language: A Holistic Approach to Networked Embedded Systems," *Proceedings of Programming Language Design and Implementation (PLDI)*, June 2003.

Hac, A., "Security Protocols for Wireless Sensor Networks," in *Wireless Sensor Network Designs*, John Wiley & Sons, 2003.

Happonen, A., *Low Power Design for Wireless Sensor Networks*, Springer, 2012.

"History of Microelectromechanical Systems (MEMS)," Southwest Center for Microsystems Education and The Regents of University of New Mexico, 2008–2010, Southwest Center for Microsystems Education (SCME), September 2013.

https://www.businesswire.com/news/home/20060123005247/en/Moteiv-Corporations-Tmote-Sky-Mote-Platform-Receives.

https://www.crunchbase.com/organization/sentilla#section-overview.

http://www2.ece.ohio-state.edu/~bibyk/ee582/XscaleMote.pdf.

https://www.slideshare.net/KailasKharse/difference-between-cisc-risc-harward-vonneuman.

Ibarra-Esquer, J. E., et al., "Tracking the Evolution of the Internet of Things Concept Across Different Application Domains," *Sensors* (Review), Vol. 17, p. 1379, doi:10.3390/s17061379, 2017.

Lee, S., et al., "Intelligent Parking Lot Application Using Wireless Sensor Networks," in *2008 International Symposium on Collaborative Technologies and Systems*, IEEE, 2008, pp. 48–57.

Madhuri, V. V., S. Umar, and P. Veeraveni, "A Study on Smart Dust (MOTE) Technology," *IJCSET*, Vol. 3, No. 3, March 2013, pp. 124–128.

Minkoff, J., *Signals, Noise, & Active Sensors*, equation (6.50), p. 13, Wiley & Sons, 1992.

Omiyi, E., K. Bür, and Y. Yang, A Technical Survey of Wireless Sensor Network Platforms, Devices, and Testbeds, *A Report for the Airbus/ESPRC Active Aircraft Project EP/F004532/1: Efficient and Reliable Wireless Communication Algorithms for Active Flow Control and Skin Friction Drag Reduction*, Lund University, March 19, 2008.

Omiyi, P., K. Bür, and Y. YangA Technical Survey of Wireless Sensor Network Platforms, Devices and Test Beds, Technical Report; Vol. EP/F004532/1 AIRBUS-01-190308), University College London, 2008.

Polastre, J., R. Szewczyk, and D. Culler, Computer Science Department TELOS: Enabling Ultra-Low Power Wireless Research, University of California, Berkeley, CA.

Rabaey, J., et al., PICORADIO: Communications/Computation Piconodes for Sensor Networks, Air Force Research Laboratory, AFRL-VS-TR-2003-1013, 2003.

Saladino, A., and S. Mitchell, DISA Services Course Executive Overview–AFCEA, June 14, 2017.

Silva, A., M. Liu, and M. Moghaddam, "Power-Management Techniques for Wireless Sensor Networks and Similar Low-Power Communication Devices Based on Nonrechargeable Batteries," *Journal of Computer Networks and Communications*, Vol. 2012, Article ID 757291, 2012.

Stollon, N., *On-Chip Instrumentation: Design and Debug for Systems on Chip*, Springer, 2011.

Tether, T., Statement to Science Committee, U.S. House of Representative Multidisciplinary Research, May 2005.

Wang, Q., and I. Balasingham, "Wireless Sensor Networks—An Introduction," *Computer Science*, DOI:10.5772/13225, 2010.

Weber, W., et al., "TinyOS: An Operating System for Sensor Networks," in *Ambient Intelligence*, Springer, December 2004, DOI: 10.1007/3-540-27139-2_7.

4

Ad Hoc Network Technology

WSN systems are configured as wireless networks that assume no a priori or static infrastructure. WSN motes are designed to be distributed in an impromptu manner with all nodes participating in the formation of a peer-to-peer network. Such a network is considered an ad hoc network, which is characterized by autonomous node status, allowing for continuous new user discovery, network-wide monitoring and resolution management, and the efficient performance of unicast, multicast and broadcast of messages to network users. In addition to WSN, other ad hoc networks exist and have been employed to support T-ISR functions. Most notable of these is the mobile ad hoc network (MANET). A MANET (pronounced ma-ney [1]) is a collection of autonomous mobile devices (e.g., handheld software-controlled radios, laptops, smart phones, and wireless sensors) that communicate to one another via wireless links under control established through cooperative network algorithms (i.e., protocols) [2].

All ad hoc networks, especially WSN and MANET systems, benefit from the enormous body of research and development of numerous technologies presented in Chapter 3. Of these, packet-switched communications is the core concept that propelled ad hoc networks to the forefront of distributed data collection and dissemination. Adapting packet-switched communications to wireless ad hoc networks meant developing responsive protocols and interfaces that take in account fundamental objectives associated with how these networks are to be used, namely:

- Wireless connectivity;
- Autonomous initialization, operation, and control;
- Minimization of energy use;
- Minimization of processing complexity (processor capabilities and memory size);
- Reliable and rapid message transmission.

Packet switching was initially developed to increase reliability (as we will describe in Section 4.1) when using wired networks that provide a static and well-defined infrastructure not unlike that of telephony systems. Wireless connectivity introduced a drastic increase of link variables, including intermittent connectivity, lost links, rapid and frequent addition and subtraction of network nodes, and loss of physical security. Ad hoc networks are designed to operate autonomously, performing self-initialization, self-organization, and in the case of link/node faults(s), self-healing where rerouting occurs as required. Incorporating such extremely valued capabilities required astute design and implementation of network functions in the areas of MAC, network routing, and operational management. As ad hoc networks avert centralized control, it is the nodes that enact processes that affect a network.

Every stage of network operations requires energy including: node initialization, self-localization, setup of wireless links, and network management (operations, monitoring and maintenance). Ad hoc nodes are designed to be self-contained and as such are dependent on local power sources. For WSN, this may be nothing more than small-volume battery packs. Ad hoc static nodes have limited processing resources, especially for nodes designed to be low-cost and low-power and to have a minimal volume profile. MANET nodes fair better than WSN nodes. For MANET, being mobile indicates dependency on either power provided specifically to the node, or, power provided via the mobile platform (e.g., unmanned and manned vehicles). As a result, MANET nodes may be designed with significantly more processor complexity than those associated with WSN; however, MANET nodes are still subject to SWaP2 constraints to allow for their mobility.

As described earlier, deployment of ad hoc networks occurs to acquire timely and accurate sensor measurements. Being wirelessly connected introduces a probability of neither occurring. As a result, adaptation of packet-switched communications to wireless networks rightfully considers preceding approaches and designs that were successfully used by systems to conduct initialization and management of the wired networks regarding message integrity, congestion, and latency throughout the network. This chapter considers principles involved to construct and operate communication networks based on packet-switching algorithms having applicability to ad hoc networks. The underlying performance modeling of packet-switched networks draws from mathematical descriptions associated with Poisson point processes. (See Chapter 5 for the underlying mathematical models.) Additionally, this chapter describes the use of the prominent packet network models, open systems interconnection (OSI) reference model, and DoD model (TCP/IP), with in-depth discussions drawing on significant literature that addresses design, development, testing, and implementation of packet networks. In addition, the chapter presents the background of MANET along with previews of designs for MANET protocols. The chapter concludes with a comparison of WSN and MANET systems and a review of ad hoc network issues and vulnerabilities.

4.1 Overview: Packet Switching

Formatting data for reliable transmission over a digital network is the raison d'être for packet switching, the central concept behind modern digital communication networks. Packet-switched network design and research began in the 1950s, with Paul Baran at the RAND Corporation (United States) [3] and Donald Davies at the National Physics Laboratory (United Kingdom) [4, 5]. Packet-switched networks were designed to accommodate networks that share network resources and whose links exhibit dynamic behavior, meaning that their network pathways are subject to congestion, intermittent link connectivity, or loss of a node. Additionally, packet-network designs assume the existence of message errors or lost messages, which would indicate a need to retransmit data. Figure 4.1 illustrates message segmentation, the process by which packet-switched architecture helps with dynamic path routing and

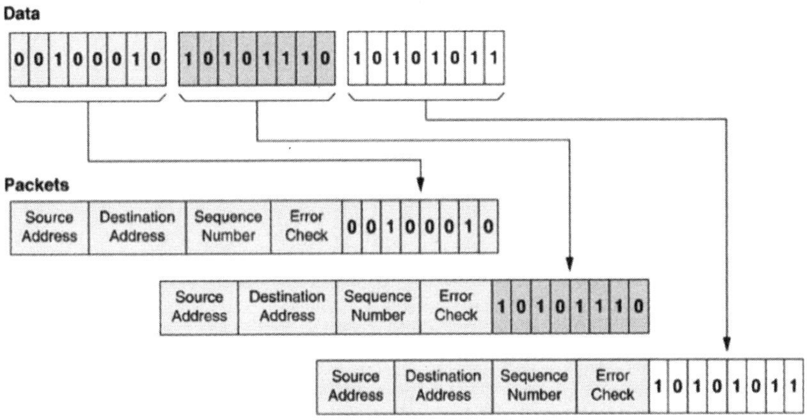

Figure 4.1 Simplified example of data segmentation of data for transmission.

recovery from message errors and losses. Actual packet formatting and processing depends upon select algorithms (encapsulated within the data link layer protocol) that are selected to match characteristics of the network.

Segmentation enables digital communication systems to send messages as a series of data packets, each shorter than the original data message. Through routing of these smaller-sized packets, from source to destination, improvements to network availability, delivery time (latency), and recovery from corrupted or dropped messages can be realized. The packetized data consists of network header information (which is protocol-driven) and data (denoted as the payload). The header provides metadata that directs the message to intended receiver(s), includes error correction coding [e.g., check redundancy check (CRC)] [6, 7], a sequence number (to enable reassembly of the message), and additional information according to network requirements and the associated protocol in use. Figure 4.1 shows a basic header appended to the segmented data.

Message errors are affirmed through agreement of both source and destination of an exception event using transport layer protocol to insert an error-correction code (e.g., CRC) and a sequence value in the message header. Corrupted messages, wherein bits have been altered, are handled via the error checking. The determination that messages have been lost makes use of the inserted sequence number, which monotonically increases with transmitted messages. Network availability is increased because network messaging is not constrained to any particular data type. Messages can relay any information (as encoded bits) regardless of type (e.g., multimedia, commands, and sensor measurements). In addition, dynamically controlled message flow and congestion mitigation is possible through packet routing. Flow control is the process of managing transmission data rates between source and destination nodes to prevent a high-rate data source from overwhelming a slow destination. Congestion mitigation and control monitor and throttle packet rates of data messages entering network paths to avoid congestive collapse of paths from excessive packet traffic.

It is important to realize that flow control is not the same as congestion control, although there is an overlap between mechanisms used by protocols that provide both services. Congestion control addresses the problem of a node overwhelming the link (i.e., the pathway between two nodes), while flow control focuses on the end node.

4.1.1 Flow Control

Confusion exists in the literature regarding the association of a particular algorithm to only one of these control mechanisms. For example, stop-and-wait algorithms, and the associated automatic repeat request (ARQ) variant, are used for both flow and congestion control. Flow control is node-based and attempts to match the source-sending rate with the destination receiving speed. A basic approach to flow control employs a simple stop-and-wait process [8, 9]. Here, the source node is enforced to wait until the destination node returns an acknowledgement (ACK) message for each transmitted message prior to sending the next data message. Unfortunately, this waiting period ties up network links with message traffic, especially if the propagation delay is much larger than the transmission delay. As a result, the source node is virtually wasting time (therefore, wasting energy) by not continuously transmitting messages. Additionally, if the ACK message is lost, the wait process halts transmission but the source transceiver remains active without data flowing.

To solve this deadlock occurrence where an ACK message is lost and the source and destination are waiting on one another to send a synchronization message, the source can employ a watchdog timer. The timer is reset and started with each message transmission. If an ACK is not received and a timeout occurs, the source can assume that the ACK message was lost and retransmit the packets assumed lost. This process ensures that data flow restarts in the advent of deadlock and minimizes the wait time for an ACK to be considered lost. A variation on the stop-and-wait flow control approach is to set up the transfer where an ACK is not required for each transfer but expected during the established time period associated with a source-based watchdog timer. Within each ACK message, the destination informs the source what the current available buffer size is. At each start of a transmit time window, the source transmits a number of frames of data. If the timer expires before an ACK is received, the source ceases transmission of data frames. The transmit uses a previous ACK message to determine if the rate of data messages needs to be adjusted downward using the destination's buffer information contained in the received ACK message. The source then resends a subset of interrupted data messages that meet the destination buffer size with the expectation that a successful ACK will arrive before timeout occurs. The process continues as the source transmits the mutually agreed number of frames (N) that the destination indicates it can handle. This provides a compromise to the basic stop-and-go, with one transfer per one ACK message. Here, the source is waiting every N frames for an ACK. Although there is N:1 improvement over the basic stop-and-wait (per message) approach, there is still a period (the watchdog timer duration) that wastes time (and energy).

Another approach to flow control is to minimize wait time through the use of a sliding window [10]. Using a sliding window, the destination stores receive data in a receive buffer and returns an ACK message to the source.

Flow control determines the number of bytes a source can transmit to the destination by having access to source and destination buffer sizes. The source cannot send more bytes than buffer availability at the destination; instead, the source node must wait to continue to send more data until all the bytes in the current transmit buffer have been received as signaled by an ACK message is received from the destination. Similar to stop-and-wait, each ACK message informs the source of the current size of the current destination buffer. If the destination buffer is full, the ACK message sets the window size to zero, and the source node must wait before transmitting more data. After the responsible application at the destination pulls data from the receiving buffer, the destination node transmits an ACK message to the source, indicating a window size equivalent to the data pulled out, and the source node can restart sending messages, not to exceed the updated window size.

4.1.2 Congestion Control

Congestion control (network congestion avoidance) is a process implemented to avoid loss of use for a link due to an excessive packet rate entering the link. This control addresses the problem of a node overwhelming a network by monitoring message traffic via the use of specific ACK messages to indicate *current* link conditions between source and destination. Congestion-control approaches share algorithms employed in operational flow control *with* information extracted from ACK messages and timers to throttle the source nodes from overloading the network pathways. Two parallel activities are employed to avoid congestion:

1. The use of routing nodes that sense data flow rates that possess a capability to redirect packets along an alternative paths to the intended destination.
2. The use of timers with information gleaned from ACK messages to throttle data flow from the source node.

Through constant vigilance of traffic flow, these parallel activities can be applied to control data flow to match the capacity afforded by the path bandwidth. As with flow control, significant research has focused on congestion control and evaluated for use within packet-switched communications. Examples of timer-based algorithms exist and have been implemented [11] in combination with network protocols to effectively implement congestion control through slow-start, fast retransmit, and fast recovery [12]. An overall network setup process to initiate network operation and maintain congestion control uses slow-start at the outset and subsequently achieves congestion avoidance using fast retransmit within the protocol, such as that realized within the transport layer protocol (i.e., the well-known TCP). When data flow monitoring notes that ACK messages are being lost, the transport layer creates a timeout, and net operations restart the slow-start mechanism and invoke the phased process again. Figure 4.2 illustrates the data flow phases, slow-start and collision avoidance.

As with other approaches to minimizing congestion, slow-start supports congestion avoidance by throttling the amount of data transmitted. This approach negotiates the connection between a source and destination by defining the amount of data that can be transmitted with each packet and slowly increases the amount of data until network capacity is reached. This ensures that as much data is transmitted as possible without clogging the network, which optimizes header overhead-to-payload bits. The source initial packet contains an estimate of the congestion window, which is determined based on the source maximum window. The destination acknowledges the packet and responds with its own window size. If the destination fails to respond, the source assumes that the data size was excessive and ceases to transmit. During successful transmissions, the source increases the window size for the next packet when an ACK is received. The window size gradually increases until the destination can no longer acknowledge each packet or until the transmitting node window limit is reached. With the limit measured, congestion control is completed, and the aforementioned flow control processes handle data rates.

Fast retransmit operates on duplicate ACK messages (dupack). As the network operates, the network management monitors transmitted messages

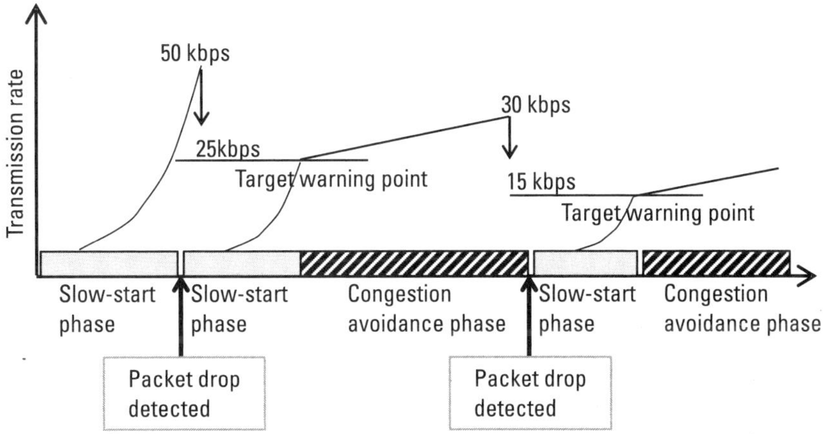

Figure 4.2 Slow-start and fast retransmit phases for congestion avoidance.

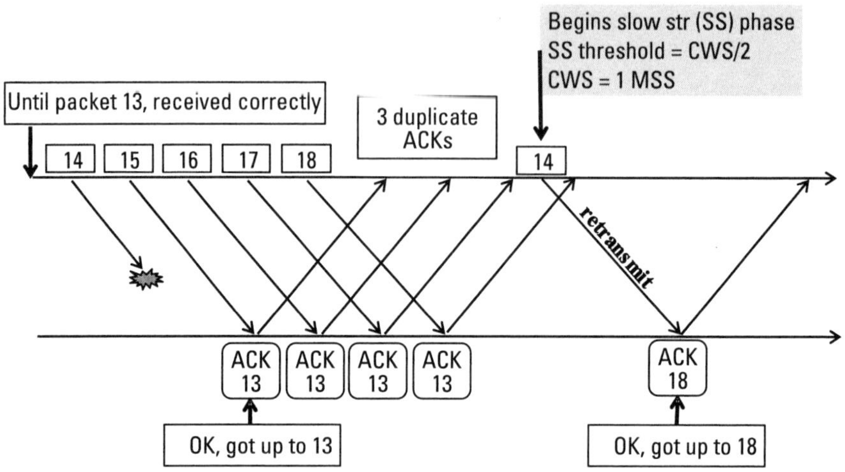

Figure 4.3 Fast retransmit timeline; example with packet 14 missing.

and associated ACK messages between source and destination nodes. If a transmitted message is dropped, the ACK messages associated with the last successfully received message will continue to resend the same ACK message (with an $m - 1$ index value, m being correlated to the lost message) until the source resends the lost message. This is signaled to the source by the existence of duplicate ACK messages. At this point the source resends the dropped message, and the destination promptly updates its ACK index to that which has been received. Figure 4.3 illustrates the timeline for slow-start and fast retransmits.

Fast recovery is an improvement over fast retransmit congestion control. Research has indicated that if congestion window size is increased beyond that with fast retransmit during collision avoidance mode that the transport protocol would operate at a faster rate than that set with fast retransmit alone. With fast recovery, the congestion window (CWS) is set to the saturation threshold plus 3 maximum segment size (MSS), versus the 1 MSS used by the fast retransmit.

4.1.3 Error Control

ARQ [13–15] represents a group of error control protocols used in the transmission of data over noisy or unreliable communication network, and applies well to ad hoc networks. These protocols reside in the data link layer (DLL) and transport layer of the OSI reference model and provide for automatic retransmission of frames that are corrupted or lost during transmission. ARQ is also referred to as a protocol using positive acknowledgment with retransmission (PAR). With these protocols, the destination sends an acknowledgment message back to the source if it receives a frame correctly. If the source does not receive the ACK of a transmitted frame before a specified period of time, a timeout occurs, and the source assumes that the frame has been corrupted, or lost, during transit. At this point, the source retransmits the frame, and the process repeats until the correct lost frame is transmitted.

Figure 4.4 illustrates error control (ARQ) protocols for packet-switched networks, including the flow control basic process of stop-and-wait, shown in Figure 4.4(a). Through receipt of an ACK message, stop-and-wait conducts error detection at a basic level. Using the stop-and-wait ARQ, the source now employs a counter, which starts with each transmission, as shown in Figure 4.4(b). If an ACK arrives during the counter countdown, the source continues to transmit the next data frame. If the counter hits a timeout before an ACK is received, the source assumes data was lost and resends the data frame (and restart the timer). The stop-and-wait ARQ mechanism does not optimally use resources. As with flow control, the source waits and is idle until an ACK is received. With a modified method, the Go-Back-N ARQ, both source and destination nodes maintain a window with the source window sized to allow transmission of multiple frames without receiving the acknowledgment of the previous ones. For flow control this approach is a variation of the stop-and-wait approach.

With Go-Back-N ARQ, the destination window enables the receiver to receive multiple frames and acknowledge them. The destination node keeps track of incoming sequence number. When the sender node sends all the frames in window, it checks up to what sequence number it has received positive acknowledgment, and if all are correctly acknowledged, the source continues to send the next set of frames. If the sender finds that it has received not-acknowledged (NAK) or has not received any ACK for a particular frame, it retransmits all the frames, starting with that associated with the first lost ACK. Figure 4.4(c) illustrates the Go-Back-N ARQ protocol. The use of the Go-back-N ARQ protocol assumes that the destination node cannot provide sufficient buffer space for the window size and has to process each frame upon arrival. This enforces the source to retransmit all frames that were not acknowledged. To selectively send only the dropped data frame(s) employs the selective-repeat ARQ protocol, shown in Figure 4.4(d), which tracks sequence numbers of the successfully delivered messages. The destination maintains the frames in memory and sends NAK only for frames missing or corrupted. The source can then selectively send only the frame for which a NAK message was received (for Figure 4.4(d), being ACK2).

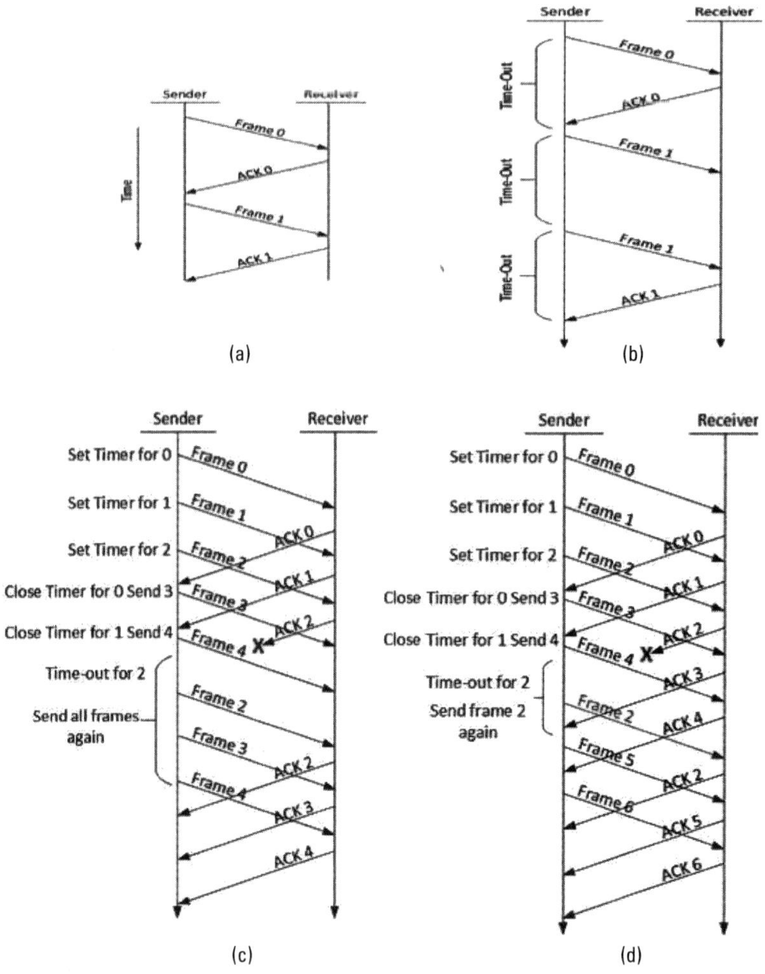

Figure 4.4 ARQ methods to approach error (dropped message) control. (a) The basic stop-and-wait process. Using stop-and-wait ARQ, the source now employs a counter (b), which starts with each transmission. (c) The Go-Back-N ARQ protocol, which assumes the destination node cannot provide sufficient buffer space and has to process each frame upon arrival. (d) The selective-repeat ARQ protocol selectively sends only dropped data frames.

In summary, using a packet-switched architecture with available protocols provides numerous advantages, including:

- Increase of network efficiency by eliminating the need to implement different networks to support different data types (voice, commands, data, and video);

- Increase of usable bandwidth as multiple logical circuits use the same physical links to send messages over the same links;
- Increase of reliability and survivability due to providing multiple paths throughout the network, origin to destination(s);
- Guarantee of accurate data delivery to intended destinations.

4.2 Basic Network Modeling Using the Poisson Distribution

Many approaches to flow and congestion control hinge on assuming that data traffic (more specifically, packet arrivals) can be represented as a Poisson process [16]. Not surprising, phone calls, traffic [17], and web hits are memory-less and are considered Poisson processes. Poisson processes exhibit independence of events with one another, with an average arrival rate of events (e.g., WSN messages) that remain constant, and the exclusion of two arrival events occurring simultaneously. The usual approach is to set up and evaluate an ad hoc network by applying the Poisson distribution mass function as a model of performance. With this model, it is the values of parameters associated with the process, namely the average rate of net messages (arrival events) and the reception time frame (denoted as timeframe), determine the probability, $P(k)$, of having k messages arrive within a time period, T, as:

$$P(k) = e^{-\frac{\text{events}}{\text{timeframe}} * T} * \frac{\left(\frac{\text{events}}{\text{timeframe}} * T\right)^k}{k!} \tag{4.1}$$

Denoting the average number of events per timeframe as λ, (4.1) becomes the more familiar form:

$$P(k) = e^{-\lambda} * \frac{\lambda^k}{k!} \tag{4.2}$$

With 4.2, here we used timeframe to define λ to indicate the duration involved when computing probability values for the net traffic throughput and other characteristics. As a simple example of the use of the Poisson distribution for network analysis, consider that the pioneering wireless packet-switched network Additive Links On-line Hawaii Area network (ALOHAnet or simply ALOHA) developed and used a medium access protocol that assumed attempted transmissions by user radio stations per any time frame (λ) followed a Poisson distribution.

To consider the throughput capability of the ALOHA network, (4.2) is used to estimate probability over a time frame, ΔT. Exactly one node transmits during a time frame, $t = 0$ to ΔT, and if no node tries to transmit during the next frame, $t = \Delta T$ to $2\Delta T$, the transmission attempt will be successful. Extending this to a subsequent frame, $2n\Delta T$ to $(2n + 1)\Delta T$, if a node attempts to transmit during this frame, and not the following frame, $(2n + 1)\Delta T$ to $(2n + 2)\Delta T$, then a message can be successfully transmitted. With ALOHA, nodes can randomly begin to transmit (no sense checking is conducted) at any time. A decision by a node to transmit data is independent of any other node. Assuming an average number of nodes (represented as λ) attempting to transmit during a timeframe, is represented as λ, we can apply (4.2) for an attempted transmission followed by an absence of transmissions as the joint (independent) probability:

$$P\{success\} = P\{1\}P\{0\} = \left(e^{-\lambda} * \lambda\right) * \left(e^{-\lambda}\right) = \lambda e^{-2\lambda} \tag{4.3}$$

Equation (4.3) presents the throughput expected for an unaltered ALOHA system. A slotted ALOHA version [18] exists and its throughput operates as

$$P\{success\} = \lambda e^{-\lambda} \tag{4.4}$$

The Poisson process can be applied to various attributes of ad hoc network design and evaluation [19]. Poisson point processes are employed in describing message arrival rates, node locations, coverage capability [20], and connectivity for ad hoc networks. This approach allows refined modeling and evaluation by breaking away from the fixed-disk assumption wherein all wireless nodes transmit isotropically, and all nodes within a RF-range successfully connect to one another. Terrain and numerous physical characteristics at the RF wavelength (e.g., multipath) affect the beam pattern associated with each node differently [21]. Noise sources and interference also work to render various connections to be fraught with errors and/or lost packets, both requiring retransmission of corrupted (lost) datagrams.

An excellent foundation in modeling connectivity of randomly distributed nodes following a Poisson point process with fixed density per unit area, was originally developed with Gilbert's 1961 work on random plane networks (currently referred to as random geometric graphs) [22]. Through application of what is known as continuum percolation theory, Gilbert's work led to a mathematical approach of how to determine network connectivity over time (generational propagation), information-carrying capability, and routing.

Discussion of the critical role of the Poisson point process in the context of network modeling is well-documented elsewhere for a randomly distributed network literature [23]. A final note concerning (4.2) to (4.4) is that numerical processing may result in issues with overflow when evaluating the λ^k and k! terms. Also, when computing the (λ^k)/k!, as well as with the $(e^{-\lambda} * \lambda^k)$/k! ratio, rounding errors may occur. To afford a numerically stable solution in using this equation, the Stirling formula for the gamma function can be used, as:

$$\Pr\{k\} = \exp\left[k\ln(\lambda) - \lambda - \ln(\Gamma(k+1))\right] \tag{4.5}$$

4.3 Standards: The OSI Reference Model

Packet-switched networks were developed to take advantage of the benefits of packetized data messaging; however, early in their development it became obvious that an overarching model of the network had to be established to manage the functionality and interfaces involved. As the growth and complexity of computer-to-computer networks expanded, a means to specify network functionality with reliance on standard services became obvious. Segregation of functions into layers was pursued to support the development of independent protocols as layered services whose functionality depends on lower layers for support. Layered functions were not constrained to depend upon one protocol to implement a layer; instead, the expectation was that a series of protocols would exist at each layer, each depending upon lower protocols for service.

In pursuit of an encompassing network model to support this layered approach, the International Organization for Standardization (ISO) began a decade-long process in the 1970s to develop general standards and methods of networking; simultaneously, work was occurring at the International Telegraph and Telephone Consultative Committee (CCITT) (with the acronym derived from the French language, Comité Consultatif International Téléphonique et Télégraphique). Results were combined and a draft network model product was revealed in 1978; with refinement and modifications, this draft became the OSI reference model in 1984. A core objective of the OSI model is support of interoperability of diverse communication systems through application of standardized communication protocols. OSI layers were defined as individual functions that operate independently of lower or upper network layers through the use of well-defined protocols. Table 4.1 presents the seven-layer OSI model, along with layer functions and nomenclature used to refer to the data associated with each layer.

Table 4.1
OSI Seven-Layer Model, Layers and Summary of Objectives

OSI Layer	Designation (Title)	Information Description	Objectives/Purpose
7	Application	Data stream	Supports application and end-user processes. Communication partners are identified, QoS identified, user authentication and privacy are considered, and any constraints on data syntax are identified. Everything at this layer is application-specific. This layer provides application services for file transfers, e-mail, and other network software services.
6	Presentation		Provides independence from differences in data representation (e.g., encryption) by translating from application to network format, and vice versa. This layer formats and encrypts data to be sent across a network, providing freedom from compatibility problems. It is sometimes called the syntax layer.
5	Session		Establishes, manages, and terminates connections between applications. The session layer sets up, coordinates, and terminates conversations, exchanges, and dialogues between the applications at each end. It deals with session and connection coordination.
4	Transport	Segment	Provides transparent transfer of data between end systems, or hosts, and is responsible for end-to-end error recovery and flow control. It ensures complete data transfer.
3	Network	Packet [1] datagram [2]	Provides switching and routing technologies, creating logical paths, known as virtual circuits, for transmitting data from node to node. Routing and forwarding are functions of this layer, as well as addressing, internetworking, error handling, congestion control, and packet sequencing.
2	Data link	Frame	Data packets encoded and decoded into bits. Furnishes transmission protocol knowledge, management and handles errors at the physical layer, flow control, and frame synchronization. Data link layer sublayers are the media access control and logical link control (LLC) layers. MAC controls network access to the data and permission to transmit. LLC controls frame synchronization, flow control, and error checking.
1	Physical	Encoded bits	This layer conveys the bit stream—electrical impulse, light or radio signal—through the network at the electrical and mechanical level. It provides the hardware means of sending and receiving data on a carrier, including defining cables, cards, and physical aspects.

The OSI reference model [24, 25] is purely a theoretical model; however, this model has served to convey network functionality, define interfaces and has aided in the presentation of numerous protocols that address each network functions. The OSI reference model helps to define terminology and to clarify protocols that define the following:

- The manner in which physical transmission media is arranged and connected;
- Interactive processes and methods that devices employ to communicate between each other;
- Algorithm(s) to ensure that devices maintain correct data flow rate;
- A useful approach for protocol designers to conceptualize network components to demonstrate how they fit together within a network;
- A process to inform devices when to, or not to, transmit data;
- How to ensure that data is passed to, and received by, the intended recipient.

Although the OSI model is a conceptual model, it continues to be used to characterize how protocols that address each layer realize a function and how layer-to-layer interfaces depend on lower-layer services [26]. Adhering to the OSI reference model fosters improvements to protocols due to the singularity of purpose (to support a single functional layer) through clear division of functionality. As a note, referring to Table 4.1, there are two conventions for transmitted information at the network layer (layer 3). If message delivery reliability is of concern, the TCP/IP protocol is used and the data messages transmitted are referred to as packets. If high-speed data transmission is desired without regard to delivery reliability, the user datagram protocol (UDP) is used, and data messages are referred to as datagrams.

Several major network and computer vendors, along with large commercial entities and governments, support the use of the OSI model [27–30]. In addition to supporting clarity of layer protocol design, a significant benefit that has resulted is that through the OSI model, a worldwide lexicon exists to describe protocol development.

4.4 Implementation Standards: TCP/IP Packet Model

For DoD, ARPANET was the catalyst that led to the DoD network model, as presented. This four-layer model differs from that originally envisioned and defined in the 1960s by those working the ARPANET program, but it is the four-layer model that continues to be referred to in subsequent discussions concerning the DoD network model [31]. The DoD layered model describes packet-switched networking functionality and interfaces are that provided as listed in Table 4.2 [32]. The DoD model, often referred to as TCP/IP, was developed in parallel to the OSI reference model, which was published by the ISO in 1984 [33, 34]. Whereas the OSI model is purely theoretical, the DoD model [35, 36] is heavily dependent on TCP/IP protocols, which resulted from intense refinement derived from deployed networks that occurred throughout a decade.

Figure 4.1 provided the idea behind packet-switched networking, segmentation of data streams into packetized groups. However, to ensure accurate delivery, smooth network operation, error detection/correction, and TCP concatenates header data to support guaranteed message delivery. The resultant packet is presented by Figure 4.5. At the TCP/IP transport layer, the data stream is accepted from the application services layer, and TCP segments the stream into appropriate packet size acceptable by the source, destination, and pathway selected by a routing protocol at the lower internet layer. TCP begins source-to-destination transmission through implementation of a three-way handshake to begin transmission of packets. TCP sends a synchronization packet from the source (client) to the destination (server), and if both are successfully sent and understood, the destination node returns a synchronization and acknowledgment packet back to the source.

Table 4.2
DoD Packet-Switched Network Model

DoD Layer	Designation (Title)	Information Description	Function
4	Application services	Data stream	Simple network time protocol Email (SMTP), TelNet (network virtual terminals), network management protocol (SNMP), hypertext transfer protocol (HTTP), file transport protocol (FTP), and miscellaneous services
3	Transport (host-to-host)	Packet	Message delivery and host management; TCP, (for reliable delivery); UDP (for speed/streaming, not guaranteed delivery); host monitoring protocol (HMP)
2	Internet	Datagram (fragments)	Unified addressing, routing, congestion and flow control; IP
1	Network access	Encoded bits	Network technologies with standardized interfaces and packet-switching aware (CCITT V24, V35, EIA RS449 MilStd-18 (RS-232C) and more)

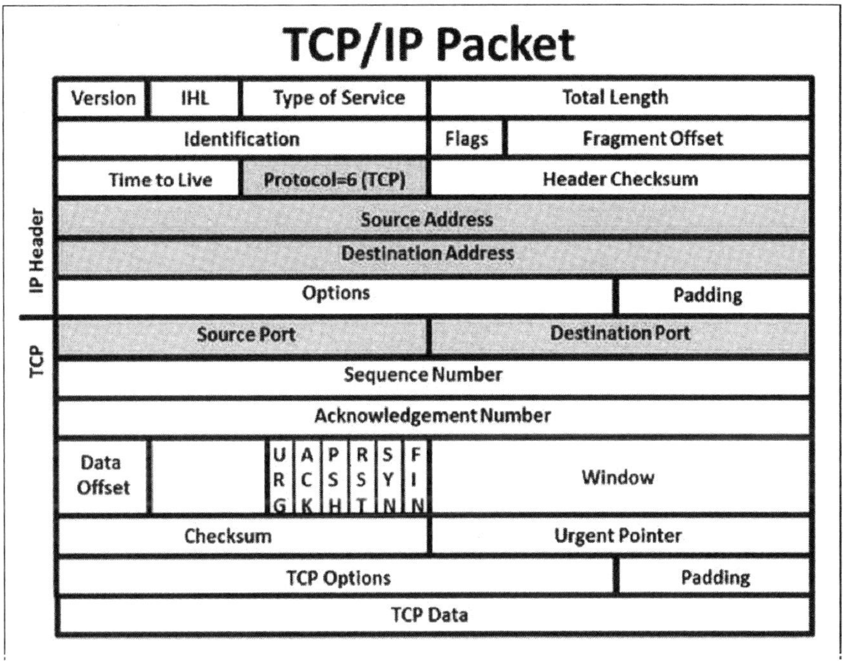

Figure 4.5 TCP/IP packet diagram; segmented data designated as TCP data.

With affirmation from a destination node, the source sends an acknowledgment packet to the destination with agreed-upon packet flags set to affirm an agreed-upon sequence number and acknowledgment values, and the pathway is open to continuous sending of packets [37]. Of note is that this three-way handshake has been exposed as causing security issues for internet servers (and users). With the client-server TCP/IP three-way handshake, unavoidable latencies occur; this gives attackers a window to flood the server by sending several connection requests in a short period. This synch (SYN) flood attack inundates the server, and the server ceases to respond. However, there have been a number of investigations and approaches as to how to remedy this exposure [38, 39].

As for delivery of packets, TCP initiates timers to ensure that reasonable elapsed times are adhered to, to provide a reasonable wait time. If a timer exceeds a preset duration, TCP will resend the packets. Sequence numbers are inserted in the header to allow for correct reassembly by the destination application services since packet-switched networks do not guarantee the time ordering of a data message. (Packets are typically received out of order due to routing, delays, and similar effects.) Initialization of the sequence values is handled, as discussed, during initialization to implement flow control to ensure matching data rate speeds between devices.

TCP provides buffering capability. Additionally, to strive for bandwidth efficiency, TCP can alter the duration to wait for acknowledgment of packets being sent to the destination. This time window is updated by either source or destination as both nodes strive to match packet rate. This sliding window process maintains a sustainable source-to-destination data rate. Checksum values are introduced via the packet header to allow for error checking and control, whereby any error will drop the expected corrupted packet, and a request to source to resend the packet will be created. Duplicate packets at the destination are monitored by TCP, with duplicates dropped when detected.

4.5 Ad Hoc Wireless Networks Standards: Cross-Layer Model

Refinement and improvements to wired networks have benefited greatly from the TCP/IP and OSI network models. Similarly, wireless networks have benefited from the TCP/IP model. Prominent examples of protocol standardization for ad hoc wireless networks include local area networks (WLANs) and wireless personal area networks (PANs) [40] that subscribe to IEEE 802.11 and 802.15 standards, respectively. However, ad hoc wireless networks possess characteristics that differ significantly from wired networks.

Notably, wireless network systems differ from wired packet-switched networks, as ad hoc wireless systems must contend with a limited energy source, variable RF propagation, unique security issues, and difficult ambient conditions. WSN, in particular, presents even greater difficulties to solve as WSN nodes (motes) have severe design restrictions, especially in the area of size, weight, and power. The network models, devised decades ago, have served satisfactorily as a foundation in the development and implementation of layered protocols aimed at ad hoc networks, but given the numerous differences between ad hoc wireless networks and the computer-to-computer networks, the need for a modified design paradigm is indicated.

Resource management is critical in the design and development of wireless network protocols with emphasis on efficient use of power and link bandwidth as critical design goals. Throughout the last decade, significant research and analyses have been completed to address what and how to modify the layer models to arrive at improved protocols for wireless networks. The basic approach has resulted in cross-layer design (CLD), wherein interactions and configurations across layers are considered and promoted to improve and/or optimize the conduct of the layer functions [41]. Reduction in overhead, which results from data sharing between the independent layers, has been evaluated through the application of cross-layer configurations to afford increased versatility, interoperability, efficient use of resources, and network maintainability. The goal behind the cross-layer design approach is to maintain functionality at the network layers but enable coordination and interaction among the various layers to further optimize the protocols' response to wireless network nodes. Key areas being worked on include improved power management [42], security [43], congestion control [44], and routing optimization [42], as well as other network management and communication functions.

Significant research has occurred to modify the data link (MAC protocols) and network (routing protocols) layers, as MAC protocols' efficiency would benefit from the collection of routing statistics, such as traffic flow, link quality, congestion, and latencies to optimize the functionality of the protocols at both levels. Providing MAC information directly to the routing task reduces internal data communications resulting in improved performance (speed and path accuracy).

4.6 Ad Hoc Network Architectures

Ad hoc networks, such as WSN and MANET, avail users of multiple solutions in forming a responsive design toward T-ISR objectives. Network nodes, especially for MANET, are mobile and as such, frequently enter and depart network proximity causing linkage to break connection(s) and require establishment of new routes, altering network topology. WSN nodes are typically static (do not change location), but autonomous behavior at the node level establishes an independence from any centralized network control. Ad hoc network topologies permit a variety of topological layouts with the two basic structures being single-hop and multiple-hop connectivity. Single-hop networks are those that have as a single link between any node and the access point (AP) without any intervening routing or relay process. Multihop topology employs a tiered architecture. Figure 4.6 presents the two topologies [45]. In Figure 4.6, the larger node (extended antenna) red represents an access point, AP, and the single-hop nodes (left) directly communicate to the AP.

In a multihop configuration (Figure 4.6(b)) the second level of network hierarchy communicates to the AP only via aggregation nodes (denoted by asterisks). Aggregation nodes achieve a level of processing (e.g., message aggregation, compression, and routing) to node data traffic and via multihop

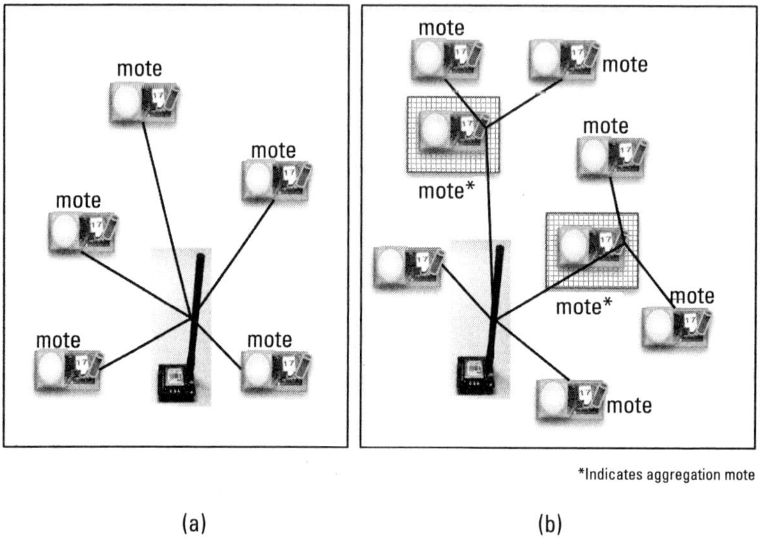

Figure 4.6 Single-hop (a) and multihop (b) network communication.

capability. These lower-tiered groups of nodes relay their data using only their assigned aggregation nodes [46–48]. The topmost node tier links nodes directly to an AP (single-hop), and a number of these nodes, designed as aggregation nodes, have a subordinate tier of nodes (cluster, or group) that obtain and send messages through the cluster head node only. Access between the AP and lower-tiered nodes is via these aggregation nodes, effectively acting as gateways (or, equivalently, routers). A tiered system allowing for reasonable RF ranges enable a significant increase in coverage (area) of the ad hoc network. Note that in the literature, aggregation node and cluster head are equivalent, with the exception of whether the clusters's head is a network-only router, or a router node that conducts additional processing and reformatting on incoming data (e.g., redundancy suppression, data smoothing, and data compression).

Figure 4.7 shows a simplified version of direct line-of-sight (LOS) and indirect [non-LOS (NLOS)] propagation paths. As with any wireless communications, there exist numerous issues apart from those associated with wired communication systems. RF fading (variable attenuation), RF-noise interference, and multipath propagation combine to cause the static link quality to fluctuate. Satisfactory, even highly reliable, link margins may disappear rapidly due to temporal propagation issues (e.g., weather and objects moving into the network vicinity). Alternately, there are times that consistently intermittent

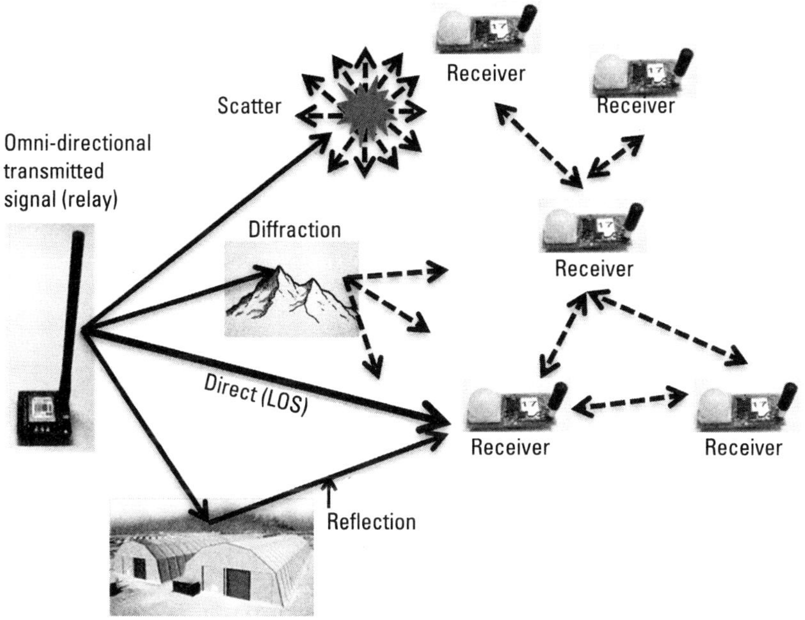

Figure 4.7 Multipath arrival of messages; direct (LOS) and indirect (NLOS) paths.

links periodically pass data messages. These unexpected propagation effects produce an anisotropic RF energy patterns, which vary with time depending on the dynamics of local propagation (e.g., soil moisture content and reflective objects moving within the RF field).

With moving transmitters and/or receivers, these behaviors take on a dynamic quality. As shown in Figure 4.7, multiple RF signals arrive at the designated receiver that include expected (direct, LOS) and unexpected (scattered, diffracted, and/or reflected) components that traveled along various pathways. Investigations have occurred to address the impact of such propagation irregularities on both the MAC and routing protocol layers [49, 50]. Existence of an unstable propagation pattern not only affects the power margin necessary for sufficient SNR to realize a communication link, but also increases the chance for MAC protocols that use the carrier-sensing algorithm to succumb to the hidden terminal problem. The hidden terminal problem illustrated in Figure 4.8 occurs when a node (A) is linked to a node (B), but does not directly communicate with other nodes (e.g., C) linked to node (B). Node A is unaware of the other nodes, and these other nodes (C included) is

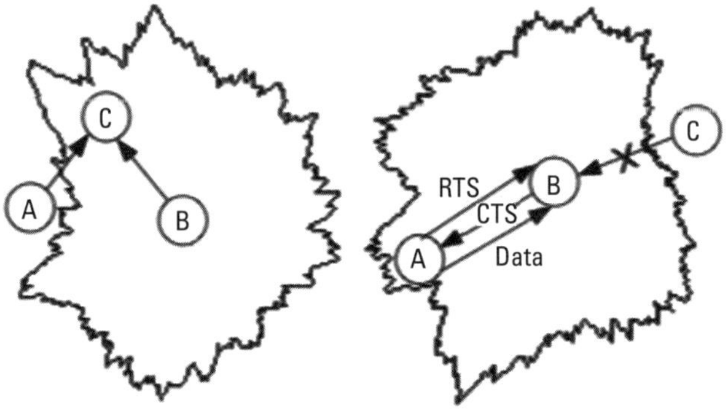

Figure 4.8 Irregular RF pattern: (a) indicates a hidden terminal with carrier sense MAC protocol, nodes A and B are unaware of each other, and (b) hidden terminal issue due to handshaking asymmetry, where note B ignores node C.

considered hidden from A. The problem occurs when A attempts to transmit a message to B simultaneously as C.

The nodes, using a well-established MAC protocol, such as carrier-sense multiple access (CSMA), verifies the absence of RF traffic before transmitting on a shared transmission medium [51]. What results are multiple nodes transmitting data packets to the common node (B) simultaneously, which creates interference preventing successful reception of any of the packets. CSMA with collision detection (CSMA/CD) does not work; packet collisions occur and accurate reception at node B occurs. Hidden nodes occur with wireless ad hoc networks occur when nodes establish connections to an access point (AP) but are unaware of one another either as a result of insufficient SNR on the RF channel or due to a configuration employing a topology where each node is within communication range of the AP but is not linked to other nodes. To overcome the hidden node problem, request-to-send (RTS) and clear-to-send (CTS) handshaking (e.g., IEEE 802.11) is implemented at the AP alongside a CSMA/CA-based MAC protocol, solving the issue seen by node C (Figure 4.8(b)).

4.7 MANET Background

The initial MANET system is a product derived from the packet radio network (PRNET) programs sponsored by DARPA in the early 1970s. The PRNET used a combination of ALOHA CSMA approaches for medium access and employed distance-vector routing to provide packet-switched networking to mobile battlefield elements in an infrastructure-less, hostile environment. It evolved to be a robust, reliable operational experimental network. The MANET form of ad hoc networks is considered because MANET networks have been deployed and operated in conjugation with WSN systems [52] and with a thorough discussion of MANET implementations and related protocols, WSN trade-space and implementations can be compared and considered.

As discussed in Chapter 3, advances in several technologies have enormously benefited WSN. These same technologies and developments contributed greatly to all forms of ad hoc wireless communications, where highly dynamic node-to-node connections are made, modified, and reformed, all without the presence of a fixed infrastructure. MANET networks, like WSNs, employ wireless linkage through dynamic, yet robust, routing. Smart phones and mobile communication devices readily support standardized wireless interfaces (IEEE 802.11, Bluetooth, and cellular 4/5G). MANET systems operate as a distributed NMS [53] in the absence of a fixed infrastructure and may include one (or multiple) access points (APs) to cellular networks, or the internet [52] (for military systems, DoDIN). MANET systems are meant to be temporary data and voice communication architectures that exhibit reliable and responsive to short-lived missions.

4.8 MANET Overview

A MANET system is comprised of an autonomous collection of mobile networked devices connected by packet-switched links that form a network architecture through which messages can be originated and delivered to select destination(s). Each MANET node acts as a MANET router that reorganizes network topology based on nodes joining or leaving the network environment. Contrary to cellular networks, where nodes are restricted to communicate with a set of carefully placed base stations, there are no base stations for MANET networks, as two MANET nodes can communicate directly using single hop (or indirectly using multihop). Additionally, MANET nodes can address a single node, provide multicasting with a number of select nodes, or operate in broadcast mode to address all network nodes.

Well-known complications in MANET systems originate directly or indirectly from instability caused by mobile nodes. The constant independent moving of nodes produces frequent failure. Additionally, activation of new links leads to increased network congestion becoming a potential source of link overload and degradation of quality of services (QoS), a widely used metric that correlates network capabilities to user satisfaction and is based on network parameters evaluated via use of a network model that possesses a priori connectivity functions such as routing. QoS performance metrics, such as connectivity, robustness, fragility, reachability, and throughput, permit the evaluation of network interoperability. A problem related to node population and mobility is connectivity, which can be excessively sparse when nodes are organized isolated groups.

Several standards have been leveraged for use within MANET systems. Bluetooth advances have made viable the widely diffused MANETs. Development and refinement of IEEE 802.15.4 (e.g., ZigBee) produced core protocols for both the ubiquitous WSN [54, 55] and MANET networks [56]. Through methodical development, IEEE 802.11 has resulted in the use of WLAN for MANET. Finally, various wireless network technologies, such as 4G (5G) over cellular networks, have been used in MANET configurations.

4.9 Routing Protocol Classification

This section provides an overview of network routing protocols employed for ad hoc systems, which, not surprisingly, are similar (if not identical) to routing methods applied to WSN systems. Routing protocols are responsible for generating routes throughout the network, maintaining links supportive of established routes, and depleting routes as connectivity changes occur. MANET routing protocols are classified according to characteristics of the protocols, algorithm attributes, and performance. Fundamental tools, such as packet casting, are invaluable to the development of routing protocol development. Final selection criteria of routing protocols depend on network metrics under all expected conditions, as well as maintaining a robust capability that allows for graceful degradation as unexpected events occur.

With design philosophy foremost in the routing protocol criteria, three MANET routing design models are considered: proactive, reactive, and a hybrid of the two approaches. For each design criteria, multiple routing protocols have been devised, evaluated, and used. Figure 4.9 presents the design philosophy behind MANET routing protocols. Proactive protocols are table-driven, which maintains current information on each network node and node-to-node connectivity. As nodes enter or leave a network, the overall connectivity map is updated and propagated throughout the network to maintain

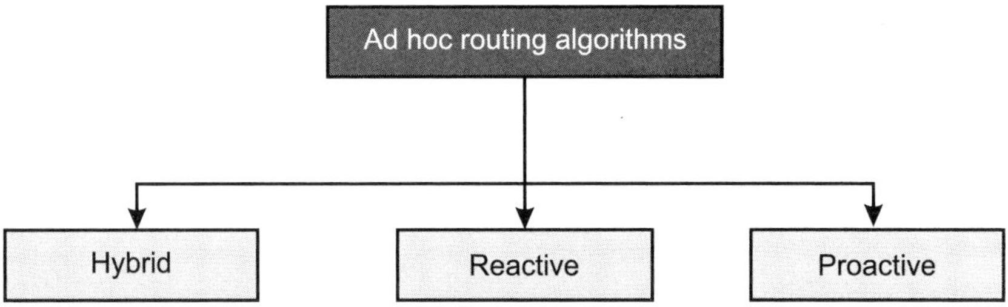

Figure 4.9 MANET protocols: design classification.

the overall network state. Onboard each node is a topology table that relates the connectivity of that particular node to the network. How the tables are formed, updated, and broadcast to the overall network is based on which routing protocol is in use.

There are several successful proactive (table-driven) routing protocols, but the following four feature most prominently in the research and deployment of ad hoc network systems [57]:

- Destination-sequenced distance-vector (DSDV): Every node in the network maintains a routing table containing all possible destinations within the network, and the number of hops to each destination is recorded.
- Optimized link state routing (OLSR): Depends upon periodic exchange of topology information. The key concept of OLSR is the use of multipoint relay (MPR) to provide an efficient flooding mechanism by reducing the number of transmissions required.
- Wireless routing protocols (WRPs): A loop-free routing protocol. Each node maintains tables for distance, routing, link-cost, and the message retransmission list. Link changes are propagated using update messages sent between neighboring nodes. Hello messages are periodically exchanged between neighbors.
- Cluster head gateway switch routing (CGSR): Employs an algorithm at the cluster level, least cluster change (LCC), to avoid excessive cluster head reselection each time the cluster subnet state changes.

Reactive routing protocols are source-dependent and consist of demand-driven algorithms. Reactive protocols generate routes only when instructed by source nodes. This process is completed once a viable route is found, or all possible route permutations are examined. Once a route is discovered and established, it is maintained by a route maintenance procedure until the destination node becomes inaccessible or the route is eliminated. Prevalent reactive (on-demand) routing protocols for ad hoc networks include the following [57]:

- Ad hoc on-demand distance vector (AODV): A DSDV-derivative with reduced broadcast net messaging; a mobility-centric protocol that implements a pure on demand route acquisition system as nodes not on a selected path do not maintain routing acquisition or participate in routing table exchanges.
- Dynamic source routing (DSR): A highly adaptive protocol in which mobile nodes are required to maintain route caches that contain the source routes of which the mobile is aware. Entries in the route cache are continually updated as new routes are learned.
- Temporally ordered Routing algorithm (TORA): Operates in a highly dynamic mobile networking environment and is source-initiated. TORA provides multiple routes for any desired source/destination via localization of control messages to a small subset of nodes near a topological change.
- Relative-distance microdiversity routing (RDMAR) protocol: Estimates the distance between two nodes using a relative distance estimation algorithm in order to limit the range of route searching in order to save the cost of flooding a route request message into the entire wireless area.

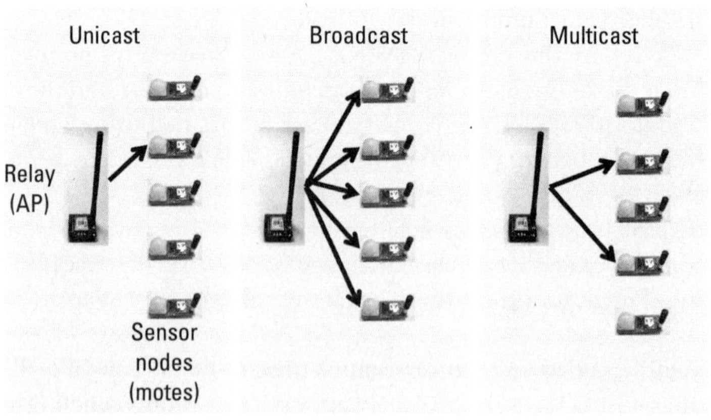

Figure 4.10 MANET protocols packet casting: unicast (one-to-one), broadcast (one-to-all), and multicast (one-to-select-subset).

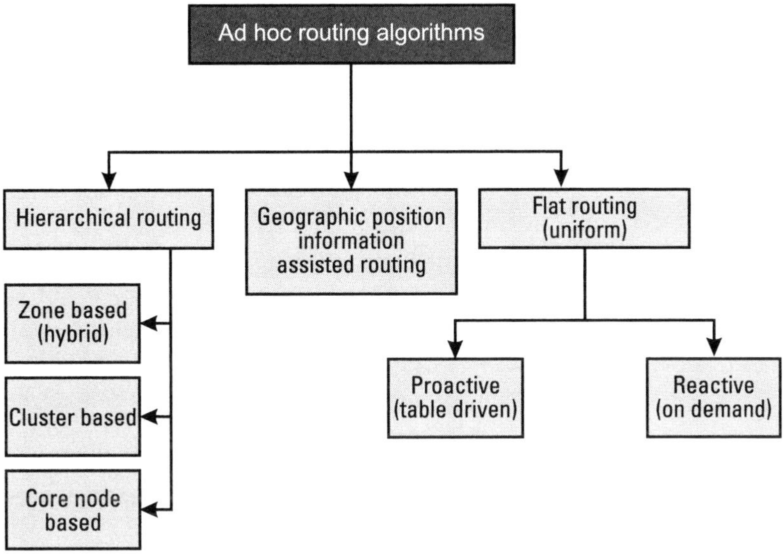

Figure 4.11 Data link protocols: network structure.

The communication mechanism inherent in routing protocols depends on casting capability and directly influences the performance of routing protocols. The availability of selective delivery of messages to network nodes is a fundamental capability that routing (and other high-level) protocols require. Figure 4.10 illustrates three casting capabilities: (1) unicast, node-to-node, (2) broadcast, one-to-all, and (3) multicast, one-to-multiple selected nodes. Having the ability to elect the casting significantly impacts overall network capability.

Viewing data link protocols from the perspective of network structure is seen in Figure 4.11, which is similar to Figure 4.9. Three network-view models are considered: hierarchical (tier-based), geographical-based (node, and node-to-node relative locations), and flat routing (uniform application of routing via proactive or reactive protocols).

4.10 WSN and MANET Comparison

T-ISR missions are acutely application-focused and require autonomous operation over extended timelines with minimal physical profile, qualities not typical of MANET components. However, MANET systems have been developed with capabilities directly applicable to the WSN design. The differences between MANET and WSN have been considered an advantage by system designers, since WSN and MANET can complement one another to fulfill T-ISR objectives.

4.10.1 WSN-MANET Commonalities

Ad hoc wireless networks, such as WSN and MANET, use an architecture that does not assume or require an existing infrastructure and employ single-hop as well as multihop (also referred to as mesh networking). A required key characteristic of WSN and MANET ad hoc networks is effective routing of links among the nodes to afford reliable delivery of messages with minimum overhead and quick reconfiguration of broken paths. WSN and MANET must be energy-efficient and capable of autonomous operation (self-organization included). Ad hoc network nodes, such as WSW and MANET, consist of active devices with computing and communication capabilities that provide remote sensing and onboard processing that supports node operation, A/D sampling of sensor signals, data compression, and formatting of data messages. In considering an aggregation node, additional processing occurs based on the designed behavior of the lower tier cluster, such as data accumulation, redundancy check, discrimination criteria, and the overall combination of lower-tier node traffic. WSN and MANFT nodes, and especially aggregation nodes, must be capable of functions that filter, share, combine, and operate with the supplied data.

4.10.2 WSN-MANET Differences

Perhaps the more obvious comparison of MANET with WSN networks is that, generally, MANET nodes move whereas WSN nodes are stationary. WSN systems are preplanned and are designed to address specific missions and provide critical measurements for use by mission operations. The middleware installed within the network nodes is tailored specifically to system requirements. This is partially due to the WSN focus on a singular or select mission objective, whereas MANET systems are set up to serve temporary but quick reaction missions where all data types can be expected and mission objectives are variable. WSN nodes usually support onboard sensors, whereas MANET nodes are typically employed as data communication nodes. The use of WSN has historically been in the remote sensing of local characteristics that are detectable by the node sensor modalities. This includes but is not limited to: weather, designated targets, signals of interest (e.g., radio traffic), and event observation (e.g., seismic activity, and forest fires). MANET nodes usually employed for data communication purposes, may or may not include a sensor.

MANET is a go-to network to allow for quick and versatile setup of communications among nodes that serve a temporary mission. For example, with natural disasters, MANET can be used to convey information and voice to orchestrate rescues and to provide logistical support and critical messaging to deployed groups. No sensing, per se, occurs. The ad hoc networks for WSN and MANET also differ in node count. To conduct complex tasks, WSN fields typically exceed 100 nodes and may require >10,000 nodes. MANET systems involve a significant number of nodes, but most systems do not exceed several hundred nodes. WSN sensors operate in large numbers, so their sensors can be designed to operate at short range to save power and nodes can be duty-cycled to prolong operational capability. Additionally, WSN data rates typically remain in the sub-megabits-per-second region. Alternately, MANET nodes have access to large power sources, and their key characteristic for performance is to reduce latency. MANET networks work with multimedia data, with data rates that far exceed that seen in WSN networks.

Finally, WSNs strive for small packaging and typically cannot be replenished (WSN motes are typically deployed where access and/or retrieval is not an option), emphasizing minimal cost per node. MANET has a human-in-the-loop (HITL) element, and equipment costs are significantly higher per node than that for WSN. As a result of these differences, WSN node requirements can be rather strict in design and implementation due to the very limited resources available to low-cost, small-package wireless units. Additionally, to push the cost per node further down, high-volume manufacturing is employed, and with such, QA processes mimic those of large volume, inexpensive microprocessor systems. MANETs, with a larger budget per node and a far lower manufacturing volume than WSN nodes, can excel at the manufacturing approach and associated QA, yielding a higher number of operational nodes per build. Table 4.3 compares characteristics associated with WSN and MANET networks that are based on recent (2017) capabilities. As these technologies evolve and even merge into a singular architecture, these differences are expected to become less pronounced with time.

4.10.3 WSN-MANET Convergence

A main concern of MANET (as with WSN) is to form viable and efficient communication links that require minimum overhead while being supportive of rapid reconfiguration if (or when) paths are disrupted. The short transmission range of diminutive WSN motes favors multihop for WSN configurations [58]. A prominent issue for WSN link reliability is not only limited power availability or smaller antenna designs to meet SWaP2 limits. WSN nodes are typically deployed on the ground, which places their electrical centers of dipole antenna (or that of a chip antenna operating using various ISM bands such as 433 MHz, 833 MHz, 2.4 GHz, and 5 GHz) within 4 inches or so off the ground, severely impeding the RF range of each node. WSN system reliability not only must overcome unreliable radio links, but the protocols used must take care not to adopt a design that provides delivery retransmission, which of course, burns more energy.

For MANET, nodes are associated with larger platforms, which support

Table 4.3 Characteristics Comparison: WSN and MANET Networks

Issues (2017)	MANET	WSN
Standards	IEEE 802.11	IEEE 802.1
Number of nodes	<WSN, 100s	Very Large, >1,000
Node movement	Decentralized	Centralized
Node functions	Nodes act both as host and router	Assigned at initialization
Interaction	HITL	Autonomous
Main purpose	Distributed communications	Sensor and information collection
Application-equipment	Comparably uniform	Application-specific
Application-specific	Uniform	Strong app dependency
Scale	Large	>>MANET
Data bandwidth	Wide	Narrow
Failure in nodes	<<WSN	Significant
Data rate	Rich multimedia	Low power, low rates
Data redundancy	No	Yes
Power	Platform-based	Very low (battery-centric)
Population of nodes	Sparse	Very dense, 10–30m spacing
Deployed by	Several organizations	Singular ownership
Application node	Mobility	Stationary
Communication mode	Unicast, omnicast, multicast	Unicast, omnicast, multicast
Routing protocols	Proactive, reactive, hybrid	Flooding, gossiping, flat routing, hierarchical, location-based
Memory constrained	>GB	kB-to-MB
Network size	Based on active users	AOI-driven
Identification	MAC address	Not unique (middleware-assigned)

larger RF antennas and/or power sources compared to those associated with WSN nodes. Links can span longer distances and operate with higher reliability. There remain issues with any wireless links whereby RF interference can result in link failure, but judicious network topology and associated monitoring (contiguously performed within LINK layer protocols) allows for quick self-healing in the event a critical link drops out.

Considering the above, there are situations where a combination of the two network types can be used to devise a hybrid network that overcomes issues associated with WSN or MANET. WSN provides persistent observation and long-term (albeit low bandwidth) capabilities whereas MANET allows for higher bandwidth and dynamic reconfiguration as required. Figure 4.12 illustrates how MANET and WSN (cluster) systems may be configured to work together [59, 60]. In this figure, cluster head nodes (CH_mote can either be a WSN node or when appropriate, a MANET node providing relay capabilities).

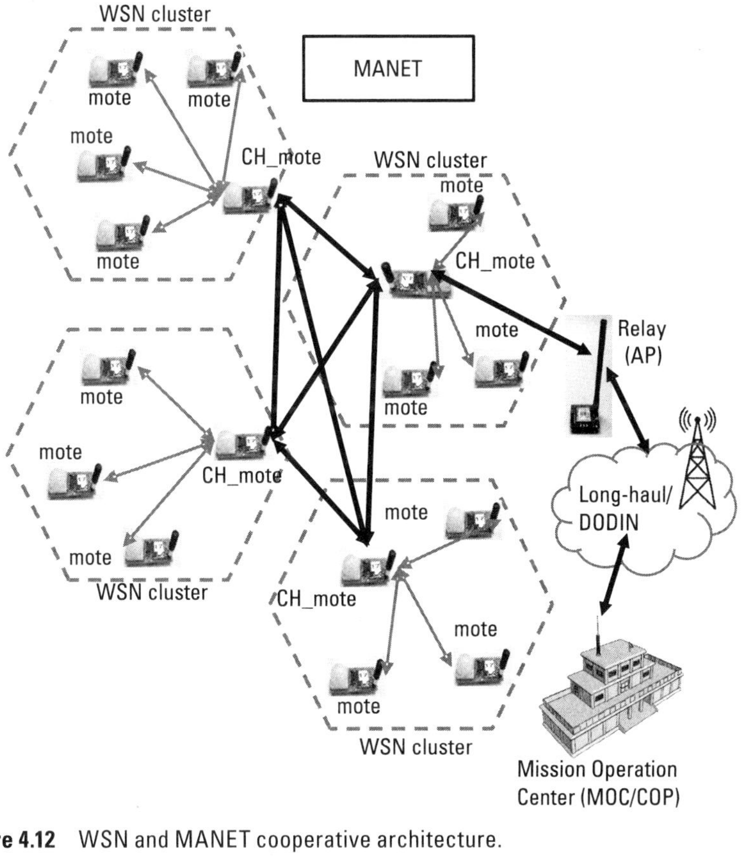

Figure 4.12 WSN and MANET cooperative architecture.

Figure 4.12 emphasizes that the integration of WSNs and MANETs establishes cross-network routing opportunities that overcome typical limitations of WSN systems, such as bandwidth and link distances, and reveal novel and cost-effective solutions to conduct T-ISR missions. Having WSN motefields connected to MANET systems supportive of Blue Force Tracking (BFT) allows a mobile group to persistently monitor areas of interest without having to be within harm's way. WSN networks can also improve overall network operability through the MANET low-latency delivery of sensor data packets. MANET-WSN hybrid networks make use of the MANET's capability to act as a backbone and enable a MANET to play the role of WSN backbone pathway. MANET can be designed to benefit from a deployed WSN system by using it to query a monitored area for alerts of arriving and departing MANET nodes. This supports conservation of overall energy expended, as discovery and advertising packets require significant and comparable amounts of energy.

A key consideration is to provide for MANET-WSN integration support while refraining from excessive packet exchanges between MANET and WSN clusters. MANET nodes ideally would remain in a low-power state and only become active when a WSN cluster (or MANET router) indicates that a shift in the underlying topology needs to be accommodated through reorganization of the net.

4.11 MANET Challenges: Issues and Vulnerabilities

Outstanding challenges remain for MANET and MANET-cooperative networks. As with most ad hoc networks, MANET networks operate autonomously without central administration to manage the nodes and/or connectivity. Being a wireless system, various malicious code attacks are possible, including viruses, worms, spyware, Trojan horses, and other attack instruments that target MANET operating system and/or user applications [61]. The core objective of MANET is to provide mobile users quick access to data, but only among network-authenticated users. As any packet-switched network, MANET depends upon the integrity and continuity of connecting nodes. Disruption by faulty (or "unauthorized") nodes can unravel a MANET network through such attacks as a denial-of-service (DoS) attack, which greatly impacts network bandwidth.

Although MANET nodes have greater RF ranges than their WSN counterparts, the transmission range is still limited. It is the task of the implemented routing protocol to maintain and update topological information, especially with MANET due to frequent changes in network topology.

MANET protocols strive to avoid unnecessary maintenance of topological information, which involves control overhead and results in more bandwidth wastage. MANET scalability can be broadly defined based on the network's ability to provide an acceptable level of service even in the presence of a large number of nodes. Having to interrogate a network frequently for newly entered users, as well as departing users, creates significant net traffic. Also, as with any wireless system, RF and data jamming is a special class of DoS attacks initiated by malicious nodes after determining the characteristics of a network physical link. Wireless connectivity is often times is plagued with poor transmission quality, an inherent problem of wireless communication caused by error sources. (See Figure 4.7.)

4.12 MANET Susceptibility and Attack Schema

MANET systems, based on wireless communications, avails its signals to both desired and unwanted listeners. An intruder can readily conduct eavesdropping to acquire critical intelligence about net traffic without actively injecting fake information. This type of attack provides the attacker with information that may reveal operational aspects of the invaded network, useful in designing or conducting a successful attack. There are numerous approaches available and used to attack an ad hoc network. For MANET (and similarly, with SWN), prominent attacks used and considered include: black hole, RF jamming/interference, flooding, DoS, wormhole, link spoofing, and session hjacking. This list is not exhaustive, but it does present a core set of possible problems that MANETs (and WSNs) must consider in their design. Beyond these attack methods, there are overarching concerns of covert emplacement, operation, and prevention of signal interception, which are addressed by systems designed for low probability of detection and intercept (LPD/LPI). Poorly designed protocols can entice an attacker with opportunities that can be exploited, such as those associated with TCP/IP in its initial implementation of the three-way handshake.

4.12.1 Black Hole Attack

The route discovery process inherent to various routing protocols opens a vulnerability to the black hole attack. In this attack, when a malicious node listens to a route request packet in the network, it responds with the claim of having the shortest and most recent route to the destination node, even if no such route exists. As a result, the malicious node easily misroutes network traffic to it and promptly drops the packets received.

4.12.2 Active Attack

Here, the intruder performs an effective violation on either network resources or data transmitted by producing routing disruption, network resource depletion, and/or node breaking. Approaches to achieving a successful attack assume the introduction of a node that produces severe RF interference or a node that joins the target network and produces excessive (worthless) network message traffic. Other approaches employed via physical attacks are removal of a node that forms a major pathway or routing junction and/or insertion of an artificial disturbance, both breaking an established link.

4.12.3 Flooding Attack

The attacker exhausts network resources, such as bandwidth, to cause node resources to be consumed, such as computational capacity or battery power. Flooding is also used to disrupt routing operation to cause severe degradation in network performance. For example, in the ad hoc on-demand distance vector (AODV) protocol, a malicious node can send a large number of routing requests in a short period to a destination node that does not exist in the network. As no node will reply to these errant routing requests (RREQs), messages will flood the whole network. As a result, node battery power and network bandwidth are consumed and eventually lead to DoS, shutting down the network.

4.12.4 Wormhole Attack

The attacker receives packets at one point in the network, and tunnels (wormholes) them to another point in the network. Subsequently, the attacker replays the intercepted packets into the network from the revised point. Routing can be disrupted when routing control messages are tunneled. With an AODV system, this attack could prevent discovery of routes and may create a wormhole when the network is using broadcasting. Wormholes are hard to detect because the path that is used to pass information is usually not part of the actual network, allowing wormholes to damage a network without the network being aware.

4.12.5 Gray-Hole Attack

Gray-hole strategies generate routing feints that lead to the dropping of messages. Gray-hole attacks have two methodologies: (1) a node advertises itself as having a valid route to destination, when in actuality it doesn't, and messages get caught in a dead end (dropped), and (2) strong receptive nodes are introduced to intercept and drop packets with a certain probability. In this case, information being transmitted is lost to the network.

4.12.6 Link Spoofing Attack

Spoofing employs malicious nodes to advertise fake links with nonneighbors to disrupt routing operations. For example, in the OLSR protocol, an attacker can advertise a fake link with two-hop neighbors about the target node. This causes the target node to select a malicious node to be its cluster head (for connective node). As a relay node, a malicious node has access to data and can manipulate either data or routing by modifying or dropping routing traffic. Line spoofing attacks can invoke a number of DoS schemes, again shutting the network down.

4.12.7 SYN Flooding Attack

The SYN flooding attack is a type of DoS attack, in which the attacker creates a large number of half-opened TCP connections with a targeted node. These half-opened connections never complete the requisite handshake to fully open the connection, and in doing so, the transmitting node continuously gets caught in a loop trying to close the connection. As with any DoS attack, the network is rendered inoperative.

4.12.8 Session Hijacking

In session hijacking, the attacker exploits an unprotected session after its initial setup. In this attack, the attacker spoofs a node's packet address, finds the correct sequence number expected by the target, and launches various DoS attacks. To succeed, the malicious node must collect secure data (e.g., passwords, secret keys, and logon names) and other information from nodes. Session hijacking attacks are also referred to as address attacks, which are very effective with flooding protocols such as OLSR.

These prominent ways that wireless networks can be compromised are known to both MANET and WSN designers through research efforts and testing. The awareness of these, and other attack modes, is why the focus over the past decade has increasingly been on ad hoc network security. Although initial WSN development considered security measures (e.g., public keying and other algorithms), most of the research was tightly connected to improving WSN-class sensors, transceivers, network reliability, and power management. Now that these facets of WSN system design have been repeatedly addressed and refined, and that WSN has been inserted into IoT and IoBT, the emphasis on secure communications and node protection has marketly been increased over the past decade.

References

[1] Moore, J., "MANET and the Art of Communication," https://fcw.com/articles/2003/08/25/manet-and-the-art-of-communication.aspx, 2003.

[2] Kochhar, et al., "Protocols for Wireless Sensor Networks: A Survey," *Journal of Telecommunications and Information Technology*, 2018.

[3] Baran, P, "On Distributed Communications Networks," *The RAND Corporation, Report P-2526*, 1962.

[4] Kirstein, P., "The Early History of Packet Switching in the UK [History of Communications]," *IEEE Communications Magazine*, 2009.

[5] Kleinrock, L., "An Early History of the Internet [History of Communications]," *IEEE Communications Magazine*, 2010.

[6] Peterson, W., and D. Brown, "Cyclic Codes for Error Detection," *Proceedings of the IRE*, 1961.

[7] Grami, A., "Chapter 10: Error Correction Coding, Introduction to Digital Communications," in *Introduction to Digital Communications*, Elsevier B.V., 2016.

[8] "DCN—Data-link Control & Protocols," Tutorialspoint, https://www.tutorialspoint.com/data_communication_computer_network/index.htm.

[9] Modiano, E., "The Data Link Layer: ARQ Protocols," MIT 6.263/16.37: Lectures 3&4.

[10] "TCP Flow Control and the Sliding Window," TCP/IP and Network Concepts for Advanced Tuning, IBM Knowledge Center. https://www.ibm.com/support/knowledgecenter/SSGSG7_7.1.0/com.ibm.itsm.perf.doc/c_tcpip_concepts.html.

[11] Internet Engineering Task Force (IETF) Working Group Request for CommNet, RFC 5681.

[12] Stevens, W., *TCP Slow Start, Congestion Avoidance, Fast Retransmit, and Fast Recovery Algorithms*, ACM Digital Library, RFC2001, https://dl.acm.org/doi/book/10.17487/RFC2001, 2001.

[13] Gerla, M., and L. Kleinrock, "Flow Control: A Comparative Survey," *IEEE Transactions on Communications*, 1980.

[14] Low, S., "Optimization Flow Control. I. Basic Algorithm and Convergence," *IEEE/ACM Transactions on Networking*, 1999.

[15] Ikegawa, T., and Y. Takahashi, "Sliding Window Protocol with Selective-Repeat ARQ: Performance Modeling and Analysis," *Telecommun Syst.*, 2007.

[16] Yin, X., "A Fairness-Aware Congestion Control Scheme in Wireless Sensor Networks," *IEEE Transactions on Vehicular Technology*, 2009.

[17] Pascale, A., et al., "Wireless Sensor Networks for Traffic Management and Road Safety," *IET Intelligent Transport Systems*, IEEE, 2012.

[18] Nakpeerayuth, S., et al., "Efficient Medium Access Control Protocols for Broadband Wireless Communications, Advanced Trends in Wireless Communications," *Advanced Trends in Wireless Communications*, 2011.

[19] Akyildiz, I., et al., "Ad Hoc and Sensor Networks, Wireless Networks, Next Generation Internet," *6th International IFIP-TC6 Networking Conference Proceedings*, 2007.

[20] Kumar, S., and D. Lobiyal, "Sensing Coverage Prediction for Wireless Sensor Networks in Shadowed and Multipath Environment," *Scientific World Journal*, 2013.

[21] Franceschetti, M., L. Booth, and M. Cook, "Percolation in Multihop Wireless Network," Lee Center for Advanced Networking at Caltech, 2003.

[22] Gilbert, E., "Random Plane Networks," *Journal of the Society for Industrial and Applied Mathematics*, 1961.

[23] Franceschetti, M., and R. Meester, *Random Networks for Communication: From Statistical Physics to Information Systems*, Cambridge University Press, 2008.

[24] Lai, R., "A Survey of Communication Protocol Testing," *Journal of Systems and Software*, 2002.

[25] Zimmermann, H. "OSI Reference Model—The ISO Model of Architecture for Open Systems Interconnection," *IEEE Transactions on Communications*, Vol. 28, No. 4, April 1980, pp. 425–432, doi: 10.1109/TCOM.1980.1094702.

[26] Russell, A., *Open Standards and the Digital Age: History, Ideology, and Networks*, Cambridge University Press, 2014.

[27] Costa, L., "Open Systems Interconnect (OSI) Model," *Journal of the Association for Laboratory Automation*, Vol. 3, 1998.

[28] Pelkey, J., *Entrepreneurial Capitalism and Innovation: A History of Computer Communications 1968–1988*, James Pekey, 2007 Chapter 12, Section 12.5.

[29] Myers, G., Tech Document 2499 Understanding Open Systems Interconnections (OSI), 1993.

[31] Chatterjee, S., "Internet of Things (IoT) Using Cross-Layer Design: Issues and Possible Solutions," 2017. https://www.researchgate.net/publication/329980937_Cross_Layer_Design_in_the_Internet_of_Things_Issues_and_Solutions.

[32] Cerf, V., and E. Cain, "The DoD Internet Architecture Model," in *Computer Networks*, Vol. 7, 1983.

[33] Zimmermann, H., "OSI Reference Model—The ISO Model of Architecture for Open Systems Interconnection," *IEEE Transactions on Communications*, Vol. 28, No. 4.

[34] Day, J., and H. Zimmermann, "The OSI Reference Model," *Proceedings of the IEEE*, 1983.

[35] MIL-STD-1777, 1983.

[36] Parziale, L., et al., *TCP/IP Tutorial and Technical Overview*, IBM Redbooks, 2006.

[37] Espina, D., and D. Baha, *The Present and the Future of TCP/IP*, 2009. https://pdfs.semanticscholar.org/1292/b32b72a303e3e19bc357779ddf4a714c4c42.pdf.

[38] Min Ma, Q., S. Liu, and X.Wen, "TCP Three-Way Handshake Protocol Based on Quantum Entanglement," *Journal of Computers*, 2016.

[39] Hsu, F., Y. Hwang, and C. Tsai, "TRAP: A Three-Way Handshake Server for TCP Connection Establishment," *Applied Sciences*, 2016.

[41] Song, L., *Cross Layer Design in Wireless Sensor Networks, A Systematic Approach*, VDM Verlag, 2008.

[42] Medagliani, P., et al., "Cross-Layer Design and Analysis of WSN-Based Mobile Target Detection Systems," in *Ad Hoc Networks*, Vol. 11, 2011.

[43] Boubiche, D., and A. Bilami, "Cross Layer Intrusion Detection System for Wireless Sensor Network," *International Journal of Network Security & Its Applications (IJNSA)*, 2012.

[44] Akyildiz, I., et al., "A Cross Layer Protocol for Wireless Sensor Networks," *IEEE 40th Annual Conference on Information Sciences and Systems*, 2006.

[45] Tsitsigkos, A., et al, "A Case Study of Internet of Things Using Wireless Sensor Networks and Smartphones," *28th Wireless World Research Forum*, 2012.

[46] Saudi, N., et al., "Mobile Ad-Hoc Network (MANET) Routing Protocols: A Performance Assessment," *Proceedings of the Third International Conference on Computing, Mathematics and Statistics*, Springer, 2019.

[47] Minha, Q., et al., "On-Site Configuration of Disaster Recovery Access Networks Made Easy," *Ad Hoc Networks*, Vol. 40, April 2016, pp. 46–60.

[48] Prabha, C., S. Kumar, and R. Khanna, "Wireless Multihop Ad-hoc Networks: A Review," *IOSR Journal of Computer Engineering (IOSR-JCE)*, 2014.

[49] Sabri, N., et al., "Performance Evaluation of Wireless Sensor Network Channel in Agricultural Application," *American Journal of Applied Sciences*, 2012.

[50] Zhou, G., Tet al., "Impact of Radio Irregularity on Wireless Sensor Networks," *MobiSys'04*, June 6–9, 2004.

[51] Ababneh, N., "Radio Irregularity Problem in Wireless Sensor Networks: New Experimental Results," *Proceedings IEEE Sarnoff Symposium*, 2009.

[52] Bellavista, P., et al., "Convergence of MANET and WSN in IoT Urban Scenarios," *IEEE Sensors Journal*, 2013.

[53] Puff, C., "Network Management System for Tactical Mobile Ad Hoc Network Segments," Master's Thesis, Naval Postgraduate School (NPS), 2011.

[54] Ouni, S., Z. Ayoub, "Predicting Communication Delay and Energy Consumption for IEEE 802.15.4/ZIGBEE Wireless Sensor Networks," *International Journal of Computer Networks & Communications (IJCNC)*, 2013.

[55] Ding, X., et al., "Link Investigation of IEEE 802.15.4 Wireless Sensor Networks in Forests," *Sensors*, 2016.

[56] Reyeset, J., "A MANET Autoconfiguration System Based on Bluetooth Technology," *IEEE 2006 3rd International Symposium on Wireless Communication Systems*, 2007.

[57] Saeed, N., "MANET Routing Protocols Taxonomy," *International Conf. Future Communication Networks (ICFCN)*, 2012.

[58] Kamal, Z., and M. Salahuddin, "Introduction to Wireless Sensor Network," in *Wireless Sensor and Mobile Ad-Hoc Networks: Vehicular and Space Applications*, D. Benhaddou, A. Al-Fuqah (eds.), Springer, 2015.

[59] Bruzgiene, R., L. Narbutait, and T. Adomkus, " MANET Network in Internet of Things System," *Ad Hoc Networks*, 2017.

[60] Cardone, G., et al., "Effective Collaborative Monitoring in Smart Cities: Converging MANET and WSN for Fast Data Collection," *ITU-T Kaleidoscope Academic Conference*, 2010.

[61] Raja, L., and S. Baboo, "An Overview of MANET: Applications, Attacks and Challenges," *International Journal of Computer Science and Mobile Computing*, 2014.

5

Basis of WSN System Performance: Theory and Application

Consider Charles Handy's description of Secretary of Defense Robert McNamara's Vietnam-era military measurement policies: "The first step is to measure whatever can be easily measured. This is OK as far as it goes. The second step is to disregard that which can't easily be measured or to give it an arbitrary quantitative value. This is artificial and misleading. The third step is to presume that what can't be measured easily isn't important. This is blindness. The fourth step is to say that what can't be easily measured really doesn't exist. This is suicide" [1]. In designing WSN (or any system), objectives must be carefully identified and defined along with meaningful metrics. Early in the development stage is when a system designer must fully comprehend and consider the problem and mission objectives to develop a responsive design. With well-defined mission objectives and top-level requirements, it is possible to flow down meaningful requirements to the constituent subsystems. In this step, supportive lower-level requirements are defined and are associated with subsystem TPMs and more importantly, KPPs, as introduced in Chapter 2. TPMs collectively impact the expected KPP values and are measured continuously throughout the development process and correlated to system-level requirements through the KPPs.

This chapter provides groundwork as to how best identify and evaluate performance parameters (TPMs and KPPs) to reflect system development progress to meet mission requirements. The chapter describes interrelationships of the WSN subsystems, which combine to influence overall system performance. Finally, to provide WSN engineers and end users background concerning how to estimate sensor performance, the chapter reviews probabilistic and statistical modeling that is fundamental to the detection and classification process.

5.1 Evaluation of System-Level Development

TPMs are defined at the basic design levels and evaluated to estimate future values of system performance as KPPs [2, 3]. How TPMs are defined and employed follows standards such as EIA-632, Processes for Engineering a System. Whereas KPPs provide a snapshot of how well subsystems and overall systems are meeting top-level technical requirements, TPM values, obtained through verification, are evaluated to estimate the probability of successfully achieving expected KPP values [4]. The TPM process confirms progress and identifies variances at the subsystem and unit levels that may jeopardize meeting a higher-level end product requirement. Assessed values falling outside established tolerances indicate a need for evaluation and corrective action. From the EIA-632 standard, characteristics and definitions of metrics used for evaluating TPM are [5]:

- *Achievement to date:* Presents the achieved value of the TPM based on estimates or actual measurements. This tracking process monitors a KPP value throughout the development phase to ensure that final design meets critical requirements (or requires modification of system objectives).
- *Current estimate:* The predicted value of the TPM. Through this estimate, risks to technical performance can be identified, tracked, and addressed early in the development process.
- *Technical milestone:* A point where TPM evaluation is accomplished or reported. This typically occurs monthly and may even approach bimonthly (two times a month) when critical systems are tested and integration begins.
- *Planned value profile:* The projected time-phased achievement projected for TPMs from the beginning of the development or when replanned as a result of a corrective projection. This is a response to the tracked values and associated revisions.
- *Tolerance band:* An envelope containing the planned value profile

indicating the allowed variation and projected estimation error. This allows a level of uncertainty and needs to be considered by other supporting parameters that may allow such uncertainty.

- *Objective:* The goal or desired value at the end of the technical effort. These, along with thresholds (described next) provide the content to the requirements verification matrix (RVM), which establishes the test's pass/fail criteria.
- *Threshold:* The limiting acceptable value that, if not met, would jeopardize the project. These values require intense evaluation and consideration as they represent the baseline system responsive to the mission objectives.
- *Variation:* The difference between the planned value and the achievement-to-date value. Variances are used to signal problem areas and possible risks. Using variances, risk charts are generated and updated throughout the development and testing phases with the system engineering team charged with risk assessment, rating, tracking, and subsequent mitigation.

Success of the tracking process requires accurate identification of TPMs. Recently, research and practice has introduced tools to help with this technical management process such as machine learning (ML) for KPP identification [6]. ML has been employed in the identification of all KPPs (and supportive TPMs) that significantly impacted system design and development to estimate software development efforts and defect prediction. On a final note, evaluation metrics consist of various types of requirement performance factors, including functional, technical, constraints, interfaces, and environment. Functional parameters indicate what is sought by an operation, with input and outputs well-defined. Constraints are limits allowed on design implementation, such as SWaP2, or on how adaptive a system can be used. Environmental factors are those associated with maintaining a working system, as well as factors that must be considered to ensure effective operation.

5.2 Developing the Baseline T-ISR System Design

The objectives-to-requirements process results in several products, including the development of CONOPS, which defines the system and interface requirements captured within a system requirements document [7, 8]. Top-level requirements are flowed down to subsystems and result in the generation of additional requirements at each level of complexity. For each system/

subsystem level, a RVM [9, 10] is generated using identified TPM values to establish success criteria to be used during subsystem and system integration and test (I&T) phases. Figure 5.1 presents a design flow path, from an identified T-ISR problem (i.e., mission) to the development of an iterated design. TPMs are used in assessing and developing a baseline system; however, in evaluating system effectiveness, operational considerations (CONOPS) must be assessed. With a description of need, mission objectives are derived and used to produce top-level requirements (where KPPs are derived and subsequently, TPMs). Other requirements are introduced at system and subsystem levels and are combined to generate expected values of performance parameters that must be met by the system to operate as desired. Specifications are derived from top-level requirements for every system element, and in parallel, CONOPS at lower levels are developed. (This should not be confused with the overarching system/mission CONOPS shown in Figure 2.3.) These elemental CONOPS exist to evaluate how well subsystems, as designed, operate both stand-alone and when interfaced with adjacent subsystems. These derived specifications lead to baseline, and later, are critical to refined designs for each subsystem. With designs completed, performance analyses are conducted through brassboard and prototype testing, mathematical modeling, and/or simulation.

In parallel with the direct evaluation of the subsystem implementation, subsystem-based CONOPS are evaluated using baseline designs. Upon evaluating technical solutions and finding the performance and CONOPS

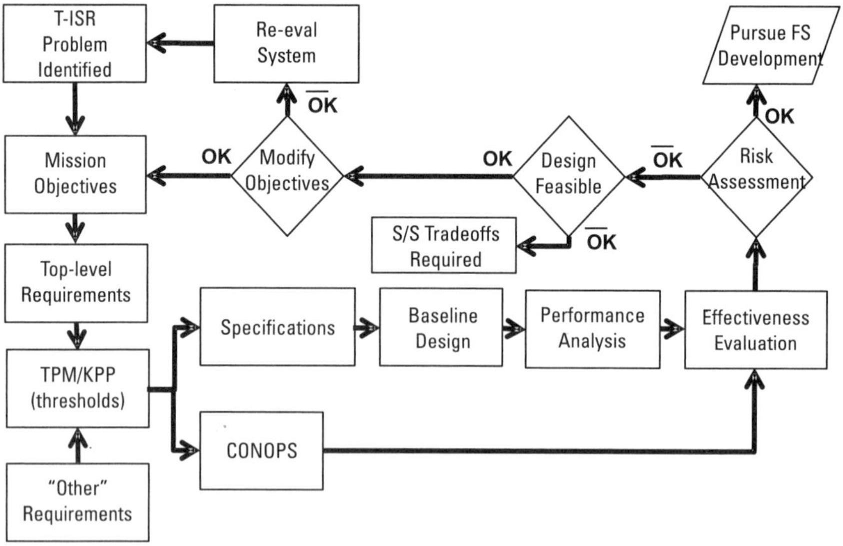

Figure 5.1 Baseline system design evaluation flow.

acceptable, the baseline design can enter the full-scale (FS) development cycle. Evaluation, which verifies and validates the design, is determined using one (or more) of the following approaches: test, analysis, simulation/demonstration, and inspection. If in an evaluation of a design, feasibility comes into question, or the design is found wanting in various key technical areas, subsystem trade-offs are performed. If the redesign of subsystems cannot accommodate the risk, a modification of system objectives is indicated and mission requirements may have to be modified. Otherwise, having exhausted design solutions, reevaluation of pursuing the program should be questioned at the top level. Previous small- and large-scale ISR programs failed to address all aspects of the system operation, and when reconsidered at the top level, these programs had to be discontinued or significantly redirected

5.3 System Engineering and Design Technical Performance

Figure 5.2 illustrates process steps from identification of operational needs to system design. It represents the downward branch of the engineering development portion of the V-diagram (Figure 2.7). Decomposition and definition occurs in this stage of system engineering to form requirements and associated TPMs (and KPPs) to generate (or iterate) design concepts to arrive at a baseline design. Figure 5.2 indicates a two-stage trade study process: (1) trades that define interfaces, which define the subsystems (trade-studies 1), and (2) trade studies (trade-studies 2) at the system level where mission objectives are derived and produce underlying requirements (KPPs). A note of caution is warranted at this point. Measures of performance (MoPs) relate directly with measurable quantifies, such as those represented by KPPs. Often, system evaluation depends heavily on MoP, whereas a full assessment of system effectiveness actually depends upon whether mission objectives are being met. Considering the latter, it is measures of effectiveness (MoEs) that should be emphasized [11]. Unfortunately, this assessment, system validation, has to wait until significant completion of the development cycle, and is typically reserved for initial system operations (IOC). This is why it is emphasized that all critical mission requirements be clearly stated, and verified as being consistent with objectives and coherent with supporting subsystem performance.

5.4 Identifying Technical and Key Performance Parameters

How do we proceed to identify and fully define system (and subsystem) TPMs (and KPPs)? There are particular indicators of performance, such as designing a system to provide a significant SNR against known target characteristics and

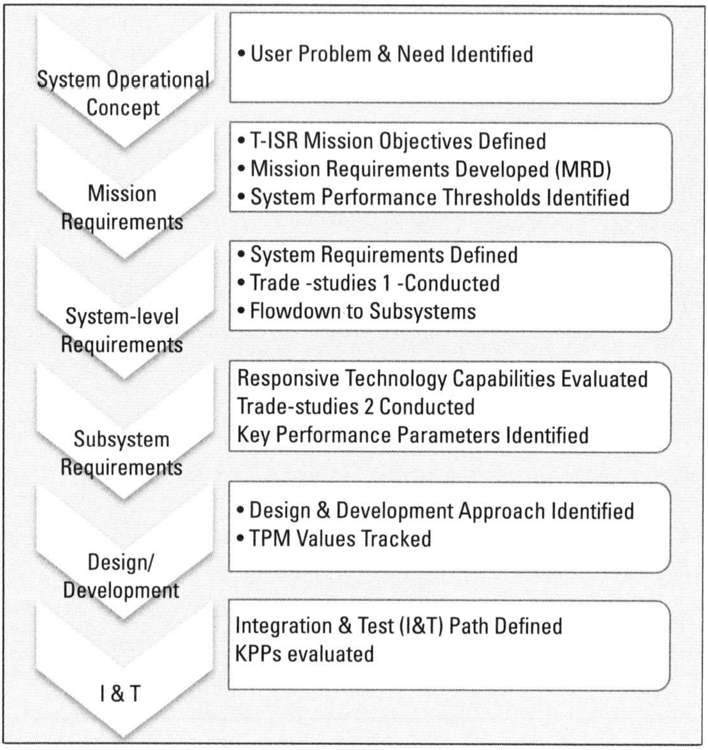

Figure 5.2 T-ISR mission objectives-to-system KPP flow.

noise sources. However, to be successful, variability and accuracy associated with assumed characteristics must be observed. The responsive system design must be capable of accommodating variability of assumed characteristics and possess operational latitude and/or design capacity to operate successfully. This is not to imply that overdesigning is desired by creating excessive capability margins. Unused margins waste resources, especially those typically scarce in T-ISR applications. However, if a characteristic is questionable, variability in performance terms should be considered and sufficient margins used to buffer unexpected decreases of performance. This is the task of conducting trade studies when performance fails to meet one or more constraints. Variability in target signatures and/or background requires that the T-ISR system be robust to aptly handle such changes. Additionally, it is not unusual that T-ISR systems are expected (and required) to address multiple mission categories. At times, these systems are required to do such simultaneously.

Our goal is to develop a system that is designed to be both robust in addressing mission objectives yet sufficiently agile to allow the system to respond to alternate T-ISR missions. In considering performance parameters, a system designer must consider the underlying mission types being addressed. From Chapter 2, we have a top-level view of objectives regarding the application of a WSN T-ISR system. Figure 5.3 illustrates objectives that overlap for typical ISR missions. As shown in Figure 5.3, ISR missions associated SIGINT and MASINT objectives share requirements. Major T-ISR mission applications and missions presented include the following:

- Force protection (FP);
- Indications and warning (I&W);
- Battlespace (damage) assessment (BDA);
- Situational awareness (SA);
- Intelligence gathering (e.g., SIGINT, MASINT).

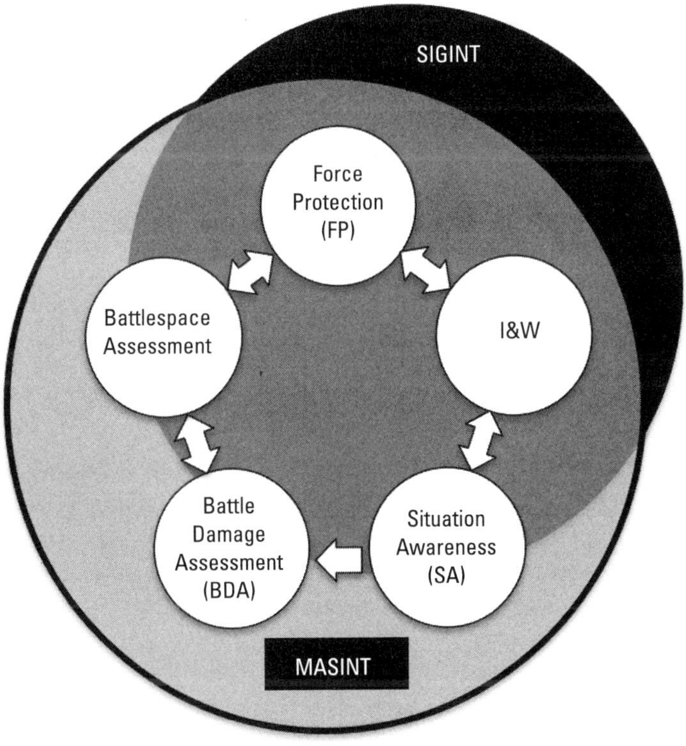

Figure 5.3 Objectives that overlap among typical T-ISR missions relating to SIGINT and MASINT objectives.

Figure 5.3 indicates that T-ISR systems must be judiciously designed and implemented to provide the performance latitude capable of supporting several missions without a need for reconfiguration and/or revision. There may be nuances in adjusting a system to respond correctly to a variant mission, but the idea is that the system should balance capability against multimission effectiveness.

5.5 System and Subsystem Objectives

A first step in the TPM/KPP definition process is to realize that the system requires supportive subsystems to achieve a responsive design. For WSN-based T-ISR systems, there are numerous approaches to defining a WSN-based system. Recall the introductions to WSN sensor node subsystems presented in Chapter 1. Not surprisingly, selected performance parameters must reflect hardware, software (firmware), subsystem-to-subsystem interfaces, and the overall behavior of the system. This involves the main subsystems discussed in Chapter 3 as follows:

- Sensor modalities (e.g., optical, RF, magnetic, acoustic/seismic, chemical-biological);
- RF transceiver;
- Microcontroller & data acquisition system, including ADC, DAC, memory, and CPU;
- Power distribution and management;
- Middleware;
- Localization (e.g., GPS receiver);
- Physical considerations (e.g., node SWaP).

As with the overall system, objectives for each of these WSN subsystems can be tracked throughout development, integration, and testing using corresponding TPMs. WSN nodes and motefields are but a small fraction of an overall T-ISR system architecture. Equally important, are external interfaces and considerations that directly contribute to system performance, listed as follows:

- Exfiltration approach and system (relay);
- Connectivity and access to long-haul communication systems (e.g., DoDIN);

- System-level concerns (i.e., security, persistence, deployment, autonomy);
- Connectivity to sophisticated sensors.

Employing WSN to T-ISR applications has been intensively investigated because of the potential to bring a networked low-cost sensor solution to solve numerous T-ISR objectives. To identify TPM/KPP parameters to evaluate WSN performance, one must consider the fundamental capability of the sensor. Certainly T-ISR mission objectives supporting subsystems as listed in Section 5.4 must be considered, but the primary goals of any ISR system are detection, classification, and tracking of targets. WSN systems have and continue to employ numerous sensor modalities, including those associated with UGS-based systems: PIR, acoustic, seismic, and magnetic. Using technologies developed and refined over recent decades, WSN nodes have integrated CMOS and thermal imagers (i.e., cameras), active sensors (e.g., laser and RF radars), and MEMS-based chemical-biological (chemical-biological) detection systems into small wireless packages. Identifying TPMs for sensors is complex and goes beyond simply characterizing an incoming signal. For each modality, signal and noise contributors must be accurately modeled, and key parameters such as range, attenuation, and noise floors must be considered in determination of appropriate TPM candidates. With optics, we consider range-to-target (R) and optical resolution ($\delta\theta$). The role of range is to exert optical energy dissipation as it propagates from a target to the sensor. Passive systems rely on reflection (e.g., visible sensors) or thermal emission by the target (infrared sensors). As the optical energy travels from target to sensor, it undergoes spreading, which is represented as a spherical wave. This results, from radiometry, with a flux that decreases as R^{-2}. For active optical systems, or laser radars, this loss can be larger depending if the target is resolved or not. With an unresolved target, where the outgoing interrogation laser illumination floods a target, there is a loss of energy for that portion of illumination not intercepted by the target. With an R^{-2} outbound loss combined with a return loss of R^{-2}, the energy acquired back at the sensor incurs an R^{-4} signal loss.

But how does this (and similar considerations) get translated into the performance of the sensor? It occurs through calculation of the SNR using performance equations for the particular sensor. For top-level consideration, we can consider those parameters that influence the SNR as a TPM/KPP for optical systems. The SNR for any sensor involves several aspects and terms associated with both signal and noise contributions. What is required is to derive, or access, performance equations (models, etc.) associated with the sensor modality to inject those characteristics measured via the performance equation, to evaluate operational performance parameters.

Verification of designs requires the development of an RVM. We know this begins with identification and evaluation of all relevant TPMs, a process that requires collaboration with subject matter experts (SMEs). Table 5.1 presents the TPMs identified for NASA's Near Earth Asteroid Rendezvous laser altimeter (NLR) sensor. Table 5.1, which exemplifies how performance and logistical/physical requirements are considered during sensor design, presents terms used by performance equations to verify that an instrument (e.g., NLR) design could operate and meet requirements imposed by the mission. In addition, Table 5.1 presents the expected (required) threshold values for each of the TPMs, which is to be verified by at various stages of integration (pass/fail criterion).

Performance parameters arise through consideration of a host of concerns associated with implementing and operating a sensor system. We have considered SNR, which is affected by range and which will be shown to impact the probability of detection (P_d). Other aspects will require attention in generating a full complement of TPM/KPPs for each sensor modality, including:

- What characteristics are associated with a target of interest to support detection, allow for tracking, and/or to conduct successful discrimination against other noninteresting objects??
- In pursuit of target discrimination, does a target signature database exist or can one be implemented given current timelines, capabilities, and/or costs?
- What is the extent of the search area/volume to be observed?
- At what range must the sensor establish reliable detection? Tracking? Discrimination?
- What integration period(s) is (are) available for target observation?
- What background or atmospheric conditions can be expected regarding line-of-sight?
- What environmental conditions must my sensor operate in?
- How much SWaP is allotted to system and subsystems?
- What data rate/data volume is tolerable?
- What is the precision and accuracy allotted to the measurement of performance values?

To get started, we consider how sensors realize the fundamental functions of detection, classification, and tracking.

Table 5.1
Example: NEAR Laser Radar Design Values (KPPs)

KPP: Maximum range, operation	327 km (maximum count)
KPP: Overall range accuracy (NLR+S/C)	< 6m
KPP: Range precision	< 0.312m
Pulse energy	15.3 mJ @ 1.064 um
Energy jitter	< 1% rms
Pulsewidth	15.2 ns (FWHM)
Pulsewidth jitter	< 0.82 nm
Wavelength spread	+/− 1 nm
Pulse frequencies	1/8, 1, 2, 8 Hz
T-0 Mask, range and steps	0–511.5 us, steps = 500 ns
Range gate, range and steps	81 ns to 42.7 us, step = 41.7 ns
TEM00 (% Gaussian fit)	> 91%
Divergence (1/e2 points)	< 235 urad
Calibration laser power jitter	< +/− 5 %
KPP: Calibration timing jitter	< 1.05 ns
Thermal control	< =/− 2 degC
Shots lifetime	> 10E+09
Effective RX Aperture, f/#	> 7.62 cm, f/3.4
Receiver spectral bandwidth	< 7 nm (FWHM)
Temporal receiver bandwidth	30 MHz, rolloff = −42 dB/octave
APD dark noise voltage	< 150 uV rms
APD hybrid responsivity (R)	> 770 kV/W-optical
Optical receiver FOV	> 2,900 urad
Threshold levels, $n = 0–7$	2^n x 16 mV
KPP: Data rates, selectable	51, 6.4 bps (bits/sec)
KPP: Boreshift shift, TX-to-RX maximum	< 345 urad

5.6 Target/Signal Detection Theory

This section reviews equations that allow us to model detection performance of sensors. These probabilistic expressions pertain to most sensor modalities, especially active sensors. The divergence in approaches for sensors occurs with downstream signal and data processing used to form track files on targets, conduct discrimination of targets of interest from background, respond to noise sources, and other data extraction processing. The underlying approach to modeling sensor performance employs statistical measurements with the application of probabilistic models. The approach is well defined in literature, and summarized here to provide clarity to underlying definitions, and to introduce/review associated mathematics used [12, 13]. With sensor performance equations, we typically see the SNR as a function of an inverse power of range to target,

$$\mathrm{SNR} \sim R^{-n} \quad (n = 0,1,2,3,4,\ldots) \tag{5.1}$$

With individual development of equations to estimate signal (S) and noise (N) terms, we see that the various modalities present unique considerations and terms. Although the system may be charged with indicating the presence of a target (or not), an innate ability to successfully operate dictates a value for maximum operating range (Rmax). This value can then be used to develop: (1) probability of detection, (2) acceptable false alarm rate (FAR), and (3) an estimation of the coverage capability by the sensor [13,14]. Chapter 9, which discuss the sensor modalities for WSN nodes, discuss the process used to provide target detection, tracking, and identification.

5.6.1 Detection via Conditional Probability Distributions

The conventional approach to detection of a signal in noise is through use of probability to determine if a signal exists, or not. The decision criteria are based upon likelihood ratio testing (LRT), which uses a probabilistic expression of an event happening based upon the occurrence of another event [14–16]. LRT formulation, which is well established, makes use of Bayes' rule for probability of causes. The problem can be stated as hypothesis testing wherein one hypothesis (H_1) is accepted when a target is present and its counterpart (H_0) occurs when only noise is present. The goal is to maximize the probability of arriving at the correct determination. A framework is provided through consideration of SNR and the use of joint and conditional distributions. Noise sources exist at the receiver, including thermal (Johnson) noise, shot noise (optical sensors), and background noise entering the receiver. Consider the measured signal envelop at the receiver, $x(t)$, as a function (F) of signal $s(t)$ and random process noise (t):

$$x(t) = F\left[s(t), n(t)\right] \tag{5.2}$$

For additive white Gaussian noise, discussed in Section 5.6.2, the function F becomes a simple summation as $x(t) = s(t) + n(t)$. Using probability expressions, we can form the conditional of H_1 as, $P[H_1|x(t)]$ and eliminate the time dependency representation for clarity, $P(H_1|x)$. This reads as the probability of H_1 occurring given the sensor measurement, x. We consider the alternate hypothesis based on the measurement as $P(H_0|x)$. We now have a simple approach to decision-making via LRT to indicate that a signal (and not noise alone) is present when

$$P(H_1|x) > P(H_0|x) \tag{5.3}$$

The objective is to equate $P(H_1|x)$ as arriving at the correct assertion, when an indication is presented that a target is present. Using (5.3), we apply Bayes' theorem to arrive at

$$\frac{P(x|H_1)}{P(x|H_0)} > \frac{P(H_0)}{P(H_1)} \tag{5.4}$$

Equation (5.4) establishes a means to arrive at a rational threshold value (TH) to test against in determining H_1 versus H_0 [14–16] using the ratio, $P(H_0)/P(H_1)$, which in many cases is assumed as 1. The left-side terms in (5.4) are known as the a priori conditional probabilities for each of the hypotheses. The terms, $P(x|H_1)$ and $P(x|H_0)$, are referred to as likelihoods. The terms in (5.3) are denoted as a posteriori conditional probabilities [15] and are computed using measurements and noise statistics. The test selects as correct the hypothesis associated with the greater of the two values, $P(H_0)$ or $P(H_1)$. This comparison is known as the maximum likelihood criterion. The correct region for declaring noise only (H_0) is denoted as R_0, and the correct region for declaring a signal plus noise exists (H_1) is referred to as R_1. Figure 5.4 provides a schematic [14–18] of what is happening with these conditional probabilities and associated hypothesis assertions using (5.4). The null hypothesis, represented by $P(x|H_0)$ is the declaration that only noise exists. The alternate hypothesis is provided by $P(x|H_1)$, which indicates that the decision is that both signal and noise exist in the measured signal.

Considering that our hypotheses are based on probabilistic estimation, a finite possibility remains that the selected assertion is incorrect. There are two errors that can occur, Type I and Type II [17]. Type I is the existence of the $P(x|H_0)$ distribution that exceeds the selected threshold, TH. In this case, noise is misconstrued as a signal, and H_1 is selected where H_0 is correct. The result is a false positive, a signal is being indicated when none is present. This is designated as a false alarm and is referred to when we read the probability of a false alarm (P_{FA}) or use the term, false alarm rate (FAR) [17, 18].

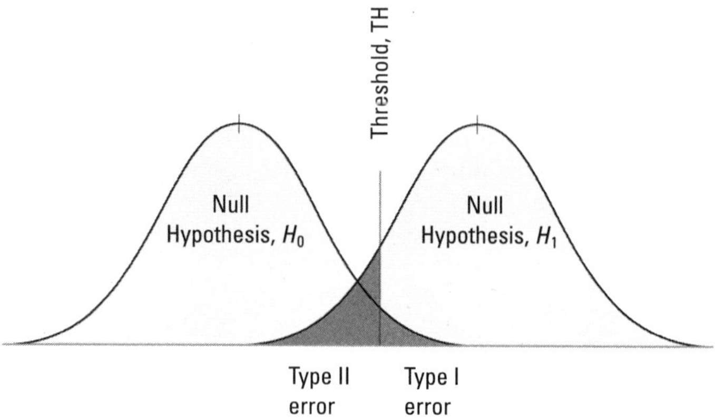

Figure 5.4 The null (H_0) and alternate (H_1) hypotheses and type errors I and II.

Alternately, a similar situation can occur when a measurement containing signal and noise does not exceed the decision threshold. In this case, denoted as a Type II error, the determination is that our measurements consist only of noise although signal is present. This represents a missed detection. Unfortunately, there are a few detractors when using either Bayes or MLC thresholds in selecting H_1 over H_0. First, initial estimates of the a priori values are required. Next, a characterization model has to be assumed for the underlying statistics. Figure 5.5 shows a Gaussian distribution for both the noise and signal + noise situations; however, usually the first (mean) and second (variance) moments differ for the noise and signal. Even a statistical distribution may differ (e.g., Poisson noise associated with a Gaussian described signal).

Finally, and perhaps the larger driving force to seek an alternative in estimating and using a threshold value, is that Type I and II errors persist. The path around this difficulty is that perhaps we should accept a set value for P_{FA} (not to exceed) and work to determine the probability of detection, P_d, using the conditional probabilities. This is the approach published by Neyman and Pearson in their 1933 paper [19] presenting the Neyman-Pearson (N-P) model. Acknowledging that Type I and II errors will persist (unless the R_0 and R_1 regions do not overlap), we can work toward allowing a value for P_{FA} and seek to maximize detection probability, P_d. Evaluating the probability (α) of for the Type I error—that is, H_1 selected when H_0 is true—can be evaluated [15], as

$$\text{TYPE I: } \alpha = \int P(x|H_0)\,dx \tag{5.5}$$

The value α is equivalent to that referred to as P_{FA}. Similarly, for the Type II error probability (β), we can use

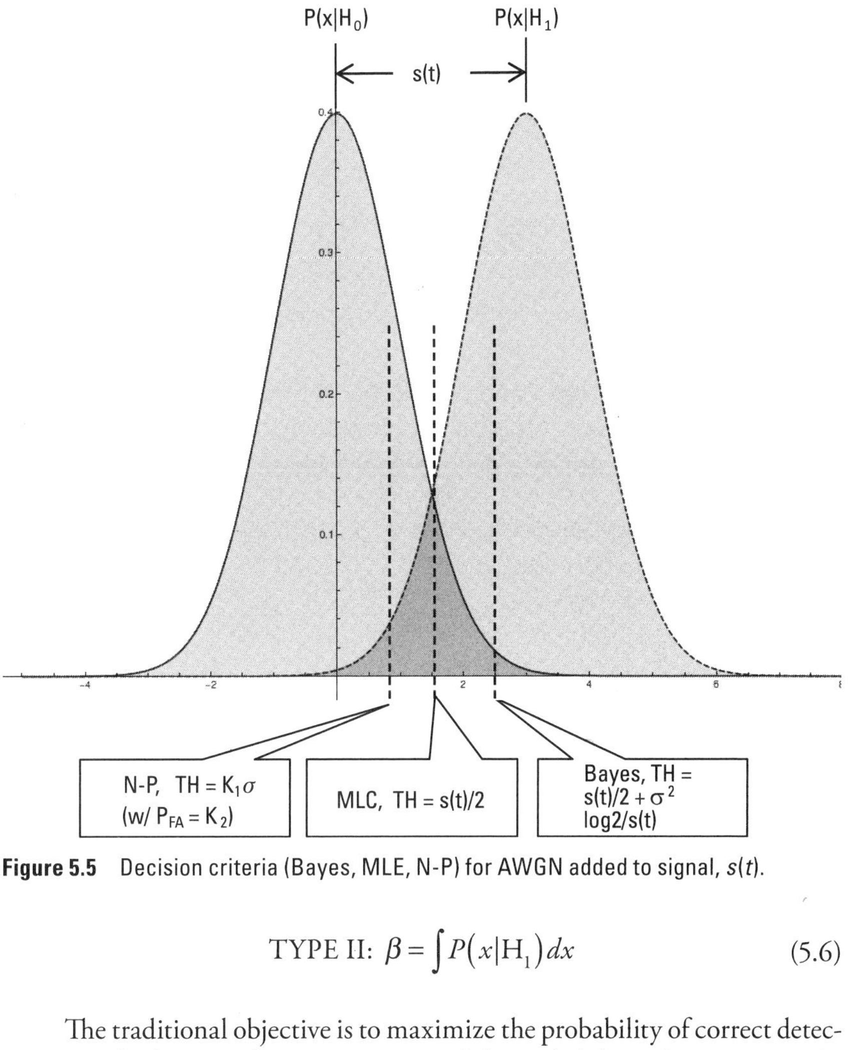

Figure 5.5 Decision criteria (Bayes, MLE, N-P) for AWGN added to signal, $s(t)$.

$$\text{TYPE II: } \beta = \int P(x|\mathrm{H}_1)\,dx \tag{5.6}$$

The traditional objective is to maximize the probability of correct detection, P_d, which comes from summation of the probabilities (correct + misdiagnosed assertions) for H_1 (or H_1)

$$P_d = \int_{R_1} P(x|\mathrm{H}_1)\,dx = 1 - \int_{R_0} P(x|\mathrm{H}_1)\,dx = 1 - \beta \tag{5.7}$$

We seek a decision threshold (TH) that maximizes P_d using the equality constraint $P_{\mathrm{FA}} = \alpha$ through the introduction of a Lagrange multiplier (λ) to maximize the Lagrangian, $L(x, \lambda)$ per [20–22],

$$L(x,\lambda) = P_d + \lambda(\alpha - P_{FA}) = \int_{R_1} P(x|H_1)\,dx + \lambda\left[\alpha - \int_{R_1} P(x|H_0)\,dx\right] \quad (5.8)$$

The product, $\lambda\alpha$, is constant, leaving integration over the R_1 region as the term to maximize by considering only positive values, or

$$P(x|H_1) - \lambda P(x|xH_0) > 0 \quad (5.9)$$

Equation (5.9) establishes the N-P threshold ($TH_{N\text{-}P}$) to choose H_1 as the Lagrange multiplier, λ, via

$$\frac{P(x|H_1)}{P(x|H_0)} > \lambda \quad (5.10)$$

To solve for the value of λ based on an accepted value for P_{FA} (α) requires

$$P_{FA} = \alpha = \int_{\lambda}^{\infty} \frac{P(x|H_1)}{P(x|H_0)}\,dx \quad (5.11)$$

Arriving at how to evaluate the correct assertion of "is a signal present or not" is a critical stage in processing sensor data. This section has allowed us to develop tools to discern the probabilities of guessing wrong, Type I and II errors [(5.5) and (5.6)]. We have also developed a means to estimate probability of detection (P_d) through assumption of noise statistics (from a priori information) and acceptance of a false alarm rate (5.11) for the selected sensor. With tolerance for a nonzero P_{FA} value, we've arrived at the means to estimate a key sensor parameter (TPM/KPP); that is, the probability of detection ($1 - \beta$).

5.6.2 Gaussian Noise Characterization

In assuming a statistical structure for the noise term, there are two major statistical distributions that are widely employed: additive white Gaussian noise (AWGN) [23] and shot noise, which is best modeled as a Poisson distribution. These statistical models are covered thoroughly by numerous references and are summarized here for convenience. AWGN, which indicates random behavior of the noise in observable measurements, is assumed by the system. Use of this model also assumes that noise, $n(t)$, is independent of the source signal, $s(t)$. This model ignores issues such as signal interference, backscattered glint, and fading channel, which are addressed specifically in the laser radar section, 9.3 in Chapter 9. The AWGN probability distribution function, pdf, for additive noise with a mean value of *zero* ($m = 0$) and a standard deviation value of σ, is given as

$$P_N[x(t)] = (2\pi\sigma^2)^{-1/2} \exp\left[\frac{-x^2}{2\sigma^2}\right] \tag{5.12}$$

With measurements assumed to be the sum of the signal and noise functions, $x = s + n$, the hypotheses tests for H_0 and H_1 respectively become

$$P(x|H_0) = P_N(x|H_0) \text{ and } P_N(x|H_1) = P_N(x - s|H_1) \tag{5.13}$$

With H_0, the observation is assumed to consist of noise only, $n(t)$. With the assumption of H_1, the observation is assumed to contain the signal and is equated to $s(t) + n(t)$, where $s(t)$ is a time series of deterministic values. Using AWGN (5.12) for the observation, $x(t)$, and (5.4), the LRT for H_1 becomes

$$\frac{P_N(x - s|H_1)}{P_N(x|H_0)} = \frac{\exp\left[\frac{-(x-s)^2}{2\sigma^2}\right]}{\exp\left[\frac{-(x)^2}{2\sigma^2}\right]} > 1 \tag{5.14}$$

Using the Neyman-Pearson criterion and assuming a $P_{FA} = K_0$, a constant, results with [15]

$$P_{FA} = K_0 = (2\pi\sigma^2)^{-1/2} \int_{\lambda}^{\infty} \exp\left[\frac{-x^2}{2\sigma^2}\right] dx \tag{5.15}$$

When integrating the Gaussian function, the error function, erf(x), is typically employed, which is defined as

$$\text{erf}(x) = \int_0^{\infty} \exp[-t^2]\, dt \tag{5.16}$$

From (5.8), set x to λ and substitute into (5.14) to obtain (using the complementary error function, erfc, definition) the calculable equation for P_{FA}

$$P_{FA} = K_0 = \frac{1}{2}\left(1 - \text{erf}\left[\frac{1}{\sqrt{2}\sigma}\right]\right) = \frac{1}{2}\text{erfc}\left[\frac{1}{\sqrt{2}\sigma}\right] \tag{5.17}$$

Figure 5.5 illustrates decision criteria for three criteria: Bayes, MLE, and N-P. Evaluating the probability of detection using the AWGN model, P_d, becomes

$$P_d = \left(2\pi\sigma^2\right)^{-1/2} \int_{\lambda}^{\infty} \exp\left[\frac{-(x-s)^2}{2\sigma^2}\right] dx = \frac{1}{2}\left(1 + \operatorname{erf}\left[\frac{s-\lambda}{\sigma\sqrt{2}}\right]\right) \tag{5.18}$$

Identifying SNR for this observation as $s/\sqrt{2}\sigma$ illustrates how SNR and P_d are (not surprisingly) directly related. This underscores the identification of TPMs for the sensor subsystem as: P_d, P_{FA} (FAR), and SNR. Within Chapter 9, SNR dependencies are developed, thereby revealing a secondary set of TPMs for sensors.

5.6.3 Poisson Noise Characterization

To address sensors subject to shot noise, such as laser and RF radars, let's consider a second probability distribution function published by Simeon Poisson in 1837. Poisson considered the number of wrongful convictions by considering the number of discrete occurrences that took place within a fixed time interval. Poisson probability theory obtained notoriety when in 1898, Bortkiewicz applied the Poisson probability distribution function (pdf) to the study of the number of Prussian army troops killed by horse kicks [24, 25]. Because Poisson distribution works especially well for event intervals, the Poisson pdf is adaptable to successfully estimating laser radars' performance due to the fundamental nature of optical receivers. Optical detectors respond to optical quanta, and in the detection process, response relates to the arrival rate of photons within an integration and read-out interval for the detector being used. For radar, electrons in transit produce shot noise behavior across a discontinuity. Poisson distribution arises when considering the number of experimental successes consisting of Bernoulli trials. As the number of trials increases, the associated binomial distribution (experiments have only two outcomes) yields the Poisson distribution.

Consider an experiment that requires a time interval Δt, with an average number of times a particular event occurs as λ (not to be confused with the variable name used previously for wavelength), then the probability of n successes in that interval, $P(n)$, becomes

$$P(n) = (\lambda\Delta t)^n \exp\left[\frac{-n\Delta t}{n!}\right] \tag{5.19}$$

From the study of Poisson probability moments, the first moment (mean, μ) is equal to the rate, λ as is the second moment (variance, σ^2). Using the expectation function E [ ], we have mean and variance as

$$\text{Mean: } E\left[P(n)\right] = m = \lambda \quad \text{and} \quad \text{Variance: } E\left[\sigma^2\ \right] = \lambda \tag{5.20}$$

Poisson statistics are employed to evaluate laser radar (and any photon counting) optical system. To develop a LRT using the Poisson probability description, we setup the null and alternate hypotheses, H_0 and H_1. Regarding optical sensors, performance is characterized beginning with the detector, which converts impinging photons to electrical current through a constant multiplier denoted as the detector responsivity, R (units are amps/watts, with input of optical power, P_o, watts). As with any receiver, there are numerous noise sources. For optical detectors, even when no optical signal enters the sensor, there exists a noise (dark) current, i_d. At this point, we can set up the signal and noise equations, using i_n (noise) = i_d and i_r (received) = i_s (signal) + i_n. We have two time series: an input optical signal (and noise) measured as is photons/sec (Q) and an output current with units of amps (coulombs/sec). To convert these time series as rate equations to apply Poisson statistics, we use the photon arrival rate parameter, λ. A quick review of photon electronics centers on the quantization of light. Energy (E, units of joules) per unit time for optical systems uses a well-established relationship, $E = h\nu$, where h is Planck's constant (6.626176×10^{-34} joule-seconds) and ν is the frequency of the optical radiation (signal). For each photon, we have a sensory input. For optical sensors, signal current results from a photon stream (photons/second, Q) impinging the detector, which translates to current (electrons (q)/second). (Unfortunately, there are two "q's" used in literature. The capitalized Q typically used to indicate the photon rate (photons/sec) and lowercase q, the electronic charge ($q = 1.60217662 \times 10^{-19}$ coulombs) used to express current.) Figure 5.6 presents the concept, showing a signal flux (photons/sec) received by an aperture directing the energy to impinge a detector array. Through the responsivity of the detector, photons are converted to a current (q electrons/sec) and are processed by the receiver electronics. The use of the photon quanta supports modeling of the signal as an arrival rate of photons in a defined interval, Δt.

Inserting the average number of photo-electrons per time interval derived from the optical input signal, the null hypotheses (H_0) becomes $i/q = i_d/q$,

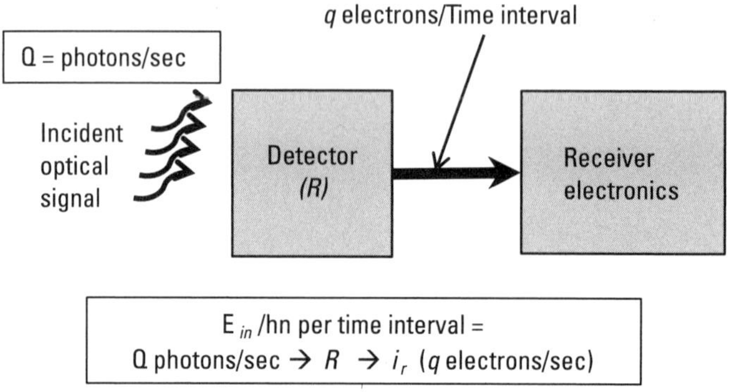

Figure 5.6 Optical input signal conversion to receiver input current (i_r).

and the alternate hypothesis (H_1) becomes $(i_d + i_s)/q$. Using Poisson pdf, the probability of k events taking place in Δt time is, by (5.19) [15],

$$P(k) = \frac{(k\Delta t)^k}{k!}\exp[-k\Delta t] \tag{5.21}$$

Setting up the LRT using the maximum likelihood approach (5.4) results with

$$\frac{P(k|H_1)}{P(k|H_0)} = \frac{\left[(i_d + i_s)\left(\frac{\Delta t}{q}\right)\right]^k \exp\left[-(i_d + i_s)\left(\frac{\Delta t}{q}\right)\right]}{\left(\frac{i_d \Delta t}{q}\right)k \exp\left[\frac{-i_d \Delta t}{q}\right]} \tag{5.22}$$

This simplifies to

$$\frac{P(k|H_1)}{P(k|H_0)} = \left(1 + \frac{i_s}{i_d}\right)^k \exp\left[\frac{-i_d \Delta t}{q}\right] > 1 \tag{5.23}$$

The decision (Bayes) threshold sets up asserting H_1 if observing k events during the interval, Δt, resulting with

$$k > \frac{\log[\lambda]}{\log\left[\lambda + \frac{i_s}{i_d}\right]} + \frac{i_s \Delta t}{q} \tag{5.24}$$

Employing the Neyman-Pearson probability model results in the decision criterion being

$$P(k|\mathrm{H}_1)=\frac{\exp\left[\frac{-i_d\Delta t}{q}\right]}{k!}\left(\frac{-i_d\Delta t}{q}\right)^k \tag{5.25}$$

As before, to estimate probability of detection (P_d), we employ an acceptable false alarm value using the N-P threshold, K_0, and solve the following:

$$P_{\mathrm{FA}}=1-\sum_{n=0}^{K_0-1}\left[\left(\frac{-i_d\Delta t}{q}\right)^n\exp\left[\frac{\frac{-i_d\Delta t}{q}}{n!}\right]\right] \tag{5.26}$$

With K_0, we decide H_1 for $k > K_0$ and H_0 for $k < K_0$. Using the $k = K_0$ value found by (5.25), we can now compute the probability of detection, P_d, as

$$P_{\mathrm{FA}d}=1-\sum_{n=0}^{K_0-1}\left[\left(\frac{(i_d+i_s)\Delta t}{q}\right)^n\exp\left[\frac{-(i_d+i_s)\Delta t}{qn!}\right]\right] \tag{5.27}$$

Using (5.26) and (5.27), we can specify an acceptable level of false alarms and evaluate the probability of detection for a sensor modeled using Poisson characteristics. Considering the decision criterion (threshold) equations, we can list TPM/KPP attributes that can be associated with each sensor subsystem, irrespective of the particular modality in use. Table 5.2 lists these parameters.

The above developed criterion selection for hypotheses testing has been based on observation of a random process. Sensors actually work on time series of measurements and can employ envelope detection, averaging, and correlation through match filtering. Figure 5.7 presents a top-level perspective of the matched-filter approach to the detection of a desired signal using a comparison of input to a threshold. In Figure 5.7, the received signal is convolved with a matched filter. If the signal contains a target signal (pulse), the matched filter attains a peak value compared to any other input since the correlation is based on receiving the expected target signal. For an integration time (or pulsewidth) of Δt, the matched filter will respond with an output signal of duration $2\Delta t$. Although the hardware implementation of the matched filter has only one such functional block, Figure 5.7 illustrates two boxes to show how a received signal is processed [18]. There is always a noise signal, and that result is shown as the bottom path. If the properly selected threshold is

Table 5.2
Sensor Subsystem KPPs

Detection KPP	**Definition**
PFA	Probability of false alarm; related to false alarm rate when expressed over a sampling interval, D_t.
P_d	Probability of detection
SNR	Signal-to-noise, generally measured as received power-to-combined noise power.
AWGN	Additive white Gaussian noise; statistical model assumed for input with a predefined mean (*m*) and variance (*s*2).
TH (l, k_0)	Threshold value, computed through assumption of *a priori* values, such as noise statistics.
Arrival rate, *l*	Poisson *a priori* statistical parameter.

used, that resultant smoother noise signal does not exceed the set threshold (TH) and no detection is declared. When a target signal is present, the input is composed of both signal and noise (shown by the summation point) and the threshold function operates against this summed signal. This illustrates SNR sensitivity to target detection versus false alarm (Type I errors). As the noise power approaches the signal power level, the threshold function will begin to consider noise-only signals as target detections (Figure 5.4). This is why we worked with conditional probabilities to optimize our selection of the threshold value [18], as presented in Figure 5.5. The presentation and use of this generalized detection theory works equally well for both passive and

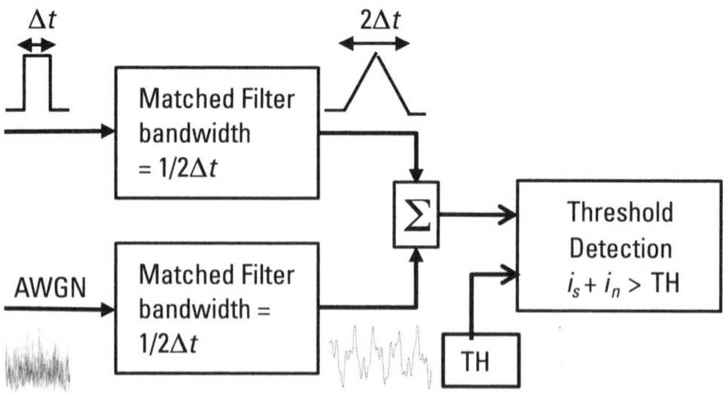

Figure 5.7 Use of threshold detection on output from a matched filter.

active sensor modalities. The equations are not specific to any particular sensor modality and were presented to indicate what parameters are critical to the detection function. Our introduction of matched filtering remains generic; however, as we implement specific sensor approaches to our T-ISR system design, the downstream processing beyond the detection function becomes increasingly specialized.

5.7 Downstream Sensor Functions

Additional functions can be applied based on the particulars of the sensor system. For example, envelope detection (Figure 5.7) operates satisfactorily for direct-detection laser radar, but coherent (homodyne) radars can extract additional SNR gain by mixing backscattered signals with the outgoing transmitted pulse (or if heterodyne, with the local oscillator) [26]. This gain produced by receiver algorithms occurs because sensor designs strive to make judicious use of a priori information to gain SNR. Additionally, detection is the initial function for any sensor subsystem. If the sole objective is to detect the presence of a target, as in a bellringer or tripwire scenario, we could discuss matched filtering and correlation processing and arrive at a reasonable end-to-end description for the sensor. However, given that we are using WSN capability to realize a T-ISR system for complex missions, downstream sensor functions, which include target tracking, reacquisition, discrimination, and target characterization, are of high interest. These downstream functions not only provide critical data concerning the target being observed but can be used to reduce data volume by sending information only for high-valued targets, and by eliminating any false alarms caused by errors in the thresholding.

References

[1] Handy, C., *The Empty Raincoat: Making Sense of the Future*, Hutchinson, 1994.

[2] Oakes, J., R. Botta, and T. Bahill, "Technical Performance Measures," *Proceedings of the INCOSE Symposium*, 2006.

[3] Roedler, G., and C. Jones, "Technical Measurement, A Collaborative Project of PSM," *INCOSE, and Industry, INCOSE-TP-2003-020-01*, 2005.

[4] "The Measureable News," *The Quarterly Magazine of the College of Performance Management*, 2016.

[5] "EIA-632, Processes for Engineering a System," Government Electronics and Information Technology Association Engineering Department, ANSI/EIA-6: 1998, Electronic Industries Association, 1999.

[6] Thorstrom, M., "Applying Machine Learning to Key Performance Indicators," *Master's thesis in Software Engineering,* Chalmers University of Technology, University of Gothenburg, Department of Computer Science and Engineering, 2017.

[7] "MIL-HDBK-520A (Supersedes MIL-HDBK-520 (USAF)," *Department of Defense Handbook System Requirements Document Guidance,* 2011.

[8] "Defense and Program-Unique Specifications Format and Content," Department of Defense Standard Practice, 2003.

[9] Levaardy, V., M. Hoppe, and E. Honour, "Verification, Validation & Testing Strategy and Planning Procedure," *Proceedings of the 14th Annual International Symposium of INCOSE,* 2004.

[10] Engel, A., *Verification, Validation, and Testing of Engineered Systems,* John Wiley & Sons, 2010.

[11] Hill, B., "Assessing ISR: Effectively Measuring Effectiveness," *Air & Space Power Journal,* 2017.

[12] Barton, D., *Modern Radar Systems Analysis,* Norwood, MA: Artech House, 1988.

[13] Cao, Q., et al., "Analysis of Target Detection Performance for Wireless Sensor Networks," *Distributed Computing in Sensor Systems: First IEEE International Conference,* 2005.

[14] Skolnik, M., *Introduction to Radar Systems,* McGraw-Hill Book Co., 1980.

[15] Minkoff, J., *Signals, Noise, and Active Sensors,* John Wiley & Sons, 1991.

[16] Oppenheim, A. V., and G. C. Verghese, *Signals, Systems and Inference,* Pearson Education Limited, 2015.

[17] Eaves, J., and E. Reedy, *Principles of Modern Radar,* Van Nordstrand Reinhold, 1987.

[18] *RCA Electro-Optics Handbook,* EOH-11, 1974.

[19] Neyman, J., and E. Pearson, "On the Problem of the Most Efficient Tests of Statistical Hypotheses," *Phil. Trans. R. Soc.,* 1933.

[20] Arfken, G., H. Hans Weber, and F. Harris, *Mathematical Methods for Physicists, A Comprehensive Guide* (Seventh Edition), Elsevier, 2013.

[21] https://llc.stat.purdue.edu/2014/41600/notes/prob1804.pdf, Purdue University.

[22] Dapaa, G., "A Common Subtle Error: Using Maximum Likelihood Tests to Choose between Different Distributions," *Casualty Actuarial Society E-Forum,* 2012.

[23] Helstrom, C., "The Resolution of Signals in White, Gaussian Noise," *Proceedings of the IRE,* 1955.

[24] Moppett, I., and S. Moppett, "Deaths by Horsekick in the Prussian Army—and Other 'Never Events' in Large Organisations," *Anaesthesia*, Vol. 71, 2016, pp. 17–30.

[25] Sheynin, O., "Bortkiewicz' Alleged Discovery: The Law of Small Numbers," *Hist. Scientiarum*, 2008.

[26] Yang, J., and Z. Zhang "A Balanced Optical Heterodyne Detection for Local-Oscillator Excess-Noise Suppression," *Proceedings of SPIE*, 2012.

6

WSN Wireless Connectivity Design and Performance

The successful operation of any WSN T-ISR system depends upon well-conceived deployment and reliable operation of a large network of sensor nodes. This necessitates appropriate and robust wireless communication at the individual mote level. RF transceivers, for use within WSN systems must be low-cost and capable of operating within the constraints available by node resources (i.e., size, weight and power). Sensor node transceiver designs must also be designed for use in harsh T-ISR environments. WSN design teams have considered using optical communication, via technology such as laser diodes and LEDs; however, issues associated with accurate and clear LOS pointing have precluded this technology from being pursued to the level available with RF connectivity.

The WSN RF transceiver provides nodes with the physical layer (PHY) foundation that enables operation as ad hoc packet-switched networks. When considering design of a WSN system, it is crucial to understand PHY layer requirements for the layout of nodes and the particular transceiver characteristics in use. While voluminous existing documentation provides in-depth details regarding the design and operation of digital wireless communication, WSN RF connectivity presents challenges that differ from those for conventional communications, including;

- The requirement for autonomous network operation—end-to-end, without manned intervention for either network operation and maintenance;
- Operation with an RF electrical center (height) at, or near (within inches), ground level;
- An emphasis on energy-conscious design capable of supporting long-endurance operation with limited energy availability;
- The need for an RF system that conforms to very small packaging to minimize the probability of discovery (clandestine missions), to support storage and shipping/handling, and to be adaptable to working with a variety of military dispersal systems (e.g., ordinance-based delivery, UAV payload dispensing, and hand emplacement by warfighters).
- Operation in harsh environments, uncommon to typical RF communication systems (e.g., blowing grass, animals) due to the proximity to the ground.

This chapter separates wireless networking capability into two distinct design and performance topics: (1) RF signal propagation and (2) RF transceiver design. In discussion of propagation, this chapter details channel modeling required to provide a realistic estimation of RF signal losses and noise characteristics. In addition, the chapter reviews applicable propagation models and transceiver design requirements and approaches and presents signal loss and noise sources. Figure 6.1 illustrates the two nodes linked through an established communication link, one designated as the data source [transmitter (TX)] and the other the sink node [receiver (RX)]. Propagation paths shown in Figure 6.1 illustrate three paths for RF signals: direct LOS, a scattered path, and a reflected path (including an Earth plane). The following sections (6.1.1 through 6.1.10) discuss models used to describe signal losses and RF noise sources for each.

6.1 WSN Link Performance: Overview of Propagation Models

T-ISR system deployments using WSN nodes require a capability to gauge network and sensor performance, but how and what tools exist to do so? This chapter provides a working description and associated mathematical models for electromagnetic (EM) wave propagation. In addition, the chapter discusses hardware implementation as it pertains to WSN node design.

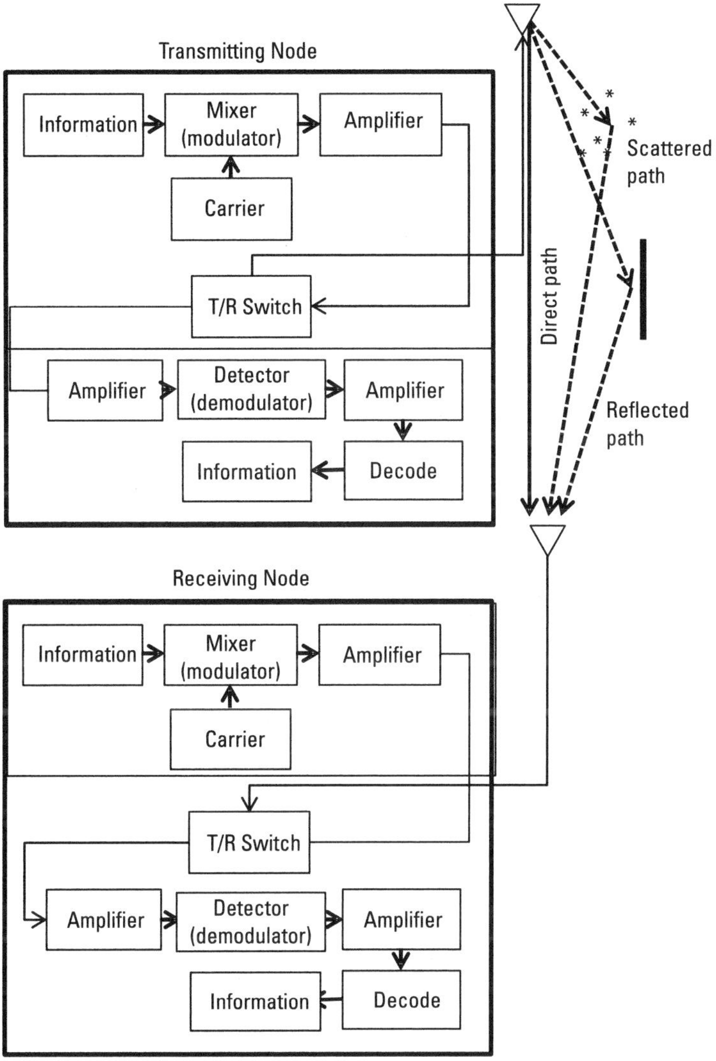

Figure 6.1 RF performance evaluation based on three path cases: direct, scattered, and reflected pathways.

Wireless communication characteristics are closely linked to environmental and physical parameters that impact the integrity of propagating RF signals. Propagation models are used to describe and assess link reliability to determine if we can expect successful transmission of packets (datagrams), and if not, what link characteristics might be improved. Given the transmitter power level, a mathematical description of propagation path(s) and receiver sensitivity values can be calculated through the use of appropriate RF models. RF models are developed through consideration of the underlying physics associated with the propagation of EM waves and the implementation aspects pertaining to the RF transceiver design and link layer protocol implementation (see discussion of MAC presented in Chapter 8, Section 8.6.1). We begin with the description of generally accepted propagation models used to describe signal loss and statistical behaviors of received waveforms. Significant effort has been expended on deriving and verifying appropriate propagation models through numerous RF measurement and analysis programs. With an appropriate propagation model, implementation issues associated with WSN RF transceivers and individual components can be considered, resulting with a responsive transceiver design. Perhaps just as important, values for figures of merit (e.g., received signal strength indicator (RSSI), bit error rate (BER), packet loss) can be estimated.

6.2 Propagation Models

Empirical measurements of RF transmitted and received signal characteristics, including transmit and received strength levels, provide guidance on how best to consider model propagation effects. With inclusion of these effects, which are responsible for the attenuation, distortion, and interference of RF signals along propagation paths, the statistical behavior of the received signal as observed can be accounted for. In the selection of a propagation model, an appropriate transfer function must be chosen based on influential RF parameters, including the transmission signal definition (frequency, polarization, modulation) and aspects of the transmission media (environment descriptors). Regarding media characterization, there is also the consideration of how dynamic behavior might alter performance. WSN transceivers, which operate within the VHF-UHF ISM bands, are constrained to transmit at a relatively low power to operate at a sustainable power consumption level. Usually WSN applications position both transmit and receive antennas (motes) at (or near) ground level. Additionally, sensor modalities must adhere to SWaP2 constraints, which further limit mote-to-mote distances. Ideally, the system design for a WSN node correlates RF system requirements to onboard sensor capabilities to avoid an imbalance in power use and/or capabilities. With short distances, sensor nodes are typically positioned within LOS to one another, which will be shown to create complex behavior for RF signal reception at the receiving mote.

To fully appreciate modeling of RF transmission behavior, propagation models are presented beginning with the basic free-space path loss (FSPL) model developed by Friis [1]. FSPL provides us the starting point to describe link capability. For our system, we must consider WSN antennas being positioned at or close to the ground, resulting in a height much lower than the RF typical antenna electrical center (e.g., for ISM dipoles, this occurs at the half-wavelength of approximately 0.5m). Defining signal loss (L) in terms of transmit (P_t) and received power (P_r) levels at the RF frequency (and at antenna terminals), and using units of decibels the relationship between transmit and receive power can be stated as

$$P_r\,[\mathrm{dB}] = P_t\,[\mathrm{dB}] - L\,[\mathrm{dB}] \tag{6.1}$$

Equation (6.1) describes a basic link budget. If using watts, loss (dimensionless) is defined as the ratio of received power to transmit power. For most of the following development and equations, we will stick with the decibel form when addressing signal loss.

6.2.1 Basic Propagation Model, Free Space (Friis Equation)

Chapter 3 summarizes the FSPL equation for free-space RF propagation (Friis propagation equation). Using (3.13) for a RF signal operating with wavelength, λ, propagating over a distance, d, between transmitter and receiver, and presenting active signal areas of A_t and A_r, respectively, we can estimate received power (P_r) in terms of transmitted power (P_t) using

$$P_r = P_t\left(\frac{A_r A_t}{\lambda^2 d^2}\right) \tag{6.2}$$

In (6.2), we relate receiver to transmitter power. Recasting (6.2) into terms of overall effective antenna gain, G_e results with [2],

$$G_e = \left(\frac{4\pi A_e}{\lambda^2}\right) \tag{6.3}$$

Equation (6.3) can be associated with radio astronomy and from thermal dynamics used to define an effective area, A_e [3]. With (6.3), (6.2) can be rewritten using effective antenna gains for receiver and transmitter, G_{RX} and G_{TX}, respectively,

$$P_r = P_t\left(\frac{\lambda^2}{4\pi d^2}\right)G_{\mathrm{RX}}G_{\mathrm{TX}} \tag{6.4}$$

Equation (6.4) allows for the evaluation of transmitted and received power levels based on antenna gains and the RF signal wavelength. Denoted as a one-ray model, (6.4) illustrates that as RF frequency increases, higher power loss can be expected at the receiver. To overcome FSPL when using higher frequencies, an approach is to employ transmitter and receiver antenna designs that increase directional gain. (Increased transmitter power would also aid in ensuring a reliable link, but at a cost of node power). Unfortunately, implementing directional antennas trades simplicity of setup to obtain precision pointing (e.g., point-to-point communications). Using (6.4) in a link equation, emploing units of decibels for all terms at an RF frequency (f) relating wavelength to the velocity of light, c using ($\lambda = c/f$), we have

$$\mathrm{FSPL} = 20\log(d) + 20\log f + 20\log\left(\frac{4\pi}{c}\right) - G_{\mathrm{RX}} - G_{\mathrm{TX}} \tag{6.5}$$

A fundamental assumption with the FSPL model is that reflection or scattering between transmitter and receiver are nonexistent. The result is that losses from multipath signals are not considered and received signal power estimates are optimistic compared to actual measurements. FSPL presents us with an ideal link loss estimate. However, for near-ground configurations, observed propagation behavior has consistently demonstrated significant loss (>10 dBm) compared to those using antenna heights >50 cm [4]. In addition to ignoring effects arising from signals traveling via multiple pathways, the FSPL propagation model does not account for signal loss due to polarization effects, signal interference, or system distortion effects. Increasingly complex propagation models and assumptions address these signal path loss mechanisms and include the influence of RF noise sources. Multipath models account for increased signal loss through superposition of relatively strong LOS with weaker NLOS signals at the receiver antenna.

Multipath behavior is statistical, and estimates of loss are best described as a random variable. Receiver power loss at different locations, although located at identical distances (d) from a transmitter, can vary considerably. A generalized formula or model to accurately describe path loss in various WSN environments does not exist. As one result, propagation models employ probability distribution functions (pdf) in estimating signal strength (i.e., signal envelop). Assuming this probabilistic model for propagation, effects such as fading, diffuse environments (e.g., rain), and scattering can be addressed and lead to results more aligned to those empirically obtained.

6.2.2 Multipath-Induced Signal Fading

If a number of reflected or scattered copies of the transmitted signal arrive at a receiver, superposition of signals occurs with the resultant instantaneous received power behaving as a random variable. Each copy of the RF signal incurs differences in attenuation, time delay, and signal phase shift. With superposition, arriving signals result in constructive or destructive interference. Typically, the result is a reduction of received signal strength, which is referred to as fading and modeled as a random process-dependent variable that modifies the received signals based on time delay, geographical position, and RF frequency. Strong destructive interference periods are referred to as deep fading, which results in temporary failure of the RF link due to a significant drop of SNR. Fading can be categorized as slow or fast fading, which refers to the rate of received signal magnitude and phase variation. Characterizing fading requires use of coherence time, which is a measure of the minimum time required for magnitude or phase variations to occur and become uncorrelated from previous values. Slow fading occurs when coherence time of the channel is greater than the time delay associated with the receiver detection process. This occurs when an obstruction blocks the main signal path between transmitter and receiver and only NLOS signals arrive at the receiver [5].

During slow fading, amplitude and phase dynamics are relatively constant during detection and signal extraction. Fast fading occurs when coherence of the channel is less than the delay associated with signal reception and processing, which results in signal amplitude and phase varying considerably during processing. Adding the temporal behavior of fading (fast and slow), accuracy in estimating signal loss becomes untenable and requires the system designer to resort to employing the propagation model best suited for the application. Using statistical models is an appropriate method to characterize path loss. Sections 6.1.3–6.1.10 discuss highly utilized fading models that reflect as-measured loss statistics, including the following:

- Two-ray (plane Earth) fading model;
- Lognormal shadowing model;
- Rayleigh fading model

- Rician fading model;
- Two-wave with diffuse power (TWDP) fading model.

6.2.3 Near-Ground Consideration: Two-Ray Fading Model

In considering WSN operation at or near the ground, an approach to propagation loss modeling considers interaction of two RF signals: the direct (LOS) and reflected ray off the ground (NLOS). This description leads to a two-ray (plane Earth) model. The two-ray propagation model, depicted in Figure 6.2, begins to approach observed data associated with near-ground propagation. Assuming antenna heights (h_{TX} and h_{RX} for transmit and receive antennas, respectively) are small compared with total path length (d), the two-ray loss determined using this model, expressed in decibels, is

$$L_{PE} = 40\log(d) - 20\log(h_{TX}) - 20\log(h_{RX}) \tag{6.6}$$

Heights for the isotropic transmit and receive antennas are in units of meters with (6.6) assuming d much larger the heights of the antennas [5, 6].

Unlike the FSPL model, the two-ray model considers the transmitter-to-receiver reflection path and ground reflection characteristics. The use of the two-ray model best correlates with observed data where the RF ground reflection coefficient approaches the ideal value, −1. Regarding WSN T-ISR applications, where antenna heights are <5 cm, test results indicate convergence of the Friis model with the two-slope approach, which is geometrically heuristic as LOS and reflected waves are virtually following similar paths. Although the rationale behind two-ray modeling appears reasonable, there are issues at higher frequencies at low heights above ground configuration. Measurements at 2.4 GHz indicate significant fluctuations in received power for short-range configurations (d < 100m) with antenna heights <50 cm [5]. There appears a critical distance, dc, which indicates a shift in the exponential

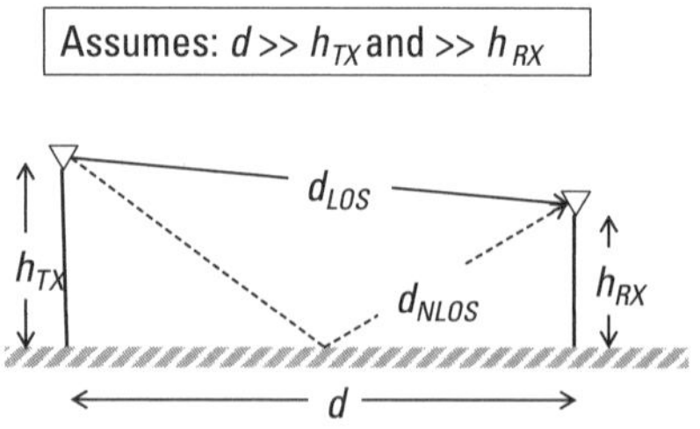

Figure 6.2 Two-ray (plane Earth) propagation model.

loss behavior with distance. Figure 6.3 illustrates the two regions of path loss for near-ground propagation, where the first region consists of ripples, where peaks and troughs exist and falloff varies as d^{-2}, while a second region ($d > d_c$) exhibits falloff without power ripples and loss varies as d^{-4}. The breakpoint [7] between these two regions occurs at d_c, given by

$$d_c = \frac{4\pi h_{\text{RX}} h_{\text{TX}}}{\lambda} \tag{6.7}$$

Nulls are representative of points where direct and reflected signals cancel while minor gains (see the Friis model) are points where signals are additive. The nulls exist in the range of falloff, before the breakpoint, d_c. The breakpoint is the position, where the transition of falloff started. After this point no nulls exist.

Figure 6.3 indicates signal ripple until a certain distance. The ripples are caused by superposition of LOS path and echo (NLOS) paths due to reflection, diffraction, and scattering. Arriving signals reach the receiver from different directions and recombine different amplitudes, phases, and time delays with regard to one another. The received signal is modeled as the sum of contributions from all paths and modeled by a multiray propagation model as shown in (6.8) [7],

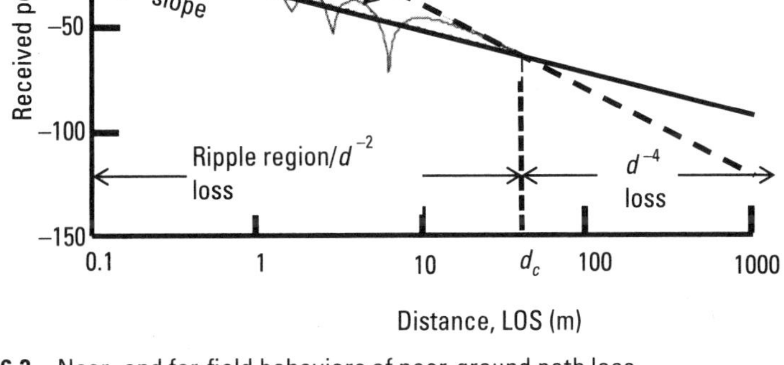

Figure 6.3 Near- and far-field behaviors of near-ground path loss.

$$P_r = P_t\left(\frac{\lambda}{4\pi}\right)\left|\sum_{i=0}^{N}\frac{c_i\,\Gamma_i\sqrt{G_i}e^{-j\Delta\phi_i}}{d_i}\right|^2 \tag{6.8}$$

with P_r and P_t representing received and transmitted power levels, RF frequency with wavelength λ, and overall antenna gain of G_i. The coefficients, c_i and Γ_i, are respectively the decaying and reflection values for path i. The path length for the individual arrival signals, d_i, associated with phase shift of the ith path ($\Delta\phi$), as

$$\Delta\phi_i = 2\pi\frac{\left(d_i - d_0\right)}{\lambda} \tag{6.9}$$

Equation (6.8) produces a periodic deep fade behavior that can be tracked (without the high-frequency fade content) by the two-ray model [7]. Although the two-ray model closely mimics measured characteristics at near-ground antenna heights within ISM RF bands (VHF-UHF), obstructions (shadowing) should be considered; these are addressed through use of a lognormal shadowing model.

6.2.4 Near-Ground + Obstructions: Lognormal Shadowing Model

Recall that the Friis free-space model also assumes an unobstructed path between the transmitter and the receiver. To address propagation loss for a wide range of environments that present random shadowing effects due to signal blockage from terrain, vegetation, and other objects, an extension to the Friis free-space (and multipath fading model) is required to account for such shadowing effects. An approach widely used is to apply a lognormal behavior to inject the randomness seen within RF ground measurement data [8]. In the far-field region of the transmitter ($d > d_0$), lognormal path losses LNS (d) for an arbitrary distance $d > d_0$ is given by (6.7) [9]. A shadowing term (χ) is introduced, which is a zero-mean Gaussian distributed random variable with shadow variation (standard deviation), σ. Fit parameters, path loss exponent (n), and shadowing variations are employed to represent the impact associated with the deployment environment (e.g., vegetation, building or tunnel interiors, and urban settings) on the RF signal propagation. A reference path loss LNS (d_0), called close-in reference distance, is obtained using the Friis path loss equation (6.5) or by field measurements at d_0. From RF measurements [8], values for n range from 1.6 (interior spaces) to 6 for shadowed urban or building obstructions (for completeness, free space, $n = 2$).

$$L_{\text{LNS}}(d) = L_{\text{LNS}}(d_0) + 10n\log\left(\frac{d}{d_0}\right) + \chi \quad \text{for } d_c \leq d_0 \leq d \tag{6.10}$$

6.2.5 Rayleigh Fading Model

When scattering occurs with numerous objects having dimensions at or smaller than the RF wavelength (λ) we again see a superposition of all arriving signals; however, underlying statistics of the paths changed the larger objects ($>\lambda$) that were used by the lognormal models. With numerous scatter centers within various propagation paths, the central limit theorem leads to a channel impulse response that can be expressed as a Gaussian process regardless of the distribution of individual scattered components. The received signal is modeled as a summation of individual contributions arising from these independent scattering centers, and the composite signal characteristic is modeled as orthogonal uncorrelated Gaussian distributed random variables [10]. Figure 6.4 presents the schematic of received signals at the receiver antenna, assuming a simple case of 2 Gaussian (probability and AWGN) distributed returns. Figure 6.4 also illustrates Rayleigh pdf and cdf (cumulative distribution functions, respectively) for differing values of the standard deviation, σ.

If the transmit signal does not directly reach the receiver, the orthogonal random variables have zero mean and uniform phase distribution (0 to 2π radians). For this scenario, received signals are comparable in amplitude and Rayleigh distribution is considered as the best fit to empirical results. Transforming the resultant envelope statistics into a radial variable ($r > 0$) and associating loss values as Rayleigh-distributed random variables having zero mean and 2nd-moment statistic, $E[r^2] = \sigma$ [11] results with

$$p_R(r) = f(r,\sigma) = \frac{r}{\sigma^2}e^{-r^2/2\sigma^2} \tag{6.11}$$

The resulting receiver signal level is a combination of the major loss component remains (FSPL), shadowing, and behavior associated with the Rayleigh fading (a smaller effect with respect to the former two mechanisms). How rapidly channel fades occur is determined by how fast the scattering centers, receiver, and/or transmitter are moving relative to the transmission path (where the use of direction cosines is required). Of note, the motion of a transmitter or receiver (e.g., MANET) may produce detectable Doppler shift in signal components and create an additional term to be considered in path loss. With static applications, where the transmitter and receiver nodes are

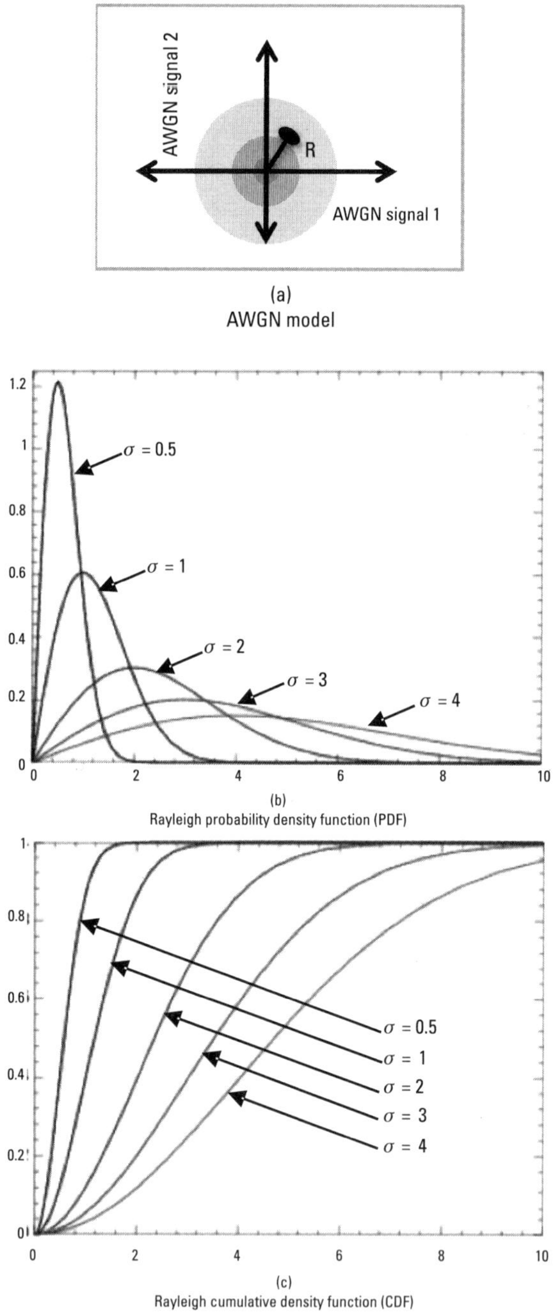

(a)
AWGN model

(b)
Rayleigh probability density function (PDF)

(c)
Rayleigh cumulative density function (CDF)

Figure 6.4 Rayleigh distribution signal schematic: (a) the schematic (model), (b) the PDF, and (c) the CDF.

stationary, the path may still inject a Doppler component produced by the motion of scatter centers (e.g., blowing vegetation or precipitation). Consideration of Doppler effects is discussed in Section 6.2.9.

6.2.6 Rician Fading Model

Assuming repeating the previous scenario set for Rayleigh fading, where numerous scattering centers exist within signal pathways, we consider the presence of a dominant signal, which is the directly received transmitted signal (LOS). In this scenario, the received signal envelope (r) follows a Rician distribution, with dominant amplitude (A). Using a modified Bessel function of the first kind, J_0, Rician loss is estimated using [12],

$$p_R(r) = f(r,\sigma) = \frac{r}{\sigma^2} e^{-\left(r^2+A^2\right)/2\sigma^2} J_0\left(\frac{rA}{\sigma^2}\right) \tag{6.12}$$

If the dominant amplitude parameter, A, is set to zero, we revert back to the Rayleigh fading model, as expected. Figure 6.5 depicts the signal schematic for Rician behavior.

Although using a probabilistic approach aids with tractability, for static applications (where the nodes are fixed and the environment is unchanging, such as an interior deployment in a building or tunnel), fading is inherently deterministic. Time fluctuations come into the picture if the local topology changes dynamically or if one of the nodes moves about; a mobile node will respond to a different topology depending on its position with respect to its linked node. Motion of objects in the pathways would also be seen as a topology change and cause similar time-of-arrival (TOA) fluctuations.

6.2.7 TWDP Fading Model

The TWDP propagation model regresses to the previous two models in specific cases. If a node-to-node link incurs multipath behavior resulting primarily from reflective objects within the RF pathway, the received signal is best characterized through an assumption of two constant-amplitude signals along with numerous, smaller-amplitude signals randomly phased with respect to one another. This is the RF environment assumed by the TWDP fading modeling [14]. The constant amplitude signals are referred to as specular components of a fading model. The TWDP-distributed envelope, R, is a summation of these constant amplitude signals with phase terms modeled as independent uniform random variables, U_1 and U_2 over the interval [0,1]. Combining these

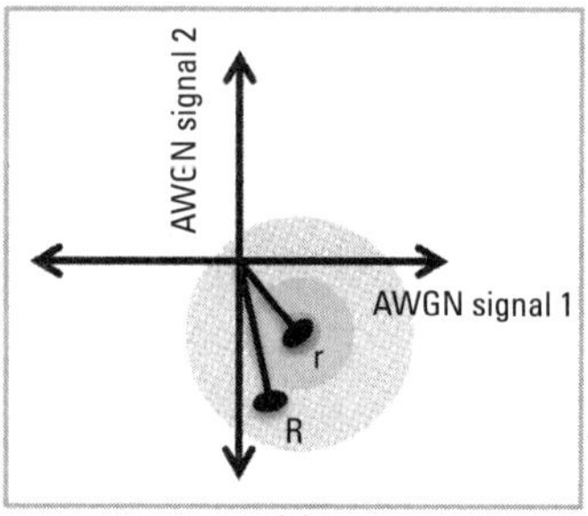

(a)
AWGN model.

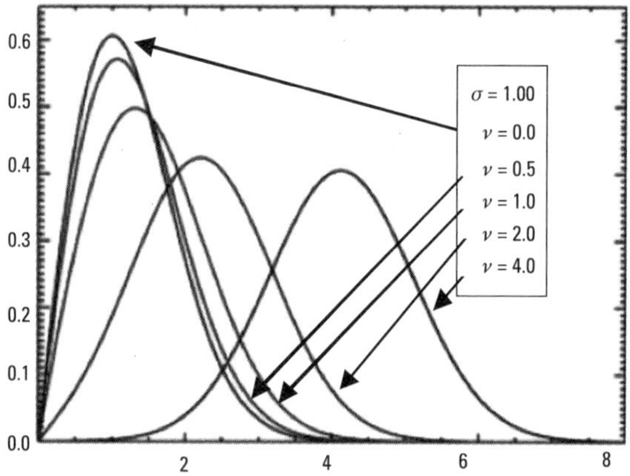

(b)
Rician Probability density function (PDF)

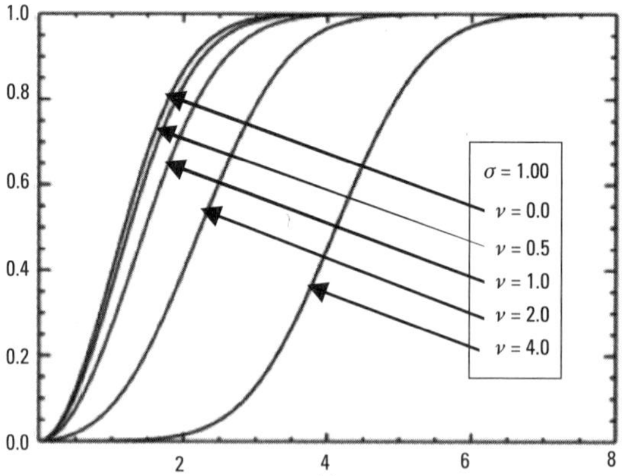

(c)
Rician Cumulative density function (CDF)

Figure 6.5 Rician distribution, appropriate signal schematic: (a) the schematic (model), (b) the PDF, and (c) the CDF.

constant amplitude and phase variables with independent, zero-mean Gaussian random variables, X and Y, with equivalent standard deviation value (σ) results with a TWDP model, as

$$R = \left\| V_1 e^{j2\pi U_1} + V_2 e^{j2\pi U_2} + X + jY \right\| \tag{6.13}$$

The pdf of the received signal envelope associated with TWDP fading is characterized using amplitudes of the two individual waves, average power P, the ratio of peak specular-to-average diffuse power (K), and disparity (Δ) between the two specular components. Average power and disparity are as follows:

$$P = V_1^2 + V_1^2 = 2\sigma^2 \tag{6.14}$$

$$K = \frac{V_1^2 + V_2^2}{2\sigma^2} \tag{6.15}$$

$$\Delta = \frac{2V_1 V_2}{2\sigma^2} \tag{6.16}$$

The K factor can vary from 0 to ∞. When $K = 0$, the envelop, R, equates to Rayleigh fading. With $K = \infty$, R corresponds to two-wave envelope fading experienced according to a transmission line with reflections. The disparity parameter varies from 0 to 1. With a disparity = 0, the TWDP model regresses to Rician fading. Figure 6.6 compares the two-wave behavior with that associated with Rayleigh and Rician distributions.

Unlike the special cases of Rayleigh and Rician fading, there is no simple, closed-form solution for TWDP fading. Instead, the exact pdf is the result of the following definite integral [13]

$$f_R(r) = r \int_0^\infty J_0(rv)\, e^{-v^2\sigma^2/2} \left(\prod_{i=1}^{N} J_0(V_i v) \right) v\,dv \tag{6.17}$$

Numerous techniques have been proposed to approximate TWDP pdf in closed form or to enable evaluation of associated statistics directly (refer to the selected bibliography for more details).

6.2.8 Selective Frequency Fading

Selective frequency fading is the partial cancelation of a RF signal due to the arrival of an signal through two separate paths, with one (or both) path(s)

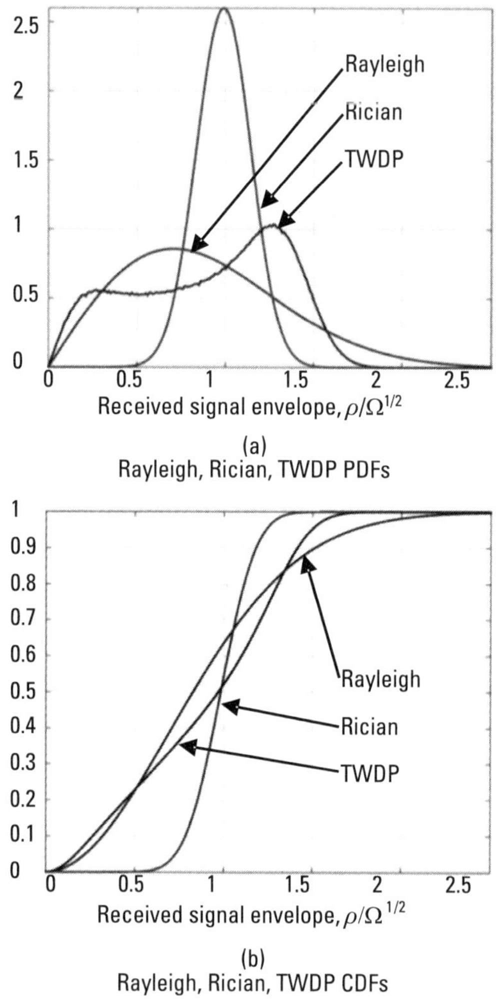

Figure 6.6 TWDP fading: (a) PDF, and (b) CDF.

having various arrival times at the receiver. Selective fading manifests as a slow, cyclic disturbance, with maximum cancelation strongest at a particular frequency. As the carrier frequency of a signal is varied, the magnitude of the change in amplitude at the receiver varies. A characteristic of channels experiencing multipath effects is a measure denoted as multipath time delay spread. Multipath time delay spread can be considered the difference between the TOA of the earliest significant multipath component (typically, the LOS component) and the TOA of the final multipath component. Multipath time delay spread can be quantified using the root mean square (rms) of the delay spreads (τ_i), which are statistical in nature. Mean channel delay $E[\tau]$, is evaluated using the power delay profile of the channel by $S(\tau)$, [14]

$$E[\tau] = \frac{\int_0^\infty \tau S(\tau)\,d\tau}{\int_0^\infty S(\tau)\,d\tau} \tag{6.18}$$

The rms delay spread, τ_{rms}, is obtained through the standard deviation of the normalized delay power density spectrum,

$$\tau_{rms} = \sqrt{\frac{\int_0^\infty (\tau - E[\tau])^2 S(\tau)\,d\tau}{\int_0^\infty S(\tau)\,d\tau}} \tag{6.19}$$

Using the mean multipath time delay spread, the bandwidth over which the channel can be assumed to be flat (spectral components are passed the channel with comparable gain and linear phase) is denoted as the coherence bandwidth, BW_c. This is the bandwidth over which the channel transfer function remains virtually constant. Coherence bandwidth is a statistical measurement of the range of frequencies likely to experience correlated amplitude fading. If the multipath time delay spread equals τ seconds, coherence bandwidth is can be approximated using

$$\text{BW}_c = \frac{1}{\tau}\ \text{Hz} \tag{6.20}$$

With frequency-selective fading, coherence bandwidth of the channel is smaller than the signal bandwidth, and signal power fades at particular frequencies. Frequency-selective fading produces a cloudy pattern to appear on a spectrogram, such as that depicted in Figure 6.7. Time is on the abscissa, frequency on the ordinate, and signal strength indicated as gray scale. Strong destructive interference is frequently referred to as a deep fade and may result in temporary failure of communication due to a severe drop in the channel SNR. [Constant (stable) frequency components are shown in Figure 6.7 as the dark, horizontal lines.]

Several diversity schemes exist that can mitigate frequency-selective fade effects, including diverse spatial reception and frequency modulation. Spatial diversity employs multiple antennas spaced a quarter-wavelength apart. The receiver continuously compares signals arriving at the antennas and passes

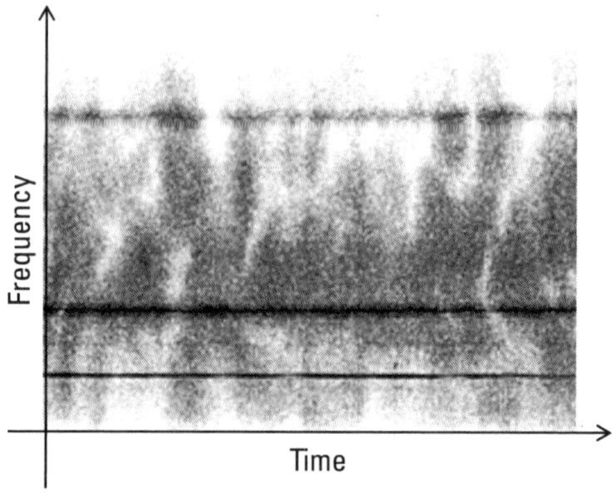

Figure 6.7 Frequency-selective time-varying fading.

the better signal. If outside the flat response region, frequency components experience uncorrelated fading. Since different frequency components are affected independently, it is unlikely the entire signal would simultaneously be affected by a deep fade. Channels having dispersive characteristics experience frequency-selective fading whereby signal energy is spread across frequencies causing transmitted symbols to interfere with symbols on adjacent frequencies. Equalizers within the receiver can compensate for the effects of intersymbol interference. Frequency diversity approaches include orthogonal frequency-division multiplexing (OFDM) and code division multiple access (CDMA) modulation. Using OFDM, a wideband signal is parsed into modulated narrowband subcarriers, which results in each subcarrier exposed to flat fading rather than frequency selective fading. With CDMA, the receiver is configured as a multiple set of received signal processors (rake processors) that processes each echo signal separately. Selective fading can also be resolved through error coding, simple equalization, or adaptive bit loading. Intersymbol interference (ISI) is avoided by introducing a temporal guard interval between symbols. If symbol duration is greater than the delay spread (by a factor of approximately 10), the channel is considered ISI-free.

6.2.9 Mobility-Induced Selective Frequency Fading

Although WSN deployments typically position nodes statically, the exfiltration relay or a moving group of nodes may be part of a WSN network (e.g., UAV-based relay, roving unique-sensor node, ground troops). This leads us to a discussion of propagation modeling relating to relative velocities between nodes to account for Doppler effects. In our discussion of propagation models, we considered a transmitted signal, $s(t)$, with frequency (f_c) and phase (ψ) of the form $s(t) = \cos(2\pi f_c t + \psi)$. Doppler effects on general RF propagation modeling are well-documented, especially for cellular telephony [15–19]. For a mobile environment, fading is categorized as large-scale fading or small-scale fading. Large-scale fading creates attenuation or path loss due to motion over large areas and results in path loss as a function of distance. Large-scale fading is characterized as a mean-path loss associated with lognormal variation about the mean.

Small-scale fading produces short-term fluctuations in received signal amplitude due to multipath. Small-scale fading is classified as flat or frequency selective fading and follows that presented in Section 6.2.8. With numerous reflective paths and no direct LOS signal component, small-scale fading exists and follows a Rayleigh behavior. When a dominant nonfading signal component is present, such as a direct LOS propagation path, the resultant small-scale fading envelope is described using Rician distribution. When a mobile node receives a large number (N) of reflected and scattered waves, the instantaneous received power is considered a random variable, dependent on the location of the antenna. We use as our model signal, a carrier signal of the form, $s(t) = \cos(2\pi f_c t + \phi)$, with the nth signal component has amplitude, a_n, phase, ϕ_b, and an arrival angle at the receiver relative to the direction of motion of α_n. The Doppler shift of this signal frequency change is based on transmitter approaching or receding as well as the and angle of the velocity vector (α_i), as

$$\Delta f_i = \pm \frac{V}{\lambda} \cos(\alpha_i) \tag{6.21}$$

Here, V is the relative velocity along the transmitter-antenna LOS. The received signal $r(t)$ can be expressed as

$$r(t) = \sum_{i=1}^{N} a_i \cos\left(2\pi f_c t + \psi + \phi_i 2\pi \Delta f_i t\right) \tag{6.22}$$

If using the quadrature approach in the transceiver (see Section 6.4.3) the in-phase $I(t)$ and quadrature $Q(t)$ components are expressed as

$$I(t) = \sum_{i=1}^{N} a_i \cos\left(\frac{2\pi f_c t}{c} \cos\alpha_i + \psi + \phi_i\right) \tag{6.23}$$

$$Q(t) = \sum_{i=1}^{N} a_i \sin\left(\frac{2\pi f_c t}{c} \cos\alpha_i + \psi + \phi_i\right) \tag{6.24}$$

A signal traveling along a propagation undergoing Doppler shifting, but that does not incur scattering, registers as an instantaneous frequency shift, Δf

$$\Delta f = \frac{d\phi(t+1)}{dt} - \frac{d\phi(t)}{dt} = \frac{V}{\lambda} \tag{6.25}$$

A signal consisting of both specular and multipath components alters instantaneous frequency time dependency and produces frequency variations greater than $\pm V/\lambda$ with nonlinear RF (square-law) detection [20]. With node-to-node velocity, the phase for the spectacular component may rotate through an angle θ, causing the received signal phase to rotate an angle $>\theta$. However, with diffuse signal occurring simultaneously with fade, the rate of change of phase is much greater than that caused by the specular Doppler. At high relative velocities (V), the instantaneous frequency shift, $d\phi(t)/dt(t)$ is significantly larger when fades compare to specular component, which increases signal errors. Coherent demodulators that lock onto and track the information signal can suppress this frequency (FM) noise and reduce the impact of Doppler shift. However, for large Doppler shifts, unmodulated signal (carrier) recovery is difficult because wideband (relative to the data rate) phase-lock loops (PLLs) are required.

6.2.10 Additional RF Path Loss Models

Evaluating alternative distributions to propagation measurements has resulted in the development and investigation of alternative distributions, such as Nakagami(-m) distribution [21, 22]. However, an issue associated with this distribution is that it is not based on physical quantities per se; rather, the application is to attempt to match measurement behavior obtained from test programs [23]. In addition to previously mentioned models, there are path loss models that have received attention in the documentation of RF propagation modeling, such as the COST231-Hata model and the COST231-Walfish-Ikegami model [24]. However, these models assume antenna placement higher than 1m (or even 10m) above the ground. As such, these models are widely used to evaluate cellular networks, but are not necessarily suitable for WSNs in most cases [25, 26].

6.3 WSN Transceiver Characteristics

Fortunately, the emergence of low-power miniaturized transceivers supports the SWAP constraints associated with WSN node design. Large-scale RF chipset production have led to low-cost WSN connectivity approaches through the use of stand-alone RF transceivers, or more recently, integrated microcontroller/transceiver SoCs. Common features of these SoC devices include the following:

- Extremely low power consumption;
- Multiple operating and sleep (power-saving) RF operational modes;
- High receiver sensitivity and frequency selectivity;
- Low output and adjustable transmit power;
- Readiness to interface to various microprocessor families;
- Software defined radio (SDR) real-time processing using specialized digital signal processing (DSP) devices and/or gate arrays (FPGA).

6.3.1 Transceiver Performance

Figures of merit for transceiver performance are related to the SNR at various points throughout the transceiver processing, starting at the receiver antenna terminals. In discussing signal losses and noise sources within a transceiver design, the use of a generic block diagram is presented in Figure 6.8. Using this figure, four major activities associated with a WSN transceiver are highlighted, namely:

- Transmit (TX) and receive (RX) RF filtering, amplification, routing, impedance matching and switching via transmit/receiver (T/R) switch. Within this segment, signals are analog, and processing operates at the carrier (interim) frequencies.
- Intermediate frequency (IF) and baseband frequency processing, which includes mixers for modulation and demodulation (envelop detection, filtering, amplification, stable frequency source [local oscillator (LO)], along with sampling circuitry associated with A/D and D/A converters. This segment is responsible for the digitalization of analog signals, a process that continues to move further toward the receiver antenna to avoid difficulties related to analog signal processing.
- Microcontroller (uC) and DSP logic, which involves interface (I/O) ports, memory capability, and associated software and firmware to implement boot logic and SDR functions. The encoding/decoding, encryption and spread/despread functions are realized within this segment.
- Power source (e.g., batteries), control, and distribution. This segment is responsible for conditioning, routing, and regulating voltage supply for all transceiver functions

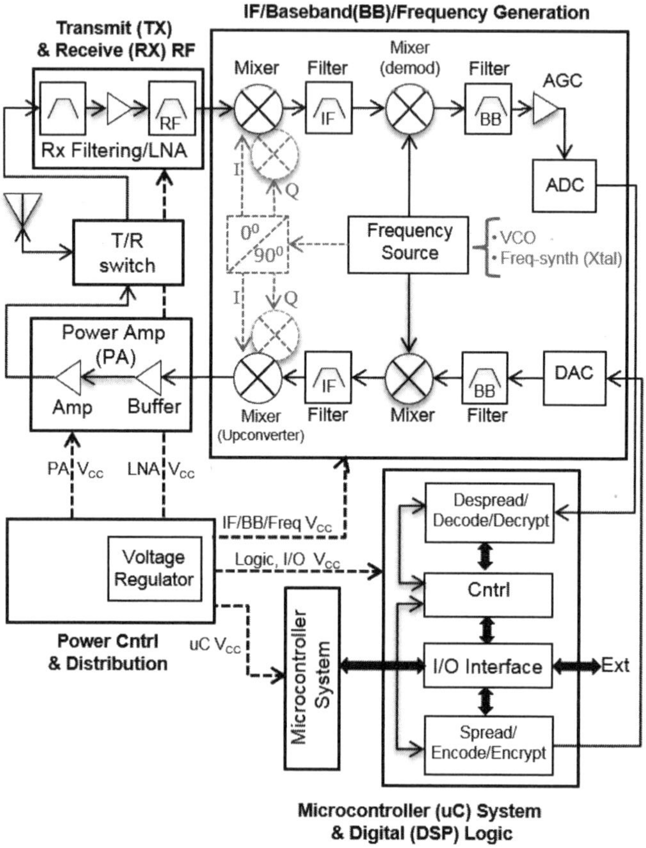

Figure 6.8 Functional block diagram for RF transceiver (dashed lines depict single-channel and quadrature sampling).

Using Figure 6.8, and starting with the TX/RX RF subassembly, an appropriately tuned and impedance-matched antenna is used for both transmission and reception of signals. Antenna design is relatively complex for WSN applications, as competing objectives exist to promote successful transmission and reception of signal. Efficient use of RF energy is dependent upon the proper matching of the antenna resonant wavelength and height above ground using a compact, miniaturized volume. Power transfer between the antenna and signal connection (transmission line) employs a balun (balanced-to-unbalanced) design to power-match transceiver circuitry with the antenna for both transmission and reception. In particular, a balun converts differential signals to single-ended signaling (and the reverse) [27]. The element before the antenna using Figure 6.8 is the transmit and receive switch (T/R switch), which shares resources and ensures sufficient isolation of transmit circuitry from that used by the receiver. Switching times for CMOS-based T/R switching structure is < 90 ns with an insertion loss (at 2.4 GHz) of 0.5 dB, and isolation between RX and TX of 25–30 dB [28]. Signals arriving at the antenna terminals (and passed through the T/R switch) are routed to a bandpass filter to reject out-of-band frequencies and to allow operation of downstream elements to work in a narrowband about the intended signal [29]. Filtering at this stage aids in maintaining linearity at the low-noise amplifier (LNA) and subsequent mixer(s). An LNA is inserted to boost the incoming signal to re-establish a link margin for signal losses and noise additions expected further downstream in the receiver.

The LNA must provide sufficient gain to establish the system noise figure. Noise figure is the difference, measured in decibels, between the receiver noise output responding to standard noise temperature (290K) and the output measured without an input signal. Thermal noise (Johnson-Nyquist noise) is determined using the product, kTB, with k the Boltzman's constant, B the bandwidth of the signal, and T the absolute temperature of the electrical load. Thermal noise power, kTB watts/Hz, is derived considering voltage variance (using the one-sided power spectral density). The received signal is routed to a down-converter mixer that converts the VHF-UHF carrier to an IF. Using an IF provides operational advantages by allowing filtering to work at a bandwidth proportionate to the lower IF frequency, versus the signal carrier frequency. Figure 6.8 shows a single-channel IF/baseband section in bold, and a quadrature-processing configuration as dimmed (gray) elements. In some systems, the RF signal is fed directly to the demodulator.

The IF signal continues and enters an IF filter, which limits system bandwidth and reduces undesired spurious signals (mixer images). A third mixer (demodulator or detector) removes transmitter modulation from the IF signal and directs the resultant baseband (BB) signal through a final filter to an ADC for digitalization. With the digitized stream, the signal is routed to a microprocessor system and/or DSP (depends upon the transceiver configuration and design). Data recovery is achieved through despreading of the signal (assumes that a spreading code using pseudorandom numbers (PRN) has been applied to the transmitted signal). Additional processing is required to decode and de-encrypt the packetized data as well. When using modulation schemes based on quadrature processing, the demodulator splits the received signal into in-phase (I) and quadrature (Q) components. At each mixer stage, the appropriate LO signal is introduced to provide frequency conversion sought after through signal multiplication. Stability of the LO is a concern because a significant phase noise could occur at this stage and produce spurious frequency components.

The formation of a data packet (datagram) for transmission occurs within the microcontroller system and DSP logic. The data stream to be transmitted is converted into an analog signal via the D/A converter (DAC). This BB signal is filtered to remove digitalization spikes from the DAC, then routed to a mixer that modulates the BB signal into an IF signal using the LO. In an FM system, the frequency source may be modulated directly. As in the receiver path, the IF is often bypassed with direct RF modulation. The final stage prior to entering the RF section is the power amplifier (PA). The PA amplifies the transmit signal to an appropriate level based upon an estimate of propagation losses and receiver capability. Depending on the modulation scheme chosen, the PA may have to be linear. For FM systems, the PA can be a class C amplifier where efficiency can be quite good. For QPSK schemes, the amplifier must be class AB and operated a few decibels below gain compression. The final subsystem in the functional train of Figure 6.8 is the power control and distribution function. This function provides power management and conditioned power to all transceiver subsystems. For WSN nodes, this subsystem is usually associated with a battery power source. Software-based (middleware) control manages power switching and conditioning to power/depower various transceiver components and subsystems to promote long-term life.

Figure 6.9 presents an example of modifying a WSN node from use of a 15-inch dipole antenna (Figure 6.9 (a)) to a chip antenna design (Figure 6.9 (b)) that includes an additional software-controlled PA. This particular mote, a joint venture of Crossbow, Ohio State University (OSU), and the University of California under auspices of the DARPA NEST program (Chapters 1 and 11), was a test bed for mote design and used by numerous demonstrations to evaluate WSN capability of WSN to T-ISR problems [30]. Simultaneous with this effort, research continued on reducing the antenna footprint while attempting to maintain a reasonable transmitter range without need to increase transmitted power. Figure 6.9 illustrates modifications to the DARPA XSM mote, from the initial experimentation model to a model used in operational evaluations by various DoD users.

Various RF chip transceivers are suitable for WSN designs and are classified by modulation schemes and system architectures. Although complex modulation such as orthogonal frequency division multiplexing (OFDM) addresses selective fading and provides effective spectrum usage, power considerations have transceiver manufacturers relying on simpler modulation, such as: on-off keying (OOK), frequency shift Keying (FSK), ultra-wideband (UWB), minimum shift keying (MSK), binary phase shift keying (BPSK), and quadrature phase shift keying (QPSK) [31].

Figure 6.9 Antenna redesign for the tactical XSM2 mote (TXSM) : (a) the orginal XSM2 mote, (b) the replacement chip antenna and power amp, (c) the overall repackaged XSM, (d) addition of a camouflage skin to produce the T-XSM, and (e) the T-XSM mote in the field.

6.3.2 Signal Loss Mechanisms and Noise Sources

Reviewing the transceiver block diagram (Figure 6.8), numerous signal loss mechanisms and noise sources can be discussed. There is significant filtering throughout transceiver processing to aid in frequency control; however, realizable filters do not have instantaneous cutoff profiles and if insufficient frequency spacing occurs, a signal envelope may enter an adjacent signal channel resulting with energy displacement from one signal to another. When the power spectrum of a signal becomes entangled with another signal power spectrum, intersymbol interference (ISI) ensues (Section 6.2.8). The imperfection associated with filtering results with ISI, which can interfere with a signal in the absence of thermal noise [32]. As previously noted, jitter in the reference LO phase causes signal loss during the detection process. Phase jitter of the LO signal when used in modulation creates out-of-band components that get rejected by the subsequent filters causing a loss of signal power. With modulation, the energy expended to transmit the carrier is energy lost from the information-relevant signal at the receiver, denoted as modulation loss.

Regarding propagation, in addition to signal losses there are atmospheric attenuation and noise sources. Atmospheric attenuation depends upon the transmitter to receiver path (distance and RF path through the atmosphere) and carrier frequency. For our application, WSN nodes are separated approximately 10–100m, minimizing atmospheric attenuation with the exception of weather effects (e.g., heavy rain), which produces both loss and noise [32]. Propagation noise also arises from background emissions, such as zodiacal RF emission and other celestial objects within the receiving antenna field-of-view. Wideband sources, including background noise and thermal noise at the antenna, also prompts bandpass filtering prior to the LNA. This prefiltering minimizes noise energy entering the receiver; however, being white noise, inevitably noise power remains and enters the receiver system. Adjacent channel interference occurs when unwanted signals from other frequency channels leak energy into the frequency band being processed. Co-channel interference results from an interfering signal energy appearing within the signal bandwidth, or from other channel users where insufficient discrimination occurs. Due to the nonlinear combination of multiple signals at the detector, intermodulation (IM) noise is generated and becomes a noise component to be evaluated and considered in the design.

6.3.3 Quadrature Sampling Advantages

Quadrature sampling is the process of digitizing a continuous (analog) bandpass signal and translating the resulting spectrum to be centered about 0 Hz. The objective to quadrature sampling is to obtain a digitized version of the analog bandpass signal, with the discrete spectrum centered at 0 Hz and not at the carrier (f_c) or IF (f_{IF}) frequency. This is accomplished by introducing a time signal, $e^{-j2\pi f_c t}$ at the appropriate mixer (Figure 6.8) to realize complex (I, Q) down-conversion. Figure 6.10 illustrates the approach to performing

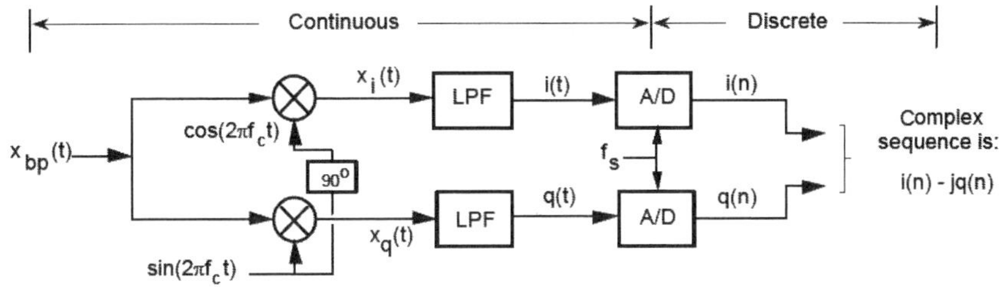

Figure 6.10 Quad sampling block diagram.

quadrature sampling leading to I/Q demodulation (Weaver demodulation) using a sampling rate of frequency f_s samples/second. The reference LO is fed to one mixer directly and a 90-degree phase-shifted version is provided to the remaining mixer where lowpass filters (LPFs) are inserted to provide anti-alias filtering. The outputs of the mixers are I and Q signals, which are injected into ADCs.

Advantages of this quadrature-sampling scheme include the following:

- A/D converters can operate at half the sampling rate of a standard real-signal sampling to realize the required conversion rate by a single (carrier signal) ADC.
- Through doubling of samples at a slower data rate, lower clock rates can be used, which saves power.
- At a given sampling rate (f_s) wideband analog signals can be easier to be process.
- Quadrature sequences make FFT processing efficient due to the wider frequency capability.
- Quadrature sequences are effectively oversampling by a factor of two, which allows for signal-squaring operations without need for up-sampling.
- Conserving I and Q signals throughout the signal processing provides knowledge of signal phase and enables coherent processing.
- Quadrature sampling facilitates the measurement of instantaneous magnitude and phase of a signal during demodulation.

6.4 Overall RF Transceiver Performance

Overall capability of a transceiver can be expressed as the maximum range that reliable connectivity is achieved throughout all expected scenarios of operation. Of course this maximum range will vary depending upon environmental conditions. Also, reliability can be quantified using one (or more) different figures-of-merit (e.g., SNR at the receiver, BER, packet loss). For WSN application, recall that a mote must communicate to its nearest neighbor but should have RF range margin to allow for mote failure and be capable of reaching the next nearest mote to enable self-healing and dynamic rerouting.

6.4.1 Minimum Received Power (SNR)

To determine the minimum power required, our communication channel is characterized as AWGN behavior, with noise appearing at the receiver input is considered thermal noise, and sensitivity can be defined using the noise figure of the receiver. Within a receiver, the SNR of the RF and analog baseband signal prior to digital signal processing is referred to as carrier-to-noise ratio (CNR). SNR is determined using energy per bit (Eb) per noise power spectral density (No) in the information-bearing signal [32], which is the digital baseband signal. The noise figure (NF) of the receiver from the antenna port to the output of the ADC is expressed as the ratio of input CNR in numeric value to the output CNR. SNR at the receiver input terminals can be used to determine the minimal received RF power required to meet the receiver noise floor. This enables us to calculate minimum required power (P_{RX}) at the receiver input. Setting the noise level at the input of a receiver to the thermal noise (kTB), the sensitivity of a receiver (minimal receiver power required, Min[P_{RX}]) can be expressed as [29]

$$\text{Min}\left[P_{\text{RX}}\right](\text{dB}) = 10\log(\text{k}T) + 10\log(BW) + NF + CNR_{\text{min}} \tag{6.26}$$

where BW is the receiver noise effective bandwidth in hertz, NF the overall noise figure of the receiver in decibels and CNR_{min} is the minimum CNR required for obtaining the required error rate. For the prevalent IEEE 802.15.4 radio transceivers used by mote designers (e.g., TI CC2420, MSP430F534X, ATMEL AT86RF230, and ATmega128RFA1), sensitivity values range from −90 to −122 dBm [33]. Minimal CNR is determined primarily by demodulation, decoding, and digital signal processing using the digital baseband signal. However, magnitude and phase frequency responses of filters in the receiver RF analog section may impact this CNR value if the bandwidth, in-band ripple, and/or group delay distortion of these filters is ill defined [34].

6.4.2 RSSI

A tool available to transceiver designers and WSN end users to estimate performance is RSSI measurement. RSSI is a metric that is measured using circuitry capability embedded in wireless (e.g., Wi-Fi) and WSN transceiver designs. RSSI is universally depended upon to aid in the evaluation of quality of service for IEEE 802.11 and 802.15.4 networks. The preponderance of transceiver chipsets and SoC have embedded RSSI measurement capability, and has been used for WSN systems. These internal RSSI measurements [variable *A* in (6.21), dBm units] evaluate received power based on distance (*d*) in units of meters, and transmitted power using an appropriate free-space loss exponent as with (5.1), *n*, as

$$P_{\mathrm{RX}}\ \mathrm{dBm} = A - 10n\log[d] \tag{6.27}$$

RSSI is not specifically a physical parameter; instead, RSSI is a relative performance metric that varies from manufacturer to manufacturer. The interpretation for RSSI is set by transceiver designers and can range report values such as 0–100, 0–127, or 0–255, depending on the manufacturer. When using RSSI, the objective is to have as close to 0 dBm as possible for best signal performance. To rank RSSI values with RF performance, we can use values ranging from high link reliability link ($A = -30$ dBm) to breakpoint between satisfactory and unsatisfactory (–90 dBm) [35]. Figure 6.11 illustrates RSSI roll-off with the distance obtained using the TXSM mote ($f = 433$ MHz) shown in Figure 6.9(e). For the environment this mote was being used, the best-fit model was a fall-off with exponential falloff ($n = -3.8$). Finally, one additional note, RSSI values (IEEE 802.11) are sampled during reception of the preamble stage of an 802.11 frame, not over the full frame.

For WSN systems, mote designs have employed RSSI as a localization tool. Localization is performed during node deployment to determine placement of individual WSN nodes to ensure network operability and sensor coverage. Although RSSI is available for use by localization process, resultant

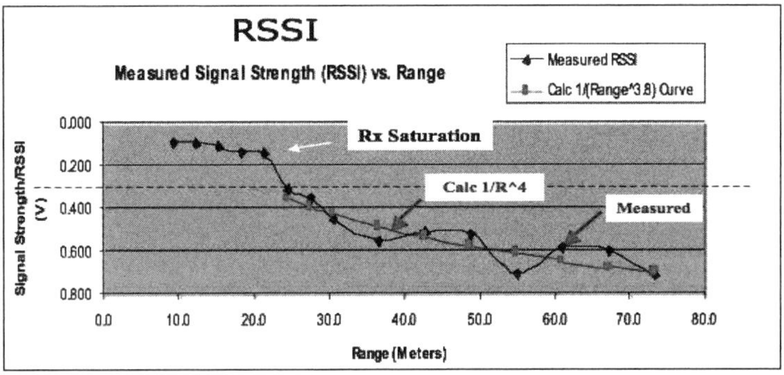

Figure 6.11 TXSM mote RSSI measurements over distances.

location accuracy may not meet the requirements of the WSN application. However, having this capability readily available with most mote designs serves as the de facto approach for initial localization.

6.4.3 Packet Loss Indication

A direct measure of reliability for a RF link is the determination if information sent was received intact and fully recoverable. Packet loss indicates an unreliable link and emphasizes that the root cause must be distinguished among various link failure possibilities, such as message collisions, insufficient RF signal power, RF interference, or cross-layer (MAC/PHY) issues. Success in identifying overlapping messages is dependent upon several elements, including MAC and physical (PHY) [36]. Our network design must consider behaviors as Jain fairness [38] and implementation of a robust NMS, which indicates tasks that address packet rate adaptation, contention, window selection, power control, and carrier sense selection.

6.4.4 Monitoring of BER

Design of the data layer protocol is directly dependent upon evaluation of SNR throughout the network to ascertain reliability of the various RF links. At this point, discussion of performance has focused on SNR as the foundation for designing and operating a viable WSN system. However, limitations exist unless link issues are sensed and fed back through the network to autonomously adjust parameters effecting SNR. Packet frames are typically lost at the receiver and not at the transmitter. This indicates that feedback using SNR measurements at the receiver side be provided to the associated transmitter. With SNR feedback, link parameters can be modified based on adjustment of parameter values involved in carrier sense, data rate, and transmission power [36]. There are concerns with this approach: (1) SNR feedback between motes requires modification to the standard 802.11 protocol, and (2) actual correlation between measured SNR and delivery probability rate is unique to individual links, and both SNR and delivery success rates are time-varying requiring a continuous update of the statistics from receivers to associated transmitters.

6.5 External RF Connectivity

Overall RF design and performance evaluation for a WSN system is not complete unless two additional RF links are considered; the GPS receiver and the two-way link for the WSN mote field to the exfiltration relay [also known as the access point (AP)]. Although (typically) used only during deployment and initialization operations for a WSN system, GPS reception must be considered based on the required localization accuracy and available power budget. Unless the node is mobile, once the setup occurs, the GPS function can be powered off. GPS chipsets and required interfaces continue to drop in cost and power requirements while providing high positional accuracy. For this subsystem, the mote has to consider interfacing to the GPS chipset and placement and connection to the attendant L-band antenna. Note that there are missions where GPS signals are not readily available, or where additional cost and power use cannot be justified. In these cases, reliance on GPS localization may be used through an external system with GPS coordinates transferred to each mote during deployment. Such GPS transfer systems have aided deployments where GPS signals are unavailable, such as inside buildings and tunnels.

To inject commands and to extract sensor and housekeeping data to a motefield requires a relay function. Relays (or APs) provide necessary connectivity to external communication architecture, including worldwide resources (e.g., DoDIN). Exfiltration employs a spectrum of wireless devices and protocols, including: military-standardized VHF communications (e.g., Low-power SEIWG), VHF-to-satellite link, and VHF-to-cellular. For T-ISR, relay design and capability is matched to the external system used allowing seamless integration into existing T-ISR (and ISR) systems using established network protocols. The Security Equipment Integration Working Group, SEIWG, coordinates and influences system architecture, technical design, and systems integration to foster interoperability of all physical security equipment to be used within the DoD. SEIWG stresses device interoperability to allow for the seamless integration of new equipment with existing systems without architectural redesign. In particular, SEIWG-005 is an interface specification (RF data transmission interfaces) for DoD physical security services, which is a specification that defines the RF communication pathway between physical security sensors and a command and control display equipment (CCDE) [37].

Numerous T-ISR sensor systems have been deployed that use a standardized communication protocol at VHF to communicate between sensor systems and control relays. A version of this capability stresses low power and is one of the communication waveforms addressed by WSN relay design to allow for insertion into DoD-related missions. Examples of T-ISR relays are presented in Figures 6.12 and 6.13, which were designed and used in the demonstration of WSN mission capabilities for NEST programs. Figure 6.12 is a relay design for aerial insertion to support WSN motes using 433 MHz/2.4 GHz communications. The deployed relay penetrates the ground and employs a

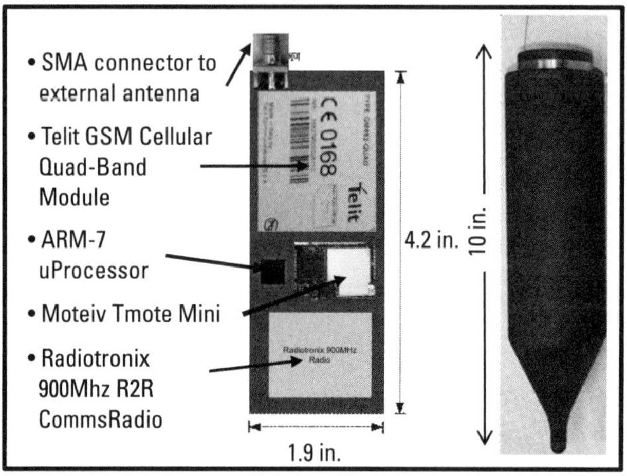

(a)

(b)

Figure 6.12 Aerial insertion WSN relay, motefield-to-LP_SEIWG: (a) depicts an aerial relay dimensions and internal board composition, and (b) shows a deployed aerial relay (with closeup) indicating how covert this element can be.

stop plate to ensure that the WSN, LP-SEIWG, and GPS antennas remain above ground level. Sufficient battery power is supplied in the relay to match mission lifetime expectation of the motefield being supported.

Figure 6.13 presents a WSN exfiltration relay design deployed within the WSN field (in this case, hand-emplaced) that connected WSN motes operating at 433 MHz to LP-SEIWG. As with the previous relay, this relay is integrated to a sufficiently capable energy source (battery bag) to match duration with

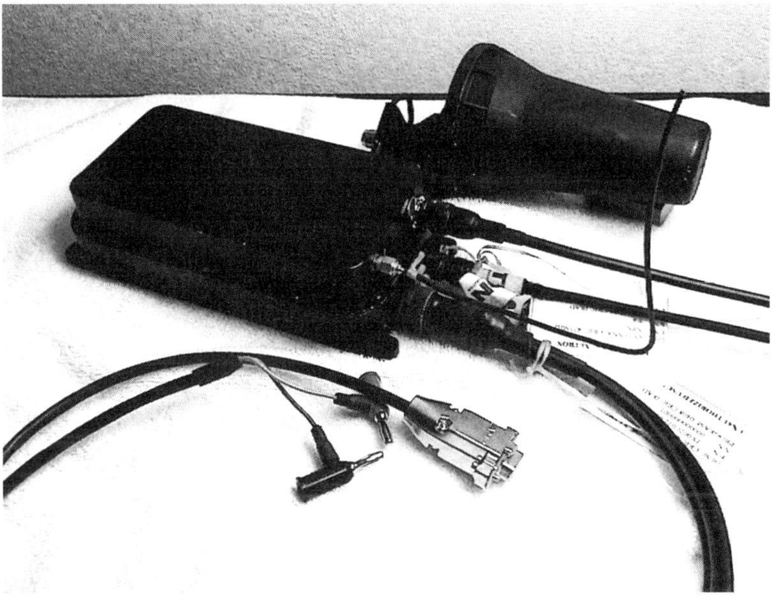

Figure 6.13 Hand emplaced WSN relay, motefield-to-LP_SEIWG communication link.

the WSN motefield. Both of these relays employed LP-SEIWG to allow for connection to sophisticated T-ISR sensors and to allow for exfiltration using a space-ground link (SGL) transceiver.

References

[1] Shaw, J., "Radiometry and the Friis Transmission Equation," *American Journal of Physics*, 2013.

[2] Stutzman, W., "Estimating Directivity and Gain of Antennas," *IEEE Antennas and Propagation Magazine*, 1998.

[3] Rohlfs, K., and T. Wilson, *Tools of Radio Astronomy* (Fourth Edition), Springer Science and Business Media, 2013.

[4] Gregory, D., "New Analytical Models and Probability Density Functions for Fading in Wireless Communications," *IEEE Transactions on Communications*, 2002.

[5] Remley, A., et al., "Radio-Wave Propagation Into Large Building Structures—Part 2: Characterization of Multipath," *IEEE Transactions on Antennas and Propagation*, 2010.

[6] Gu, Q., *RF System Design of Transceivers for Wireless Communications*, Springer Publishing Company, Incorporated, 2010.

[7] Yusof, K., J. Woods, and S. Fitz, "Short-Range and Near Ground Propagation Model for Wireless Sensor Networks," *IEEE Student Conference on Research and Development (SCOReD)*, 2012.

[8] Wang, D., et al., "Near-Ground Path Loss Measurements and Modeling for Wireless Sensor Networks at 2.4 GHz," *International Journal of Distributed Sensor Networks*, 2012.

[9] Viswanathan, M., *Wireless Communication Systems in MATLAB*, Second Edition, 2018, https://www.gaussianwaves.com/wireless-communication-systems-in-matlab/.

[10] Sabri, N., et al., "Performance Evaluation of Wireless Sensor Network Channel in Agricultural Application," *American Journal of Applied Sciences*, 2012.

[11] Sklar, B., "The Characterization of Fading Channels," *IEEE Communications Magazine*, 1997.

[12] Puccinelli, D., and M. Haenggi, "Multipath Fading in Wireless Sensor Networks: Measurements and Interpretation," *ACM*, 2006.

[13] Kritsis, K., et al., "A Tutorial on Performance Evaluation and Validation Methodology for Low-Power and Lossy Networks," 2018, *IEEE Communications Surveys & Tutorials*, Vol. 20, No. 3, 2018, pp. 1799–1825, doi: 10.1109/COMST.2018.2820810.

[14] Goldsmith, A., *Wireless Communications*, Cambridge University Press, 2012.

[15] Boeglen, H., et al., "A Survey of V2V Channel Modeling for VANET Simulations," *2011 Eighth International Conference on Wireless On-Demand Network Systems and Services*, 2011.

[16] Patel, C., et al., "Comparative Analysis of Statistical Models for the Simulation of Rayleigh Faded Cellular Channels," *IEEE Transactions on Communications*, 2005.

[17] Balachandran, K., et al., "Channel Quality Estimation and Rate Adaptation for Cellular Mobile Radio," *IEEE Journal on Selected Areas in Communications*, 1999.

[18] Abdi, A., and M. Kaveh, "A Space-Time Correlation Model for Multielement Antenna Systems in Mobile Fading Channels," *IEEE Journal on Selected Areas in Communications*, 2002.

[19] Feukeu, E., K. Djouani, and A. Kurien, "Compensating the Effect of Doppler Shift in a Vehicular Network," *Africon*, Pointe-Aux-Piments, 2013, pp. 1–7, doi: 10.1109/AFRCON.2013.6757685.

[20] https://en.wikipedia.org/wiki/Rayleigh_fading.

[21] Zedini, E., I. Ansari, and M. Alouini, "Performance Analysis of Mixed Nakagami-m and Gamma–GammaDual-Hop FSO Transmission Systems," *IEEE Photonics Journal*, 2014.

[22] Stefanovic, H., and A. Savic, "Some General Characteristics of Nakagami-m Distribution," *1st International Symposium on Computing in Informatics and Mathematics*, ISCIM, 2011.

[23] Oestges, C., et al., "Experimental Characterization and Modeling of Outdoor-to-Indoor and Indoor-to-Indoor Distributed Channels," *IEEE Transactions on Vehicular Technology*, 2010.

[24] Lee, S., and L. Choi, "ZeroMAC: Toward a Zero Sleep Delay and Zero Idle Listening Media Access Control Protocol with Ultralow Power Radio Frequency Wakeup Sensor," *International Journal of Distributed Sensor Networks*, 2017.

[25] Tang, W., et al., "Measurement and Analysis of Near-Ground Propagation Models under Different Terrains for Wireless Sensor Networks," *Sensors*, 2019.

[26] Ikegami, F., T. Takeuchi, and S. Yoshida, "Theoretical Prediction of Mean Field Strength for Urban Mobile Radio," *IEEE Trans. Antennas Propagat.*, 1991.

[27] Grini, D., "RF Basics, RF for Non-RF Engineers," *MSP430 Advanced Technical Conference*, Texas Instrument, 2006.

[28] Kidwai, A., C. Fu, and J. Fully, "Integrated Ultra-Low Insertion Loss T/R Switch for 802.11b/g/n Application in 90 nm CMOS Process," *IEEE Journal of Solid-State Circuits*, 2009.

[29] Ismail, A., and A. Abidi, "A 3–10-GHz Low-Noise Amplifier with Wideband LC-Ladder Matching Network," *IEEE Journal of Solid-State Circuits*, 2004.

[30] Madhuri, V., S. Umar, and P. Veerave, "A Study on Smart Dust (MOTE) Technology," *IJCSET*, 2013.

[31] Zhao, B., and H. Yang, "Design of Radio-Frequency Transceivers for Wireless Sensor Networks, in *Wireless Sensor Networks: Application–Centric Design*," *Analog Integrated Circuits and Signal Processing*, Vol. 79 No. 2, 2010, pp. 319–329, doi: 10.1007/s10470-014-0267-3.

[32] Sklar, B., *Digital Communications Fundamentals and Applications*, Prentice Hall, 1988.

[33] https://en.wikipedia.org/wiki/Comparison_of_802.15.4_radio_modules.

[34] Gu, Q., *RF System Design of Transceivers for Wireless Communications*, Springer, 2006.

[35] Dolha, S., P. Negirla, and F. Alexa, "Considerations About the Signal Level Measurement in Wireless Sensor Networks for Node Position Estimation," *Sensors*, 2019

[36] Giustiniano, D., D. Malone, and D. J. Leith, "Measuring Transmission Opportunities in 802.11 Links," *IEEE/ACM Transactions on Networking*, 2010.

[37] SEIWG-005, https://www.0cq.osd.mil/ncbdp/nm/pse0g/0bout/seiwg.html.

Selected Bibliography

Jain, R., D. –M. Chiu, and W. Hawe, "A Quantitative Measure of Fairness and Discrimination for resource Allocation in Shared Computer System," *ACM Transaction on Computer Systems*, September 26, 1984.

Nakagami, D., "The m-Distribution, a General Formula of Intensity of Rapid Fading," in *Statistical Methods in Radio Wave Propagation: Proceedings of a Symposium*, June 18–20, 1958, pp. 3–36; W. C. Hoffman (ed.), Pergamon Press, 2006.

Nakagami, D., and P. Viswanath, *Fundamentals of Wireless Communication*, Fourth Edition, Cambridge, UK: Cambridge University Press, p. 31.

Young, W. F., et al., "IEEE Radio-Wave Propagation Into Large Building Structures—Part 1: CW Signal Attenuation and Variability," *IEEE Transactions on Antennas and Propagation*, Vol. 58, No. 4, April 2010.

7

Localization

As wireless sensor networks are deployed to provide persistent observation of areas and/or targets-of-interest within monitored area(s), an awareness of physical location for each of the sensor nodes is required to uniquely provide position of targets being monitored. There are exceptions when the goal is to acquire ambient or event information (e.g., temperature and humidity), but for T-ISR missions, accurate and precise estimation of target state vectors (location, velocity, and direction) is required. For T-ISR, sensor data must be capable of being referenced to a global coordinate frame such as the World Geodetic System (WGS84). WGS 84 is the standard U.S. DoD definition of a global reference system for geospatial information and is the accepted reference system for the U.S. Global Positioning System (GPS). The definition for WGS 84, an Earth-centered, Earth-fixed terrestrial reference system and geodetic datum, depends upon a consistent set of constants and model parameters that describe Earth size, shape, gravity field, and geomagnetic flux. WGS84 is compatible with the International Terrestrial Reference System (ITRS). ITRS describes procedures for creating geocentric coordinates reference frames suitable for use with geodetic measurements on or near the surface using the SI system of measurement.

A target state vector (SV) is defined as a collection of estimated positions (x, y, z) using a Cartesian coordinate system. Included with this vector are associated time tags (t) and target velocity (Vx, Vy, Vz), which provides direction along the x, y, z coordinates. The SV is expressed as a column vector (or transpose of a row vector) as (7.1).

$$SV = \begin{pmatrix} x \\ y \\ z \\ Vx \\ Vy \\ Vz \\ t \end{pmatrix} = \begin{pmatrix} x & y & z & Vx & Vy & Vz & t \end{pmatrix}^T \tag{7.1}$$

Localization is the determination of the absolute, or relative, position of individual sensor nodes that are uniquely defined through a generally accepted coordinate system. WSN deployments result in a random distribution of locations for deployed sensor nodes, which indicates a need of localization upon completion of deployment. In addition to allowing correlation of a positional SV to observed targets, many WSN systems (including those not associated with T-ISR) depend upon node location to realize location-based processing. Examples of location-based processing of WSN data traffic includes network management, intelligent sensor data processing (e.g., discrimination algorithms), data aggregation, system status, and housekeeping evaluation.

The need to perform localization was recognized early in WSN development and addressed through a multitude of approaches to provide an estimate of node (and access point) position [1–4]. Although use of a an appropriate GPS receiver for each node is a direct approach to obtaining positional information, cost, complexity, and/or power requirements may not be met by including a GPS receiver in the node design; this spurred investigation and evaluation of alternative approaches to localization. An alternate approach, relative localization, has been investigated and implemented where position data from a GPS-localized node (denoted as an anchor, or beacon node) is transferred to nodes that do not support GPS reception. Additionally, numerous range-based and proximity-based localization algorithms have been studied, implemented, and evaluated, as categorized in Figure 7.1.

Localization activities, and associated hardware for T-ISR systems, must avoid excessive power consumption, node complexity, or emissions that readily reveal the presence of sensor nodes. All but the GPS transceiver approach emits detectable signals; however, if designed properly, the localization phase during WSN initialization is short-lived and could be timed during local (AOI) quiescent periods. It can be argued that normal operation of WSN systems emits RF signals; however, the RF transmitted power associated with node-to-node communication is relatively low (1 mW, 0 dBm) [5, 6] and duration is on

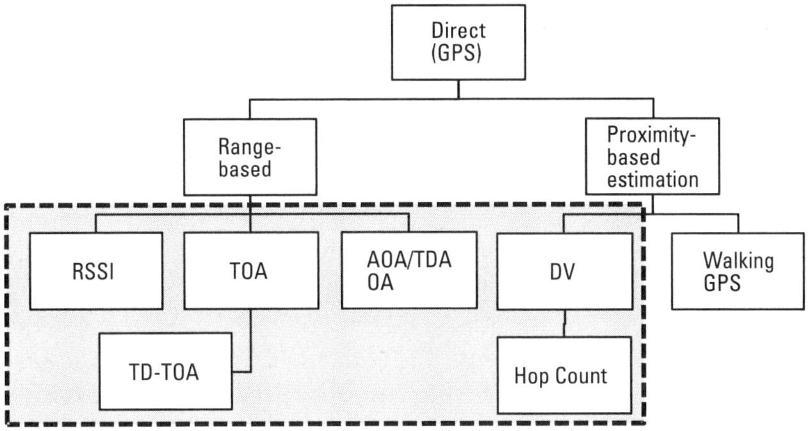

Figure 7.1 Localization methodologies (etymology).

the order of seconds. Exfiltration relay(s) emit RF signals as well, which emit stronger signal energy than that associated with individual nodes. However, relay SWaP constraints are not as stringent. This permits relay designs that incorporate larger energy sources (see Chapter 6) and allow for sophisticated antenna designs to realize narrowbeam RF transmission or space-to-ground links (SGLs) to be directed skyward. The two broadly classified localization categories used to determine position indirectly from geodetic localization are range-based transference and proximity-based estimation [7]. The approach used by both of these localization approaches is to apply updates to known coordinate values to reflect the location of the node to be localized. Range-based systems make use of RF signals to estimate distances and direction between nodes. Success has been obtained with other approaches to localization that depend upon RF signals including solving for distance via RSSI, application of trilateration with time of arrival or angle of arrival (Section 7.3.2), and time difference between signals to estimate the propagation angle (Section 7.3.1).

A final consideration in the design and implementation of a localization process is the goal of matching accuracy and precision requirements to capability. Where reasonable accuracy and precision in locating a node for a planar geometry (e.g., building interior) can be satisfied by two-dimensional (2-D) solutions, T-ISR missions occur in rather harsh terrains where topology deviates from a flat surface; this requires accuracy and precision thresholds to be satisfied in three dimensions. Without GPS, to obtain a 2-D localization solution unambiguously requires three fixed coordinate nodes, but for a 3-D solution, a minimum of four fixed coordinates is required [8].

7.1 Geolocation (Navigation Satellite Constellations)

Directly measured localization (2-D or 3-D) to sensor nodes is available through use of positioning and timing data signals derived from the ~92-spacecraft Global Navigation Satellite System (GNSS). These signals are available anywhere on (or near) the Earth where unobstructed LOSs of multiple (≥4) navigation satellites exists. GNSS is a collection of navigation satellite constellations that includes: the U.S. NAVSTAR GPS, the European Galileo System, the Russian Global'naya Navigatsionnaya Sputnikovaya Sistema (Russian Global Navigation Satellite System, GLONASS), and the Chinese BeiDou Navigation Satellite System. Figure 7.2 illustrates the various L-band frequency bands for GNSS [9]. The GNSS radio-frequency bands are denoted as RNSS, and two bands in the region allocated to the Aeronautical Radio Navigation Service are denoted as ARNS. L-band frequencies are uses, as there is reasonable penetration through clouds, fog, rain, snow, and vegetation at these wavelengths. However, in a dense and persistent environmental concentration, such as heavy rain or a dense forest canopy, attenuation of GNSS signals can occur to a level meeting the required accuracy (or even solution) of GNSS position fixes is not possible. However, for most outdoor applications, depending on viewing angles to GNSS spacecraft, signal levels are sufficient to allow for reasonably accurate position determination.

Most GNSS systems (and GNSS augmentation systems) use similar frequencies within the L1 band. As one result, multisystem receivers exist

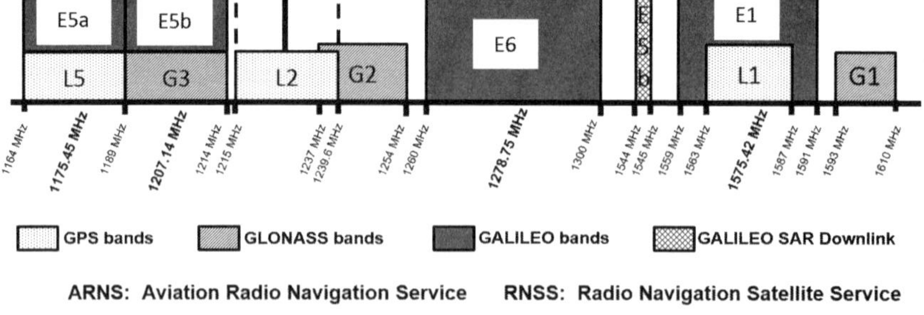

Figure 7.2 GNSS navigation frequency bands.

that are capable of using any one of the GNSS navigation systems [10]. As GNSS signals propagate through layers of atmosphere, variable delays at the receiver are introduced. This statistical delay of GNSS signals produces errors throughout the GPS receiver process resulting in noisy and/or offsets of the measured position. Depending upon the satellite-to-receiver geometry, time of day, or solar activity, the ionosphere may introduce relatively large delays (up to 16 ns, or, 5-m errors). The troposphere contains the weather elements (e.g., precipitation and wind), which produces up to 1.5-ns (0.5-m) additional positional error. Given these errors, a solution to mitigate resultant atmospheric positional error is through the use of differential GNSS. Differential GNSS involves the participation of additional, stationary receivers that monitor regional GNSS signals. These GNSS reference stations receive GNSS signals but unlike GNSS users, stationary receiver algorithms implement reverse-solution equations. Instead of using timing signals to calculate position, they use the known satellite position to calculate timing errors to form an error correction factor. This allows stationary receivers to correlate satellite measurements to a fixed, highly accurate local reference capable of measuring residual timing errors, which in turn, forms corrections transmitted to GNSS users.

7.2 GPS Overview

The GPS, originally NAVSTAR, is a satellite-based radio-navigation system owned by the U.S. government. Operated by the U.S. Air Force, GPS is one of the GNSS assets that provide geolocation and time information to a GPS receiver anywhere on Earth. The primary measurement used by navigation satellite systems is the timing of bit transitions in the navigation signal. Precise positioning requires subnanosecond measurements of bit edges to reliably resolve to 1-ft errors. To obtain the necessary measurement of bit edges and to implement effective multipath rejection requires a GPS receiver to have a wide bandwidth, which is accommodated through the use of pseudorandom noise (PRN) modulation of the carrier.

Position and timing data is extracted and processed via military GPS receivers tuned to two carrier signals in the L-band, referred to as L1 (1,575.42 MHz) and L2 (1,227.60 MHz). A third frequency, L5 (1,176.45 MHz) is used in conjunction with aircraft receiving systems as safety-of-life transportation and other high-performance (accuracy) applications. Performance through processing L5 signals occurs due to the combination of high radiated power and increased chip rate (narrower correlation peak) compared to L1 [11]. Multiple frequencies and codes are employed to remove first-order effects caused by ionosphere refraction, which is dispersive. Civilian GPS use is constrained to that provided by L1, L2, and L5 signals, which are not encrypted. Military use can gain access to encrypted code on L1 and L2 and operate using the latest codes, which are currently the M-coded L1C and L2C signals. Timing information, encoded on the carriers, allows GNSS receivers to continuously determine the time signal reception. The signal contains the data that a receiver uses to compute locations of the satellites and adjust for accurate positioning. The receiver uses the time difference between the time of signal reception and the broadcast time to compute the range from the receiver to the satellite. When the receiver knows the precise position of itself with respect to each satellite, it translates its own position into an Earth-based coordinate system, which results in latitude, longitude, and altitude.

For the U.S. NAVSTAR GPS, the model and spatial reference coordinate system employed is dependent upon WGS-84 definitions. WGS-84, managed by the National Geospatial-Intelligence Agency (NGA), provides a reference coordinate frame, an ellipsoidal Earth gravitation model (EGM), a description of the World Magnetic Model (WMM), and a list of local data transformations. (A transformation, or more specifically a geodetic datum transformation, is a change in a coordinate based on to what geodetic datum a coordinate is referenced.) WGS 84 geodetic coordinates are generated through the use of a reference ellipsoid. The WGS 84 coordinate system origin serves as the geometric center of the WGS 84 ellipsoid with the +*Z*-axis as the rotational axis of this ellipsoid of revolution. The +*X*-axis is defined as the intersection of the International Earth Rotation and Reference Systems Service (IERS) Reference Meridian (IRM) [12] and the plane passing through the origin and normal to the +*Z*-axis. The IRM is coincident with the Bureau International de l'Heure (BIH) Zero Meridian (epoch 1984.0) with an uncertainty of 0.005 arc-seconds. The +*Y*-axis completes the right-handed, Earth-centered Earth-fixed (ECEF) orthogonal coordinate system.

7.2.1 GPS Codes

There are three types of code associated with GPS carrier signals:

- The course acquisition (C/A) code;
- The precise (P) code;
- Navigation message.

The C/A code can be found on the L1 channel with a sequence that repeats every 1 ms. The C/A code is accomplished through generated

PRN-determined code. The carrier transmits the C/A code at 1.023 Mbps (million bits per second) with a chip length (physical distance) between binary transitions (+1 and −1) of 293m. The C/A code provides time according to the satellite clock as to when the signal was transmitted (with an ambiguity of 1 ms), which is easily resolved as this corresponds to a distance of 299.7 km). C/A is a course code that is appropriate for initially locking onto the signal. Each satellite has a different C/A code, so that they can be uniquely identified. The P code is identical on both L1 and L2 channels and is provided to attain precise positioning.

P code conveys basic information such as the satellite clock time or transmission identical to C/A information except that it provides 10 times the resolution. Unlike C/A code, a process known as antispoofing can encrypt P code. P code is designated using a PRN-generated code for each GPS satellite, which is divided into seven-day segments and designed to repeat every 267 days. The carrier transmits P code at 10.23 Mbps using a chip length of 29.3m. Navigation message data, available from the L1 channel, is transmitted at 50 bps. The message is a 1,500-bit sequence requiring 30 seconds to transmit. A navigation message includes information regarding broadcast ephemeris (satellite orbital parameters), satellite clock corrections, almanac data (crude ephemeris for all satellites), ionosphere information, and satellite health status. Figure 7.3 presents a pictorial representation for the coding of carriers with the C/A and P codes.

To perform geodetic datum transformations requires access to a NGA- and DoD-approved geographic translator. A datum is a point, a line, or surface used as a reference in surveying and mapping. A geodetic datum is a mathematical model of the Earth used to calculate the coordinates on any map [13]. Useful to mission operations in handing of intelligence, information, and/or engagement activity arising from WSN observations is the Mensuration Services Program (MSP) Geographic Translator (GEOTRANS), an application program that enables conversion of geographic coordinates among a wide variety of coordinate systems, map projections, and datums.

7.2.2 GPS (GNSS) Chipsets for WSN

With the availability of worldwide GNSS signals and ample margins of SWAP2, WSN sensor node designers may consider the addition of a GPS receiver module unless the application would be subject to questionable connectivity with GNSS constellations (e.g., emplacement within tunnels and/or building interiors). Focusing on T-ISR, only GPS signals are designed to be pulled from the GNSS constellation, whereas other applications (civilian and

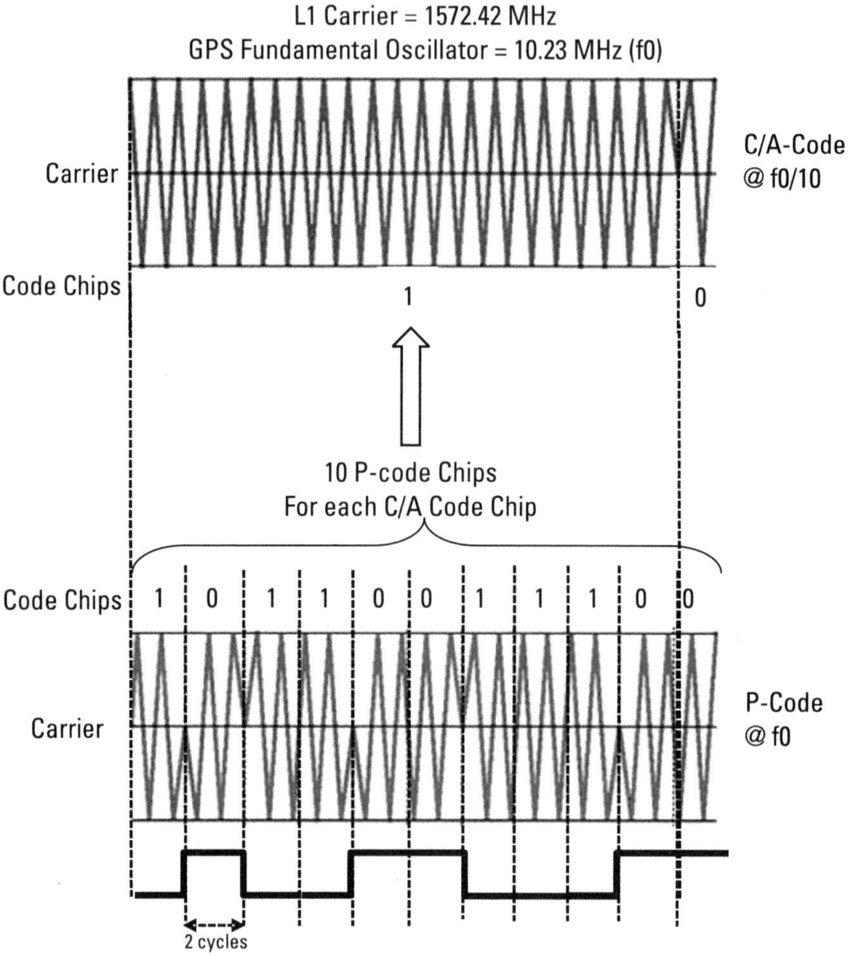

- Carrier shown as 2 cycles per P-code chip; actually 154 cycles per P-code chip.
- C/A Code modulated on 90° phase-shifted version of L1 (quadrature component).
- Codes are modulated onto carrier via multiplication with code states; each shift from 1 to 0, or 0 to 1 causes carrier phase shift of 180o
- Repetition rates: P-code = 7 days, C/A code = 1 ms

Figure 7.3 Coding the GPS carriers: C/A-code and P-code are illustrated.

public works) may consider the entirety of the GNSS signal coverage. For this reason, T-ISR WSN chipsets and modules only focus on GPS-based systems.

7.2.3 GPS Chipset Performance

Characteristics and capability are two aspects of any chipset to be explored in the design of a WSN node. Certainly issues associated with SWAP2 constraints, discussed in Section 7.2.2, must be considered. Measures of accuracy, update rate, receiver sensitivity, interference/jammer resistance, multiconstellation operation, and ease of integration within a WSN node are all top concerns to any WSN designer; these concerns are presented in this section.

7.2.3.1 GPS Chipset Accuracy and Sensitivity

GPS receiver accuracy depends upon numerous variables, including receiver design, time of day, signal strength, process time, and multipath effects. Measured accuracy from relatively low-cost GPS chipsets results in <1.89m horizontal position error [14]. Improved accuracies, at the centimeter accuracy (<30 cm), have recently arrived using receiver designs (e.g., RTK) that simultaneously process the L1 and L5 signals [15, 16]. Real-time kinematic (RTK) positioning is an approach used to enhance precision of position data derived from GNSS. The basis of RTK is to consider the phase of the GPS (GNSS) signal carrier wave in addition to the data content of the signal. RTK positioning relies on a single reference station or an interpolated virtual station to provide real-time corrections, providing centimeter accuracy [17]. For GPS in particular, the RTK approach is referred to as carrier-phase enhancement, or CPGPS [18]. Receiver sensitivity is another key measurements of GPS chipset capability. Using (6.20), the receiver sensitivity, and substituting the value for kTB product with $T = 290$K and $B = 1$ Hz, we obtain (7.2) for the minimal receiver sensitivity, S_{RX}.

$$\text{Min}\left[S_{\text{RX}}\right](\text{dBmHz}) = -174\ \text{dBmHz} + NF + CNR_{\min} \tag{7.2}$$

Of note are the units for the power and sensitivity equations. For (6.20), units are watts-normalized against 1 mW (dBm). Regarding satellite communications, the parlance is expressed as CNR. Sensitivity units, associated with CNR, employ units of decibels relative to one milliwatt-hertz not decibels relative to one milliwatt as CNR is measuring a power density. Cumulative effects of in-band interference can increase the noise floor and degrade performance. Sensitivity, as expressed in (7.2), is defined as the lowest satellite power level at which a GPS receiver is still able to track a signal and process a position fix. To help GPS receivers to meet minimal signal power levels, significant gain values are achieved through the implementation of several cascaded LNAs at the receiver input. As shown in Figure 7.4, the first LNA establishes the receiver noise figure. Using multiple amplifiers with noise figures n_i (i = 1-to-M) and amplification gains g_k (k = 1-to-M), (7.3) provides the noise input to the system.

$$NOISE_{\text{total}} = 10\log_{10}\left[n_1 + \sum_{i=2}^{M}\frac{n_i - 1}{\prod_{k=1}^{i-1} g_k}\right] \text{dB} \tag{7.3}$$

To determine overall gain, individual gains are summed, as shown by (7.4).

$$GAIN_{\text{total}} = \sum_{k=1}^{M} g_k \text{ dB} \tag{7.4}$$

GPS receivers specify two variants of sensitivity: acquisition and signal tracking. Acquisition sensitivity represents the lowest power level at which a GPS receiver is able to achieve a position fix. Signal-tracking sensitivity is the lowest power level at which a GPS receiver is able to track an individual satellite. Unfortunately, this lowest power level is difficult to predict, as there are numerous variables involved. An improved concept of sensitivity is expressed as the lowest signal power level to obtain a minimum BER. As BER correlates with CNR, sensitivity can be measured by validating the CNR reported by the receiver using a known input power level. This relationship is shown using (5.5) and (5.6), which express Type I and Type II errors. Recall that Type I is the hypothesis of a target being declared when it is not present (i.e.,

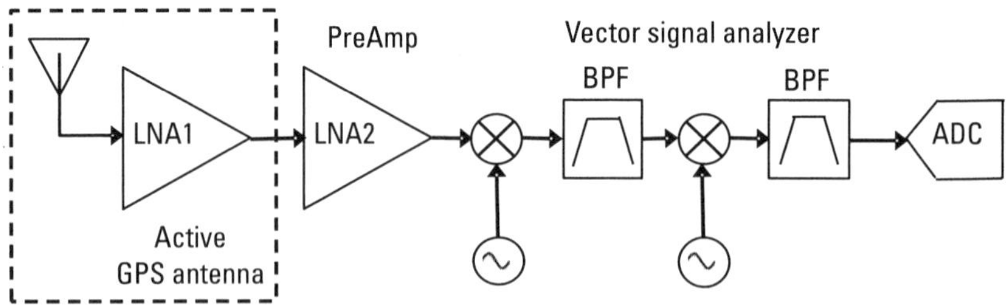

Figure 7.4 Cascaded LNA stages.

false alarm, where Type II is the declaration that a target doesn't exist when it does.) For digital communications, BER is the summation of Type I error (α) and Type II error (β) probabilities, as [19]

$$BER = \frac{\alpha + \beta}{2} \tag{7.5}$$

Consider the equations developed in Chapter 5, where we defined the probabilities associated with the observed signal against two hypotheses: H_0 for noise only, and H_1 in the case in which both signal and noise are present. Working with AWGN, we have a noise pdf described as

$$P_n(x) = \frac{e^{-x^2/2\sigma^2}}{\sigma\sqrt{2\pi}} \tag{7.6}$$

Using this AWGN pdf for (7.5), where (5.16) (α) and 5.17 (β), the BER equation can be expressed in terms of energy per bit (E_b) and noise power density (N_0) as

$$BER = \frac{1}{2}\text{erfc}\left(\sqrt{\frac{E_b}{N_0}}\right) \tag{7.7}$$

A complementary error function, erfc(x), is often used, as defined by (7.8) using the error function, erf(x).

$$\text{erfc}(x) \equiv 1 - \text{erf}(x) \tag{7.8}$$

Furthermore, regarding (5.15), documentation expresses (7.7) using the Q-function definition [20], defined as

$$Q(x) \equiv \frac{1}{2}\text{erfc}\left(\frac{x}{\sqrt{2}}\right) \tag{7.9}$$

This results in (7.7) expressed using the Q-function as,

$$BER = Q\left(\sqrt{\frac{2E_b}{N_0}}\right) \tag{7.10}$$

Figure 7.5 presents the typical performance behavior of BER as a function of bit energy, expressed as E_b/N_0, generated using [21]. These operational curves, also known as waterfall curves, are used in the evaluation of GPS receivers beginning with BER testing. Testing is not focused on the durations between bit errors, but with the determination of the level of background noise that produces excessive bit errors. A waterfall curve plot is the end result of the testing and is a significant factor when comparing and selecting wireless devices. Through observation of the probability distribution, testing can be conducted effectively with tradeoffs between the amount of test time that is acceptable and that required to attain the desired measurement confidence level and accuracy. When testing is complete, the confidence level and accuracy of the measurements can then be applied to the waterfall curve for reliable, quantitative comparisons [22].

The CNR value for a satellite is estimated by GPS receiver chipsets through the calculation of BER based upon the navigation message content. With a clear view of the sky, typical CNR values for GPS vary from 30 to 50 dB-Hz. For most GPS receivers, the minimum CNR required to achieve a position fix (acquisition sensitivity) varies from 28 to 32 dB-Hz [23]. Using (7.10), a relationship between sensitivity, as measured using the BER and CNR Eb/N0, is established. As a result, SNR is defined by (7.11), which is a tool employed to evaluate BER as a function of CNR for various modulation

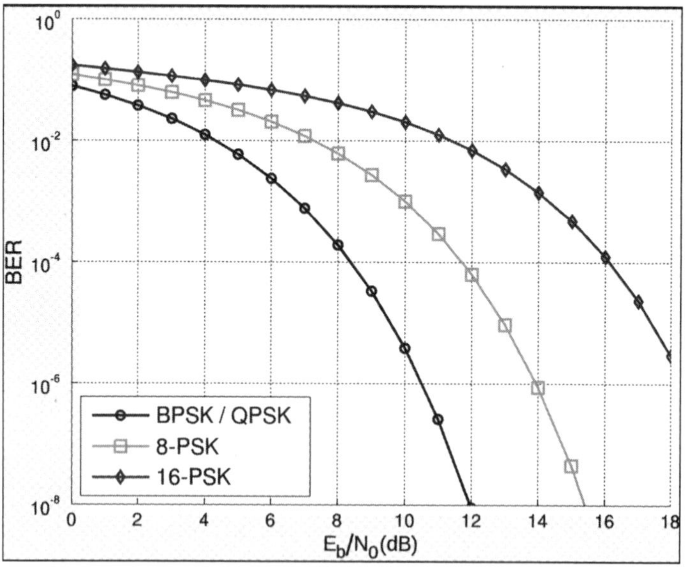

Figure 7.5 GPS performance curves; BER as a function of CNR.

schemes (e.g., BPSK, QPSK, MPSK, and QAM) [24]. As a final point, SNR and the normalized SNR (signal to noise per bit, E_b/N_0) are related through link spectral efficiency (bits/sec/Hz), the ratio of channel data rate (f_b) to channel bandwidth (B) [21, 24]:

$$SNR = (CNR)\left(\frac{f_b}{B}\right) = \left(\frac{E_b}{N_0}\right)\left(\frac{f_b}{B}\right) \tag{7.11}$$

7.2.3.2 Number of Channels and Update Rate

When multiple GPS satellites are simultaneously in view, the number of channels a GPS receiver accesses to process affects the time expended to generate the first fix. GPS receivers are designed to operate anywhere from four to hundreds of channels simultaneously, with 12–14 channels producing acceptable tracking capability. Since a GPS chipset does not anticipate which satellites are in view at any point in time, the more frequencies that are received and used, the faster a position fix can be produced. (There are receiver systems that maintain a GPS almanac based on location and can predict which spacecraft are in view over time, but several receiver systems do not do so.)

Another characteristic of GPS receivers of interest, especially for moving users, is the update rate. The update rate of a GPS module is how often recalculations are performed and how often updates to position are generated. The standard for most chipsets and GPS modules is 1 Hz. However, fast-moving platforms (e.g., vehicles and UAVs) may require faster recalculation rates (such as 5–10 Hz) requiring high-performance microprocessors to handle incoming position navigation messages. For T-ISR applications, the GPS receivers are designed to conserve power. After GPS signal lock, power savings is realized through the implementation of chipsets that can be isolated and/or chipsets that can shut down extraneous channel processing.

7.2.3.3 Constraints on GNSS Chipsets (SWaP and Costs)

Recalling that WSN sensor nodes have acute constraints on SWAP2, inclusion of a GPS receiver capability has largely been avoided. However, as with technologies cited in Chapter 1, GPS chipsets (SoC and module technologies) have improved significantly in accuracy, sensitivity, and SWAP2. Regarding cost, the integration of RF components into CMOS has led to end-to-end CMOS radios, which reduces cost. CMOS process improvement is an area that GPS chipset/module manufacturers have successfully addressed. The cascade of reduced feature dimensions from 90 nm to <45 nm have been realized and resulted in the reduction of both power consumption and cost. On average,

GPS chipsets consume 30 mA at 3.3 VDC including the LNA located at the input (antenna terminals).

In reference to packaging, WSN node designers must also account for the L-band antenna for a GPS receiver. The preponderance of GPS antennas for WSN purposes are realized using chip antennas rather than the omnidirectional ceramic patch antenna (dielectric resonator), which requires a large ground plane or the larger (and more costly) helical and dipole (whip) designs [25–27].

7.3 Range-Based Transference

To correlate a WSN field to geodetic coordinates, an absolute measurement of position in the system coordinates is preferred, requiring no transformation processing. This can be performed either using anchor nodes (also referred to as beacons), where the nodes are located in the worldwide coordinate system, or where each node locates itself directly in the worldwide coordinates, as in the case of using GPS receivers. The use of anchor nodes (also known as beacons) is pursued. With anchors, network nodes are localized through the use of transference, where the relative range (Δx, Δy, Δz) from an anchor or geodetic data for a GPS receiver is used to update the coordinates of a node being positioned.

Numerous techniques exist for estimating range from an anchor node using a variety of approaches: RF propagation, acoustic propagation, and determination of network hierarchy and nearest neighbors through the application of the likelihood estimation [28, 29]. The direct transference of GPS can also be pursued where a GPS receiver obtains geodetic solutions for a node being emplaced and transfers that information (coordinates) to the node for future reference (e.g., walking GPS [30]). Aggregate approaches position sensors via multiple measurements available from numerous sensor nodes in contrast to individual localization techniques that operate with information describing a single sensor node and a set of landmarks.

7.3.1 RSSI Approach

RSSI, described in Chapter 6, can be used to locate a node through the use of SNR and (6.26) with reasonable estimates for transmitted (P_{TX}), received power (P_{RX}), and the path (e.g., free-space attenuation exponent, (n). To use this approach, the correct propagation equation, power levels, and initial distance (*d*) value between transmitter and receiver nodes must be estimated. However, understanding the propagation characteristics is difficult. The existence of several nonideal situations typically occurs, as explored in Chapter 6. A basic concern with RSSI measurement integrity is that the algorithm and circuitry designed to use the baseband voltage reflects RSSI. Where accuracy is required, more capable circuits that operate over the frequency range of the transceiver may be required.

Multipath fading is dependent on the environment and is in many cases dependent on the frequency of the signals being transmitted. If a transceiver employs multiple frequencies and appropriate filtering, the impact of frequency-dependent multipath fading are mitigated but frequency-independent multipath effects (e.g., ground reflection) remain [31]. Another difficulty with RSSI-based localization is that transmit power may not be accurate enough. Without knowing the precise transmitter power, confidence in path loss is relatively low. A final concern: given that the distance using the propagation equation leads to a measure of distance only, precise positioning in three dimensions is not possible. Recall that to refine measurement of position based on RSSI requires multilateration with three anchor points. For three-dimensional precision, four anchor points are required [5]. Techniques using a combination of RSSI with other ranging measurements may lead to reliable location estimates; however, designers and end users must consider that nonuniform propagation environments render RSSI methods unreliable and inaccurate.

7.3.2 TOA Approach

TOA measurements are used to estimate distances from anchor (reference) points by measuring one-way propagation times (or differences) of RF signals. Each measured TOA quantity presents a circle with its center at the receiver and the source on the circumference for a two-dimensional solution and a spherical surface for a three-dimensional solution as illustrated in Figure 7.6(a, b). Not surprisingly, with noise-free measurements, three or more sources are required to provide a distinct intersection point that represents the source position. For two-dimensional position estimates, a minimum of three sensors is necessary. With less than three sources, it is possible that there will be no intersecting points, which precludes determination of a unique positioning solution. Hence, a minimum of three sensors is required to obtain the intersection, and these can be represented as a set of circular equations, based on the optimization criterion; the source position can be estimated with the knowledge of the known sensor array geometry.

Techniques based on TOA send a single packet from one node to another containing time of transmission. Assuming perfect clock synchronization between the nodes, distance traveled can be calculated. An advantage of localization using TOA is that it is more accurate than RSSI and not affected by channel fading. Unfortunately, however, time synchronization between nodes is relatively difficult [8]. Timing circuits, oscillator stability, and thermal insensitivity associated with clock distribution within a node all impact the ability to perform a high level of synchronization. Another, fundamental

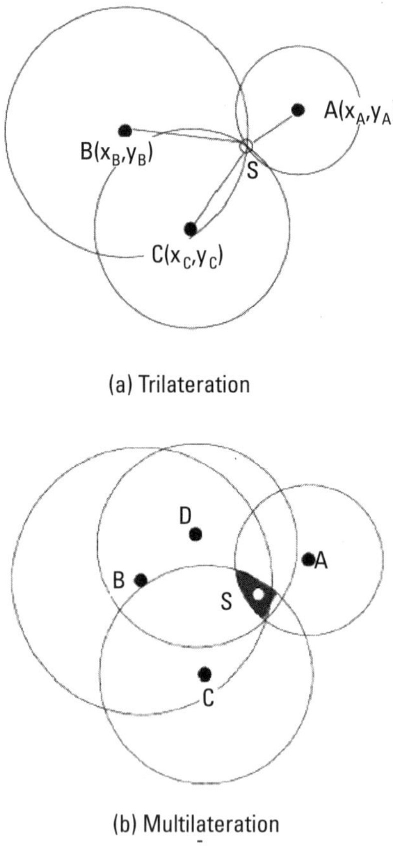

Figure 7.6 TOA configuration multiple anchor (beacon) sources: (a) trilateration solving of a transmitting source position, and (b) multilateration (>3 receiving nodes) to ascertain transmitter source location.

difficulty lies in the high propagation speed of wireless signals. The free-space propagation velocity of RF signals occurs at the speed of light, requiring high accuracy regarding all time measurements to avoid large uncertainties (e.g., 1-m accuracy requires an overall time error ≤ 3.3 ns).

An approach applied to circumvent high-accuracy timing measurements is the use of signals that travel much slower than RF, such as acoustical (ultrasonic) waveforms. However, recall that for T-ISR, audible emissions that reveal the presence and/or location of a WSN system are to be avoided.

7.3.2.1 Clock Synchronization

WSN systems strive to provide distributed-information processing tasks like data sampling, data fusion, and other time-based tasks. Every node in the network has an autonomous clock. These individual clocks drift from one another due to imperfections with oscillators, aging, the thermal environment, and other temporal variations. It is essential to calibrate these imperfections from time to time to achieve network-wide time coherence. A plethora of algorithms for clock synchronization exist and are well-documented in the literature [29, 32, 33].

7.3.2.2 Time Difference of Arrival

A variant of TOA is time difference of arrival (TDOA), which does not require the transmission times that a ranging signal was sent from a beacon (anchor node). Instead, TDOA processing requires calculation of the difference in time for a ranging signal received by anchor node pairs from the target node being positioned, along with the associated propagation coefficient (typically, the free-space speed of light). Once the ranging signal, $s(t)$, is received by a beacon pair, a difference is formed using arrival times to calculate the difference in distances (Δd) between the target node and two anchor nodes, which is the product of propagation velocity and time difference. This difference can be calculated in two dimensions, using known positions of a beacon pair, which are ($x1$, $y1$) and ($x2$, $y2$). Equation (7.12) is used to determine the difference in distance as

$$\Delta d = \sqrt{(x2-x)^2} - \sqrt{(y2-y)^2} - \sqrt{(x1-x)^2} - \sqrt{(y1-y)^2} \qquad (7.12)$$

Using nonlinear regression, (7.12) can be converted to a hyperbolic form. With a sufficient number of hyperbolas calculated, the position of the target can be determined using the loci intersection. Figure 7.7 illustrates a simplified example: A target node requires support from three beacons (B_1, B_2, and B_3) for the two-dimensional case; a signal is sent from the target at an

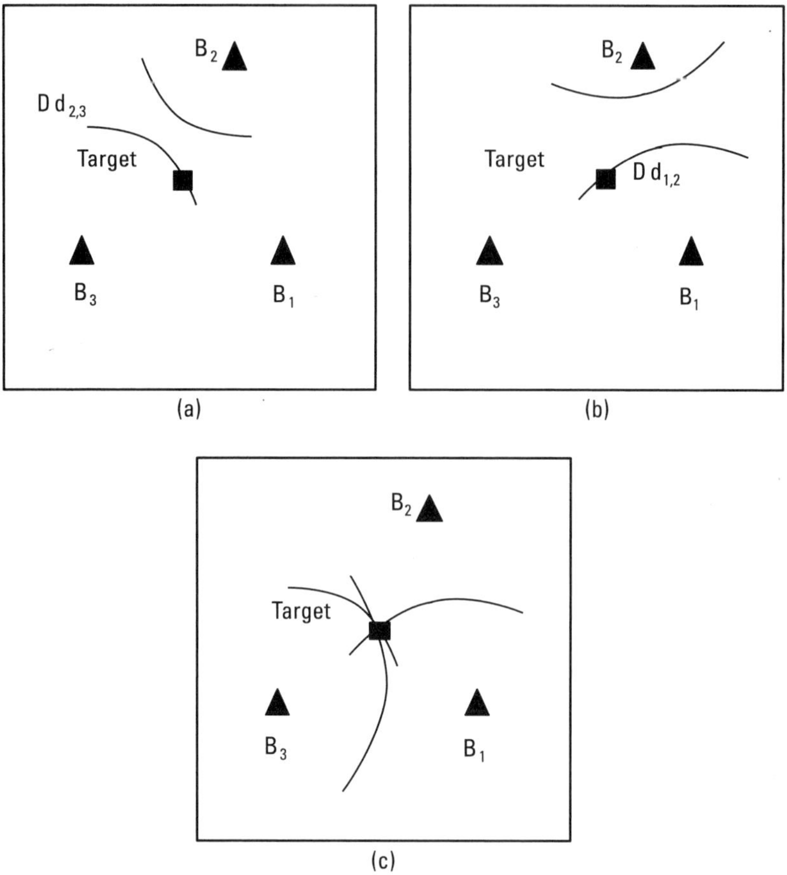

Figure 7.7 TDOA configuration with multiple beacons: (a) setup where two beacons, 2 and 3, transmit and indicate a target position, (b) a second beacon pair, 1 and 2 transmit and indicate a second location for the target, and (c) combining locii from beacons 2 and 3 with those from 1 and 2 to provide a best estimate of target node location.

unknown time, which is received by the first beacon pair, B_2 and B_3 at time, t_1. A difference in distance (Δd) is calculated and results in a hyperbolic loci of possible locations for the target node position. This hyperbola presents two branches making discerning the correct intersection difficult [Figure 7.7(a)]. However, an approximate location of the target can be estimated with the second beacon pair, B_1-B_2 [Figure 7.7(b)], and a branch can be discarded. In this case, the top branch is discarded and the beacon pairs indicate an interception [Figure 7.7(c)] [34].

The typical systems engineering approach to measuring TDOA values of a RF signal arriving at two receivers, at separate locations, uses generalized cross-correlation (GCC) [28]. GCC methods cross-correlate a prefiltered version of the received signal at two receivers that have been filtered with the proper frequency response, $H(f)$, and then cross-correlated, integrated, squared, and processed through a peak detector. GCC is conducted via a series of time shifts, τ, between two received signals, $s_1(t)$ and $s_2(t)$ until a peak value is obtained. The time delay, τ_{max}, associated with the correlated magnitude peak is the estimate for TDOA. Prefiltering is implemented to emphasize signal frequencies presenting prominent strength and to attenuate noise power prior to the signals entering the correlation process. The choice of the frequency response, $H(f)$, is important, especially when multiple delays occur from a multipath environment. If the various signal delays are not sufficiently separated, the spreading of the first multipath component function may overlap subsequent signal delays, making estimation of the peak (TDOA) difficult. The frequency response is designed to ensure that the correlation maximum occurs with cross-correlation of the received signals achieving narrowband spectrum, thereby enhancing TDOA estimation accuracy.

To clarify the above, consider a ranging signal, $s(t)$, from the target node, which is transmitted and received by two receivers following propagation channels having interference (amplitude attenuation, A_i) and noise, $n_i(t)$. A model for the time-delay of the ranging signal reaching two beacon sensors, $s_1(t)$ and $s_2(t)$, is

$$\begin{aligned} s_1(t) &= A_1 s(t-\tau_1) + n_1(t) \\ s_2(t) &= A_2 s(t-\tau_2) + n_2(t) \end{aligned} \tag{7.13}$$

Using this model assumes that both the received signals and noise are stationary random processes allowing temporal and statistical sampling to provide equivalent results. This assumption renders all marginal and joint density functions of these processes independent of the choice for the time origin [35]. The model also assumes uncorrelated noise processes (n_1, n_2). Finally, observation of the signals is assumed to occur during a finite duration, T. We are interested in the peak output obtained by the correlator, which occurs when the delay, τ, best correlates signal $s_1(t)$ with signal $s_2(t)$. With focus on the TDOA as represented by the time difference, $\tau = \tau_1 - \tau_2$ (here, we assumed $\tau_1 > \tau_2$), we can ignore the attenuation coefficients and rewrite (7.13) [31],

$$\begin{aligned} s_1(t) &= s(t) + n_1(t) \\ s_2(t) &= s(t-\tau) + n_2(t) \end{aligned} \tag{7.14}$$

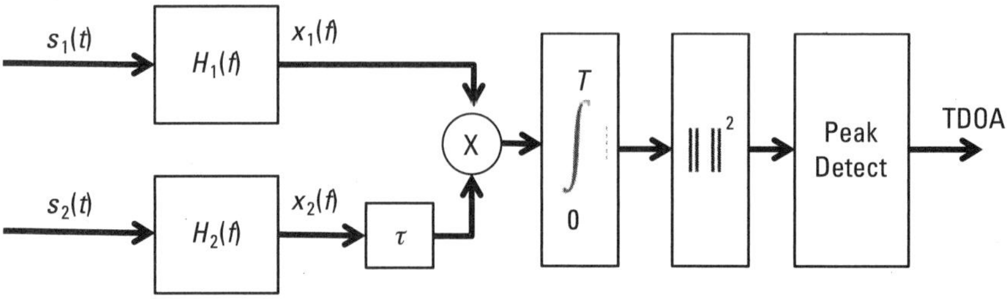

Figure 7.8 Cross-correlation process for TDOA estimation.

A cross-correlation and auto-correlation process ensues, and is depicted by Figure 7.8. The cross and autocorrelation products are summed forming the cross-correlation of the two received signals, as

$$R_{s_1s_2} = R_{ss}(t - \tau) + R_{n_1n_2}(t) \tag{7.15}$$

Accurate estimation of TDOA requires the use of time delay estimation techniques that provide inherent resistance to noise and interference along with an ability to resolve multipath signal components. Also to be considered is that cross-correlation functionality can be obtained by implementing the inverse Fourier transform of the estimated frequency domain cross-spectral density function. In particular cases, frequency domain processing is preferred because signals can be filtered prior to computation of the cross-correlation function. Cross-correlation requires synchronization among receivers but does not impose any requirement on the signal transmitted by the transmitter. The accuracy and temporal resolution capabilities of TDOA measurements improve when separation of beacon receivers increases because this distance increases better separates TOA of the signals. TDOA of multiple signals that are not separated by more than the width of their cross-correlation peaks (whose location on the time-delay axis corresponds to TDOA) usually cannot be reliably resolved by conventional TDOA measurement techniques [31].

7.3.3 Angle of Arrival

Positioning of sensor nodes without direct GPS assistance using anchor nodes (beacons) can also be performed through the use of angle of arrival (AOA) measurements. AOA methodologies, processing necessary, accuracies, and considerations are well documented. With AOA, anchor nodes are required to transmit omnidirectional signals to provide signals for the target node to determine bearings relative to the anchor locations. Measurement of AOA is achieved through the determination of arriving RF signals incident on an antenna array (or rotational position of a asymmetric antenna pattern). The AOA measurements are calculated using TDOA among individual elements of the target node array. This can be thought of as beamforming, but in reverse. With beamforming, the signal from each element is weighed to steer gain of an antenna array. In AOA, the arrival delay for each element is measured and converted to an AOA measurement.

Consider, for example, a two-element array spaced apart by one-half the wavelength of an incoming RF wave. If a wave is incident upon the array at boresight, it will arrive at each antenna simultaneously. This will yield a 0° phase-difference measured between the two antenna elements, equivalent to AOA = 0°. If a wave is incident to the array at broadside, then a 180° phase difference will be measured between the elements, corresponding to a 90° AOA [36]. Figure 7.9 illustrates the geometric explanation behind AOA.

As with TDOA, positional accuracy suffers in the presence of multipath reflections. The target receiver cannot differentiate signal strength variation due to the varying amplitude of the transmitted signal and/or signal strength variation resulting from the anisotropic receiver antenna pattern. One approach to dealing with the problem is through the use of an omnidirectional antenna at the receivers. By normalizing signal strength received by the anisotropic antenna with respect to the signal strength received by the omnidirectional antenna, the impact of varying signal strength can be significantly addressed and removed. A second antenna and associated AOA process does increase

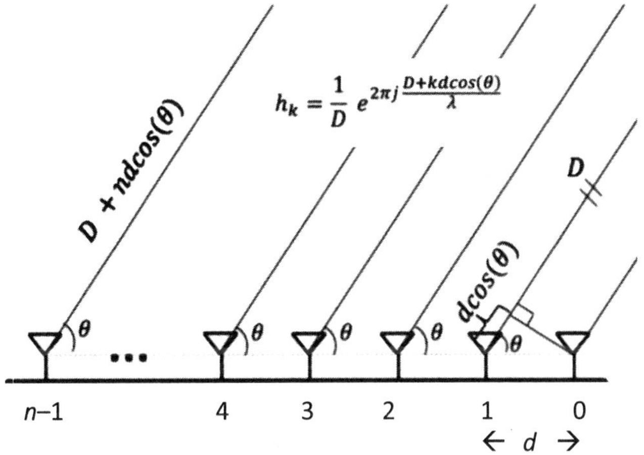

Figure 7.9 Geometric explanation in measuring phase (angle) for AOA positioning.

complexity of the node design; however, this complexity is not as significant as that required to implement scaled antenna arrays and associated processing.

To summarize AOA position transference using anchor nodes, at least two noncollinear anchor nodes are required to localize a target node if its orientation is known (i.e., use of a magnet or magnetometer to establish an origin for bearing measurements). If orientation is not available, at least three anchor nodes are needed to discover both the location and the orientation. Similar to localization using TDOA measurements, AOA localization is susceptible to measurement noise and additional problems arise if target node cannot receive ranging signals from a sufficient number of beacons. Other variations of the AOA approach have AOA measurements exchanged among neighbor nodes, and the relative AOA with respect to each beacon (even multiple hops away) can be calculated using geometric relations among the nodes. This enables the use of triangulation for localization.

7.3.4 Distance-Vector Hop Count

The distance vector hop count (DV-HOP) localization approach begins with anchor nodes (beacons) flooding their established locations to other nodes in the network being deployed. Messages are propagated throughout the mote field, hop-by-hop, with accumulated hop count (HC_{ij}) inserted into the transmitted message [2, 5]. Each node maintains an anchor information table and counts the least number of hops that the node is located from an anchor. When an anchor receives a message from another anchor, it estimates the average distance of one hop using locations of both anchors, and the hop count. With this information, the anchor node transmits the determined value through the network as a correction factor. Upon receiving a correction factor, a nonanchor node is able to estimate its distance to anchors and conducts trilateration to estimate its location.

The DV-HOP algorithm is performed using four phases [37]. During the first phase, an anchor (B_i) floods the field with a message containing its coordinates (x_i, y_i) and a variable, $HC_{i,j}$, that signifies the number of links (hops) between the sensor node (j) that received this message and the transmitting anchor (i). Sensors store and exchange hop counts for anchor nodes. In the second phase, the B_i anchor node computes hop length as HOP_i, using (7.16),

$$HOP_i = \frac{\sum_{j=1,i\neq j}^{M} \sqrt{\left(x_i - x_j\right)^2 + \left(y_i - y_j\right)^2}}{\sum_{j=1,i\neq j}^{M} HC_{i,j}} \tag{7.16}$$

In (7.16), M is the number of anchors in the network, the subscript j identifies other anchor nodes, $HC_{i,j}$ is the distance in hops between anchor i and anchor j, and (x_i, y_i) and (x_j, y_j) represent coordinates for anchors i and j. After computing average hop length, anchor i floods the network for other anchors and sensors. An unknown sensor node maintains only the average hop length flooded by the closest beacon to it. The unknown node uses the received average hop length to compute the distance, d_j, between the target node and anchor j using

$$d_j = \left(HOP_i\right)\left(HC_j\right) \tag{7.17}$$

where HC_j represents a hop counter that the target node maintains for anchor j. With this DV-Hop algorithm, the distance between anchor nodes and unknown nodes are used as the average hop distances [3, 37, 38]. In the third phase, a sensor uses a least square (LS) technique to trilaterate its position. The objective of this LS problem is to minimize summation of the square of errors between the measured distances from the reporting sensor to each anchor obtained according to (7.17) and the distance between the relative anchor and the estimated position of sensor. With the final phase, localization error ϵ_i is calculated as,

$$\epsilon_i = \sqrt{\left(x_i - x_j\right)^2 + \left(y_i - y_j\right)^2} \tag{7.18}$$

where (x_i, y_i) and (x_j, y_j) are coordinates of anchor nodes and target node, and accuracy (a_i) can be calculated using communication range, R,

$$a_i = \frac{\epsilon_i}{R} \tag{7.19}$$

Use of the DV-HOP transference method does not provide accurate results if applied to a sparse topology.

7.4 Special Localization: Walking GPS

Walking GPS is a technique [30] that avoids having to implement a GPS receiver into each sensor node, yet still depends directly on GPS to inject a node position as the node is emplaced. Node deployment works by having the person performing the hand emplacement of sensor nodes carry a GPS-enabled device that periodically broadcasts its location to sensor nodes. Implementation is through connection of a mote to a GPS device, which efficiently solves wireless connectivity between motes being deployed by using the same node design. A walking GPS receiving middleware component runs on sensor nodes being deployed in the network. This component receives location data that is broadcast by GPS-connected mote and infers position from packets received. This architecture relegates complexity derived from the interaction with the GPS device to a single sensor node, which reduces code size and data memory for the sensor node. Through this decoupling, a single GPS-connected mote is sufficient for localization of an entire sensor network.

GPS positions are represented by latitude and longitude, which are angular measures from the equator North to South and the prime meridian East to West, respectively. Due to the relatively small size of a WSN network ($<1 \times 1$ km^2), incorporating node messages with the full global (i.e., GPS) coordinates is inefficient. This inefficiency arises from the size of GPS packets for passing locations. As the more significant measure of the location is expected to be identical for all sensor nodes within a field, reduced overhead is realized using a local coordinate system. (GPS coordinates use 11 bytes out of 29 bytes, the payload size of a TinyOS packet.) Additionally, a local coordinate system that adopts linear units is better suited for WSN than a global coordinate system.

Inefficiency is also avoided from the computational costs encountered when aggregating data (e.g., triangulation of several GPS signals for positioning a target). A local coordinate system employed with walking GPS methodology establishes its origin (denoted as the reference point). The distance (D) between the reference point (GPS coordinates λ_1 and ϕ_1) and another point is specified as pseudo-GPS coordinates λ_2 and ϕ_2, using a distance in the local coordinate system, as

$$D = \sqrt{\left(F_{\text{lat}} * (\phi_1 - \phi_2)^2\right) + \left(F_{\text{lon}} * (\lambda_1 - \lambda_2)^2\right)} \tag{7.20}$$

The terms, F_{lat} and F_{lon} are conversion factors that represent distances for 1-degree changes in latitude and longitude, respectively. The unit of measure is meter/degree, and the (7.20) parameters are: a = 6,378,137m, b = 6,356,752.3142m, with height above the ellipsoid.

$$F_{\text{lat}} = \frac{\pi}{180}\left(\frac{a^2 b^2}{\sqrt{\left(a^2\cos^2\phi + b^2\sin^2\phi\right)^2}} + h\right) \tag{7.21}$$

$$F_{\text{lalont}} = \frac{\pi}{180}\left(\frac{a^2}{\sqrt{\left(a^2\cos^2\phi + b^2\sin^2\phi\right)^2}} + h\right)\cos\phi \tag{7.22}$$

The influence of h on conversion factors is minimal ($h = 200$m is typically assumed). For a GPS location specified by λ_2 and ϕ_2, the x and y coordinates of this point are provided via the local coordinate system. The x-axis is oriented in the East-West direction whereas the y-axis is aligned in the North-South direction. Values for the variables in equations (7.20) to (7.22) are obtained using a GPS receiver.

Walking GPS localization has two distinct phases. The first phase occurs during deployment of sensor nodes. The carrier (e.g., soldier or vehicle) performing the manual deployment has a GPS-connected mote that periodically transits its location to the deployed nodes. The GPS-connected node acts as an anchor (beacon), and nodes infer their location based on the information provided in the broadcast message. This is when the walking GPS solution takes place. The second phase is during system initialization. If a sensor node does not obtain a location, the node will request from a neighboring node their location information. The location information received from neighbors is used within a triangulation procedure by the requesting node to infer its position. This second phase enhances robustness of the walking GPS scheme. Results from the use of walking GPS have demonstrated a very good match between locations reported by the tracking application and actual deployment of nodes. Walking GPS localization is applicable only to manual deployment of sensor nodes; however, there remain numerous WSN deployments where manual deployment is the only viable solution, and walking GPS localization provides a methodology for these situations.

References

[1] Nasipuri, A., and K. Li, "A Directionality Based Location Discovery Scheme for Wireless Sensor Networks," *WSNA'02*, September 28, 2002, ACM, 2002.

[2] Mao, G., and B. Fidan, "Localization Algorithms and Strategies for Wireless Sensor Networks," *IGI Global*, 2009.

[3] Chariabi, Y., "Localization in Wireless Sensor Networks," Master's Degree Project Stockholm, Sweden, report IR-RT-EX-0523, 2005.

[4] Paul, A., and T. Sato, "Localization in Wireless Sensor Networks: A Survey on Algorithms, Measurement Techniques, Applications and Challenges," *Journal of Sensor and Actuator Networks*, 2017.

[5] Chee, Y., "Ultra Low Power Transmitters for Wireless Sensor Networks," Doctoral dissertation, University of California, Berkeley, Technical Report No. UCB/EECS-2006-57, 2006.

[6] Yusof, Y., M. Islam, and S. Baharun, "An Experimental Study of WSN Transmission Power Optimization Using MICAz Motes," *2015 International Conference on Advances in Electrical Engineering (ICAEE)*, 2016.

[7] Mesmoudi, A., M. Feham, and N. Labraoui, "Wireless Sensor Network Localization Algorithms: A Comprehensive Survey," *International Journal of Computer Networks & Communications (IJCNC)*, 2013.

[8] Garg, V., and M. Jhamb, "A Review of Wireless Sensor Network on Localization Techniques," *International Journal of Engineering Trends and Technology (IJETT)*, 2013.

[9] https://gssc.esa.int/navipedia/index.php/.

[10] Soderholm, S., et al., "A MultiGNSS Software-Defined Receiver: Design, Implementation, and Performance Benefits," *Ann. Telecommun.*, 2016.

[11] Leclère, J., R. Landry, Jr., and C. Botteron, "Comparison of L1 and L5 Bands GNSS Signals Acquisition," *Sensors*, 2018.

[12] https://wiki2.org/en/IERS_Reference_Meridian. International Earth Rotation and Reference Systems Service (IERS) Reference Meridian (IRM).

[13] https://earth-info.nga.mil/GandG/coordsys/datums/index.html.

[14] William J. Hughes Technical Center, WAAS T&E Team "Global Positioning System (GPS) Standard Positioning Service (SPS) Performance Analysis Report," Federal Aviation Administration GPS Product Team, Report #96, 2017.

[15] Dabove, P., and V. Di Pietra, "Single-Baseline RTK Positioning Using Dual-Frequency GNSS Receivers Inside Smartphones," *Sensors*, 2019.

[16] Moore, S., "Super-Accurate GPS Chips Coming to Smartphones in 2018," *IEEE Spectrum*, 2017.

[17] Wanninger, L., "Introduction to Network RTK," IAG Working Group 4.5.1, Accessed: 3 March 2020.

[18] Mannings, R., *Ubiquitous Positioning*, Norwood, MA: Artech House, 2008.

[19] Minko, J., *Signals, Noise, & Active Sensors (Radar, Sonar, Laser Radar)*, John Wiley & Sons, 1992.

[20] https://en.wikipedia.org/wiki/Q-function.

[21] http://www.montana.edu/aolson/ee447/EB%20and%20NO.pdf.

[22] https://www.eetimes.com/confidence-in-waterfall-curves-guides-noise-analysis-in-wireless-system-test/#.

[23] https://www.ni.com/en-us/innovations/white-papers/08/gps-receiver-testing.html#section--455231815.

[24] Goldsmith, G., *Wireless Communications*, Cambridge University Press, 2005.

[25] Palihawadana, A., "GPS & GPS Antenna Designing Literature Review," *GPS Antenna Design*, 2019.

[26] Falade, O., et al., "Single Feed Stacked Patch Circular Polarized Antenna for Triple Band GPS Receivers," *IEEE Transactions on Antennas and Propagation*, 2012.

[27] Boccia, L., G. Amendola, and G. Di Massa, "A Shorted Elliptical Patch Antenna for GPS Applications," *IEEE Antennas and Wireless Propagation Letters*, 2003.

[28] Mao, G., B. Fidan, and B. Anderson, "Wireless Sensor Network Localization Techniques," *Computer Networks*, Vol. 51, No. 10, pp. 2529–251.

[29] Chepuri, S., G. Leus, and A. van der Veen, "Joint Localization and Clock Synchronization for Wireless Sensor Networks," *IEEE 2012 Conference Record of the Forty Sixth Asilomar Conference on Signals, Systems and Computers*, 2012.

[30] Stoleru, R., T. He, and J. Stankovic, "Walking GPS: A Practical Solution for Localization in Manually Deployed Wireless Sensor Networks," *29th Annual IEEE International Conference on Local Computer Networks*, 2004.

[31] Ahmed, H., et al., "Estimation of Time Difference of Arrival (TDOA) for the Source Radiates BPSK Signal," *IJCSI International Journal of Computer Science Issues*, 2013.

[32] Chepuri, S., G. Leus, and A. van der Veen, "Joint Localization and Clock Synchronization for Wireless Sensor Networks," *IEEE Signals, Systems & Computers, Asilomar Conference*, 2012.

[33] Wu, Y., Q. Chaudhari, and E. Serephin, "Clock Synchronization of Wireless Sensor Networks," *IEEE Signal Processing Magazine*, 2011.

[34] https://sites.tufts.edu/eeseniordesignhandbook/files/2017/05/FireBrick_OKeefe_F1-.pdf.

[35] Cooper, C., and C. McGillem, *Methods of Signal and System Analysis*, Holt, Rinehart, and Winston, 1967.

[36] Peng, R., and R. Sichitiu, "Angle of Arrival Localization for Wireless Sensor Networks," *Sensor and Ad Hoc Communications and Networks (SECON)*, 2006.

[37] Chen, H., et al., "An Improved DV-Hop Localization Algorithm for Wireless Sensor Networks," *2018 13th IEEE Conference on Industrial Electronics and Applications (ICIEA)*, Wuhan, 2018, pp. 1831–1836, doi: 10.1109/ICIEA.2018.8398006.

[38] Pachnand, G., "Modified DV-Hop Algorithm for Localization in Wireless Sensor Networks," *Indonesian Journal of Electrical Engineering and Informatics (IJEEI)*, 2014.

8

WSN Middleware-Based Functions

By focusing on data provided through a large distribution of sensor nodes, WSN systems are considered data-centric communication networking. T-ISR mission applications are unlikely to request sensor observations unless a node signals the MOC that an event of interest (based on preset conditions) is unfolding. This allows for increased system robustness by decoupling data from sensor nodes and applying a combination of processing procedures that can address objectives by solving varying combinations of quality and resource requirements [1]. This data centricity continues from a sensor node to nearest-neighbor nodes (~10 nodes), area nodes (~100 nodes), and to the entire motefield (>1,000 nodes), using a rough order of magnitude of motes. To meet mission objectives, WSN systems continuously monitor and gauge status of subsystems with an ability to dynamically balance the overall system to maintain responsiveness by focusing on three system requirements:

1. Successful collection of sensor data associated with T-ISR mission objectives;
2. Timely routing of observation and status data to mission centers for analysis and reaction;
3. Sufficient persistence to maintain an effective system through sufficient power availability and successful security [2].

Fundamental to successfully designing a WSN responsive to a T-ISR mission is to understand the particular objectives; however, it is just as important to embed system capability that fosters sensor performance, operational adaptability, robust network formation, and efficient real-time data processing. Chapter 5 presents a comprehensive set of T-ISR mission objectives that depend upon successfully meeting these requirements. In correlating mission characteristics to node and system design, we can consider approaches that realize these key system functions through trade-offs among hardware, system design, and most notably, the algorithmic code that seamlessly integrates WSN elements, middleware. Middleware provides adaptive and efficient utilization of system resources. Through such utilization, standardized system services can be wielded by a diversity of applications, creating a runtime environment that supports and coordinates multiple applications [3]. With realization of these characteristics, WSN can accomplish the T-ISR key functions discussed throughout this text.

8.1 WSN Foundation: Middleware, Services, and Resources

Ideally, middleware inserts a software layer that removes complexities associated with code generation and operates with heterogeneous elements such as a microcontroller system, ADC converters, radio chipsets, and network protocols. WSN middleware provides those requisite special services associated with remote sensing based on pervasive computing applications that make use of a wireless sensor network and embedded RTOS. Adopting modular design, middleware is reconfigurable and introduces a granularity that decouples tasks, services, and applications through a service-oriented architecture (SOA). SOA is a software paradigm employed during code development to provide functionality to applications using a communication protocol over a network. SOA services are elemental units of functionality that are accessed remotely, thereby enabling users to quickly combine functionality in an ad hoc manner to form applications built purely from existing services.

WSN resources are required to be dynamic, as sensor nodes may leave or join a network through intentional mobility of a sensor node or due to failure of network link or nodes. This dynamic network environment emphasizes an underlying approach to WSN middleware to monitor and continuously update resource management, which enables the services to be provided as required and benefits system efficiency. Optimal management of resources is emphasized for WSN systems so as not to deplete the availability of node resources such as battery power, link capacity, and/or microprocessor memory.

8.2 WSN Middleware Virtualization

In computing, virtualization refers to the act of creating a virtual (rather than actual) version of something, including virtual computer hardware platforms, storage devices, and computer network resources. Virtualization is an established method of logically dividing system resources provided by distributed processors among multiple applications. Virtualization creates an environment that supports efficient sharing of resources, services, and networks. Virtualization WSN virtualization of the system can integrate different hardware, applications, and network capabilities to monitor and manage the network resources. Figure 8.1 presents a middleware virtualization model that intervenes between the application level and the network and physical layers. Figure 8.1 depicts middleware supporting several critical functions associated with management of the network, sensors, power, and access to the various services provided through virtualization [4]. How does middleware virtualization benefit WSN node and system designs? Virtualization can emulate a physical sensor, a network formed from a collaboration of various node designs, or a conglomerations of nodes associated with disparate tasks where virtualization enables subgroups of nodes execute multiple applications, simultaneously. This capability supports the development and test phases as well as investigating how best to construct various interfaces.

8.3 WSN Middleware-Enabled Capabilities

Once deployed WSN systems collect observations that provide critical data required by command decision-making, which ultimately equates to supporting mission success. How well the system operates depends upon the technical and implementation attention provided during selection of node design, middleware design, and overall WSN architecture. The need for WSN middleware is fundamental to implementing WSN node algorithms. Table 8.1 presents the functional capabilities associated with WSN operation acquired through middleware. Middleware makes available high-level sensing tasks that enable sensor nodes to coordinate and delegate tasks based upon individual characteristics of the nodes (e.g., sensing modalities, node location, and energy reserves). Designing and implementing such real-time middleware is a nontrivial task. Nodes are small-scale, low-cost devices, with limited energy, which translates into restricted resources, microcontroller performance, sensor performance, wireless communication bandwidth, and radio transmission range. Finally, to keep costs per node low, quality assurance factors are considered.

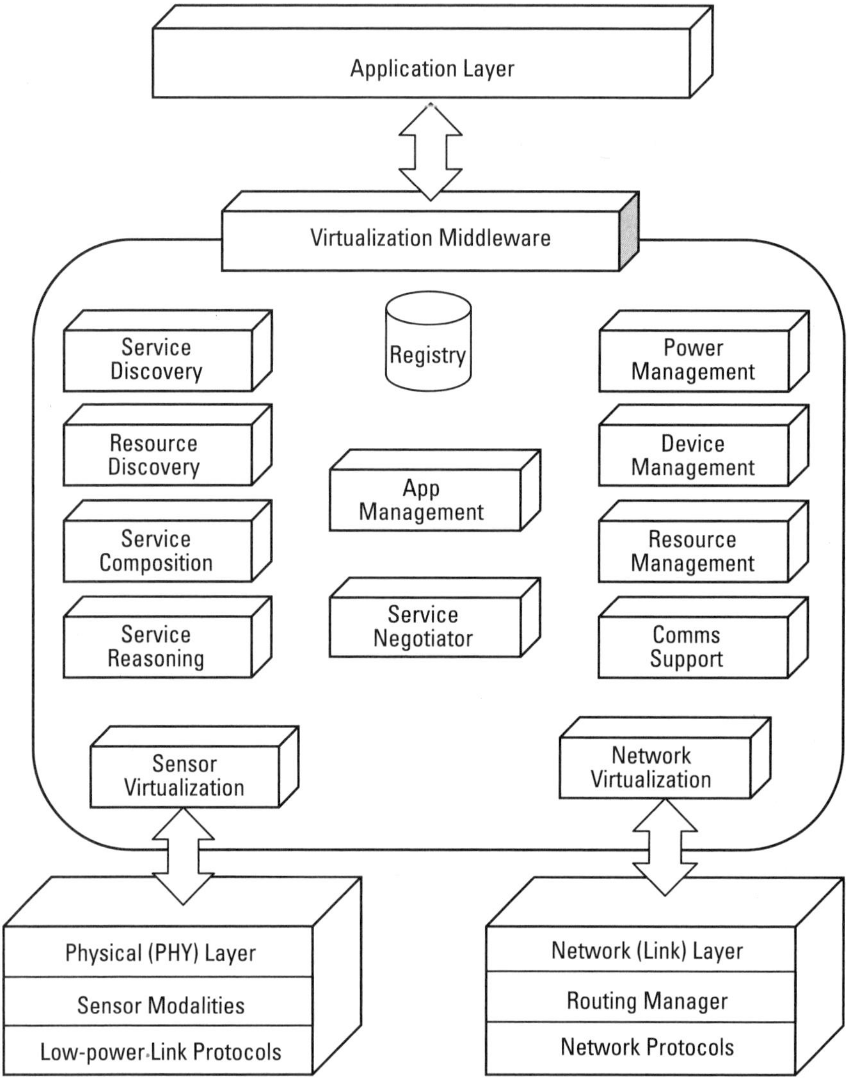

Figure 8.1 WSN middleware virtualization model.

Node mobility (e.g., MANET systems), node failures, and environmentally produced disruptions result in highly dynamic scenarios for WSN, causing frequent changes of network topology. Although multihop and mobile architectures can transport information across partitions, rerouting of links to

Table 8.1
Middleware-Enabled Functions for WSN Nodes

NMS	Network formation (routing protocol) Nearest neighbor statusing Scalability Self-healing	MANET management Link tables Network (links) statusing Store and forward
Localization	Individual node location (relative, absolute)	Network relative locations Hop counters
Transceiver	Active: transmit Active: receive Inactive Sleep	Code/Decode/ECC datagrams Retransmission process RTS/CTS
Processor	Programming upload (OTA) Code image version monitoring Grenade timer RTOS and services Data processing Target tasking	Detection Tracking and track files (multiple targets) Discrimination Classification Identification Aggregation
Sensor	Active Standby Power off	Threshold set, CFAR process Operating parameter table preset & update
Power management	LPL protocol	Senor activity monitoring and control
I/O	Test port enable/disabled	Diagnostic interface
Security	Crypto key management Command authorization Command verification	Command ack Node key/group key (cluster) Tamper monitoring and action

accommodate absent nodes and/or disrupted paths may result in unbounded delays and may potentially establish unidirectional links. WSN systems must be designed to respond to these link failures typically done within middleware to conduct monitoring and dynamic rerouting. The functions of identifying a failure and resolving network outage depend on middleware composed of real-time energy-efficient algorithms. During determination of overall status and/or pathway rerouting, network efficiency does decrease; however, autonomy is maintained.

8.4 Persistent Monitoring

Persistent monitoring can be described as a two-stage activity. The first stage occurs when the WSN system is operating in a reduced power mode, which conserves power as the system monitors assigned AOI awaiting appearance of a potential target. In the second stage, a potential target has been initially detected, and the WSN system activity increases as designed, to perform the detection function to validate the existence of a target of interest. The entire system may not be activated, but based on the target activity as observed by the system, additional node and network services are brought online to validate the existence of a target of interest.

8.5 WSN Functional Requirements

Key functional capabilities of a T-ISR system were enumerated in Table 8.1. Figure 8.2 illustrates node functions that directly access and process sensor information: detection, tracking, discrimination/classification, and identification. In addition, Figure 8.2 illustrates reacquisition (update track files), aggregation (compression of message traffic with data from other nodes), and algorithms capable of multiple target and/or closely-space object (CSO) tracking, where the tracking function detects multiple objects and invokes algorithms relating to multiple targets and/or CSO. Along with timely reporting, these key functions describe the overarching requirement of any T-ISR system (not just WSN): to find, classify, and track targets. There remain numerous other functions that a WSN design must perform; however, we can consider these functions supportive to the fundamental functions, which include power management, network management, and security. The need for well-designed middleware is evident.

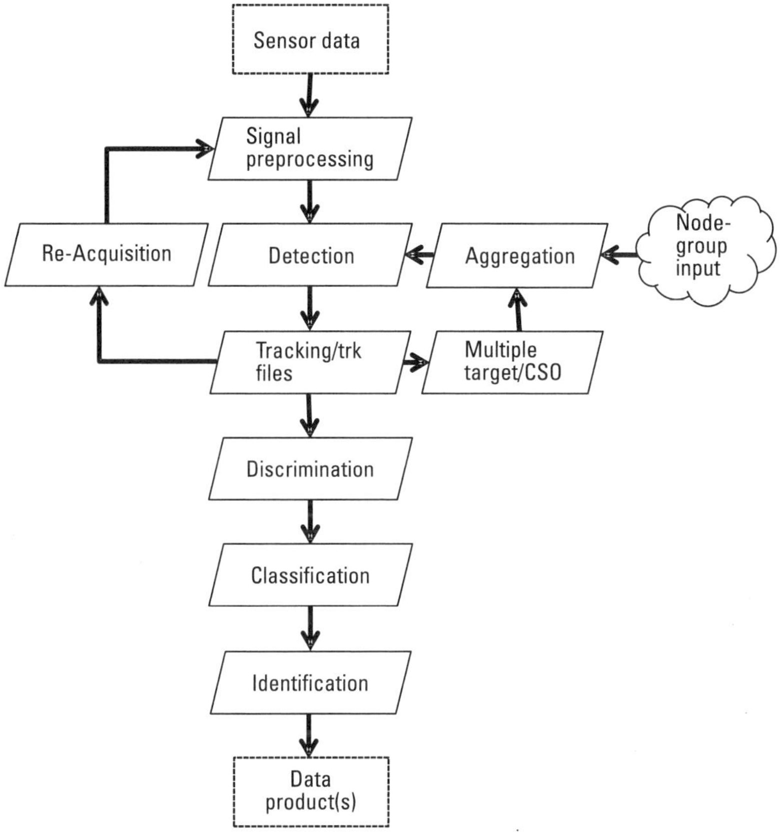

Figure 8.2 Middleware-enabled target processing functions.

8.5.1 Detection Function

Detection begins with an event that signals a WSN system of a potential target. Key requirements associated with a detection function are: allow no misses (objects entering the AOI, yet go undetected) and minimization (or ideally, absence) of false alarms. As discussed in Chapter 5, detection methods are derived using estimation theory and are closely related to hypothesis testing. Recall that performance metrics for detection include evaluation of probabilities for detection (P_d), false alarm (P_{FA}), and the probability of a miss (P_m), where a target was present but the system did not generate the expected alert response. The goal is to maximize P_d and minimize P_{FA} given the performance level of sensing, computation, and communication. Unfortunately,

P_d and P_{FA} may be positively correlated. Additionally, driving false alarms can mean a loss of detections that should have been made, thereby increasing the probability of a missed target (P_m). As presented in Chapter 5, we deal with the two probabilities, P_d and P_{FA}, as P_m is provided to us as

$$P_d = 1 - P_m \tag{8.1}$$

Regarding probabilities of detection, the system design process reviewed in Chapter 5 begins with specification of an acceptable P_{FA} while maximizing the best possible P_d. A means of boosting detection probability without simultaneously increasing the FAR can be accomplished in certain situations. For active sensor modalities (e.g., laser and RF radars), coherent integration of the backscattered signal can be implemented using a matched filter (recall Figures 5.7 and 6.8) arrangement that preserves phase information via quadrature sampling. Assuming a nonfluctuating target and background noise that is independent, sample-to-sample (e.g., AWGN) integration of observations results in a linear gain proportional to sampling size, N. With incoherent integration and the above assumptions, gain increases as the square root of the sample size ($N^{-1/2}$) [5].

Increased observation time can be realized through a number of approaches, but the prominent (and perhaps easiest) approach is to employ a large number of sensor nodes to prolong the observation window of potential targets within the AOI. Certainly the decision about the number of nodes to deploy must take into account the mission AOI as well as sensor effective range capabilities in the operating environment. This consideration directly influences deployment design as WSN system designers attempt to balance coverage with node count and as a function of onboard sensor and processing capabilities. The deployed number is also selected to support link and hardware reliability as well as longevity. Link failures are resolved through MAC rerouting using alternative pathways, and having nodes in reserve supports self-healing. Given a deployment pattern and M-deployed sensors, intrusion distance can be characterized. Intrusion distance is the total distance a potential target can enter a monitored AOI before being detected. Detection of a possible target is based on the number of triggered nodes, M_{trig}, and because of such, introduces a WSN-based lag time between actual intrusion and signaling of a detection event, T_D [6]. Allowable system latency (T_{sys}) is calculated as elapsed time measured between actual target incursion and the associated time stamp of notification to an end user (e.g., mission operations center).

T_{sys} is heavily dependent upon the value for T_D; however, other factors enter the overall latency equation, such as node and group processing times, detection message propagation, exfiltration relay to backbone communication architecture delays, and alert message delivery/recognition by the end user (MDC). Previously, time lags associated with these other factors dominated, but given vast improvements in worldwide communications and recent processor speeds, the decision time directly associated with by the sensor nodes (T_D) has become the predominant factor in estimating overall latency. To fully meet T-ISR system objectives, WSN may need to depend upon multiple sensor observations (M_{obs}) and data fusion at higher network levels. Higher-processing levels is defined by the combining of observations (time-history of measurements and associated data) from a number of nearest neighboring nodes. Multinode observations provide major advantages. Sensors employed on WSN nodes are not required to provide extended range capabilities. Also, having an array of sensors wirelessly connected inherently extends sensor coverage beyond that associated with a single node. Finally, sensors do not have to perform detection over a large field-of-view (FOV).

8.5.1.1 Active and Passive Modality Nodes

With random distribution of sensor orientation during WSN deployment, sensors with narrow FOVs are pointed randomly; however, the AOI is amply covered as neighboring nodes provide coverage where a single node may be blind. Through this leveraging of sensor technology and associated processing comes power and cost savings. Energy savings is also achievable for particular mission configurations by using active sensor modalities (laser and RF radars). To balance the increased energy use by active-sensor nodes, the increased sensing range must provide efficient energy coverage of the AOI. This can occur when extended sensing, beyond that of the passive modalities, is required but not at a significant duty cycle. This hybrid WSN depends on passive sensor nodes to provide the bulk of persistent monitoring and to act as a bellringer function for activation of active sensors. Conceivable scenarios suited for such a WSN are those presenting an AOI that emphasize long linear dimensions (or areas).

The hybrid system balances energy use (and costs) through use of a tiered approach. Passive sensing is used to discern when a target enters an AOI, whereas active sensing can provide precise target location and improved classification. This minimizes the active sensor duty cycle while reducing the number of passive motes required for the monitoring of a large AOI. For example, mixed passive-active node deployments can be effective for missions that involve roadways, pipelines, waterways, terrain features, or building perimeters. Such a hybrid sensor motefield would have passive sensors provide control and operation of the active sensors through autonomy rules, as depicted in Figure 8.3. Figure 8.3 portrays the use of microlaser motes within a mote-field consisting of passive sensor motes, such as PIR. The range of the active system far exceeds that of the passive node (100:1); however, power consumed by the active system is also a few orders of magnitude more. (Costs are more too, albeit at a lower, 10× ratio.) The hybrid approach balances longevity of the passive node against accuracy (localization of the target) and significant

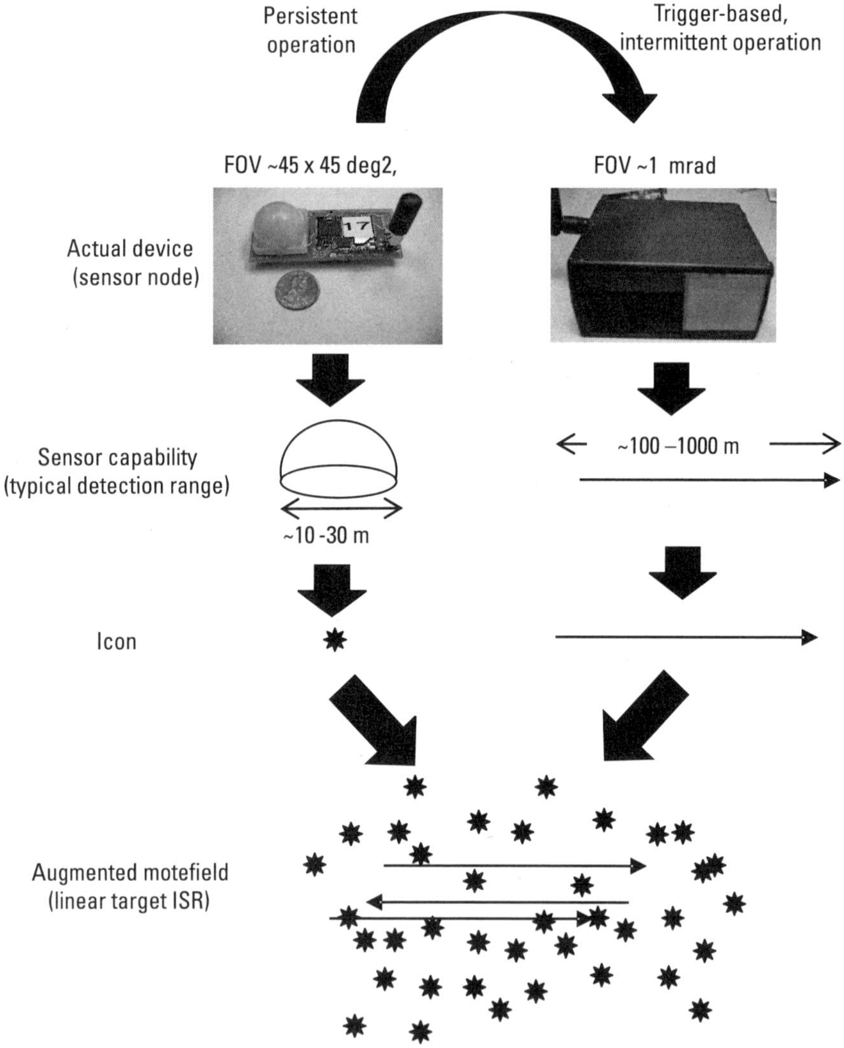

Figure 8.3 Hybrid sensor mote field showing the use of a persistent sensor node used to trigger an active sensor node upon object detection and qualification.

sensor range (10^3:1) of the active nodes. The passive nodes perform persistent surveillance role as bellringer to activate active-sensor nodes.

8.5.1.2 Constant False Alarm Rate Processors

A final consideration of a system composed of distributed sensors, homogenous or heterogeneous nodes, is that multiple sensors provide multiple observations for processing. Appropriately designed detection algorithms can make use of these multiple observations to increase performance either through direct fusion of measured data (i.e., incoherent, or time-delay integration of measurements) or through use of data constructs as constant false alarm rate (CFAR) processors. The basis of CFAR is to automatically detect a target in nonstationary noise and clutter background while maintaining a predetermined probability of false alarm. For target detection, a threshold is set adaptively based on the noise power level estimated from background observations. Usually, the noise power level is not known a priori, and a fixed-threshold value may be employed. Unfortunately, a constant threshold may increase the probability of a false alarm simultaneously with a marked decrease of probability of detection. CFAR sets the threshold according to current conditions, through the use of background values made by adjoining measurements, as shown in Figure 8.4. Through this approach, the threshold can be updated via a weighted moving average of a number of observations, versus using a single measurement, or use of a fixed a priori threshold. In this way, it is possible to avoid outliers and measurements of background that deviate significantly from the average background.

Assessment of CFAR detection has been extensively documented in open documentation for both homogeneous and nonhomogeneous backgrounds (e.g., multiple-target and clutter-edge cases) [7, 8]. In analyses, a centralized

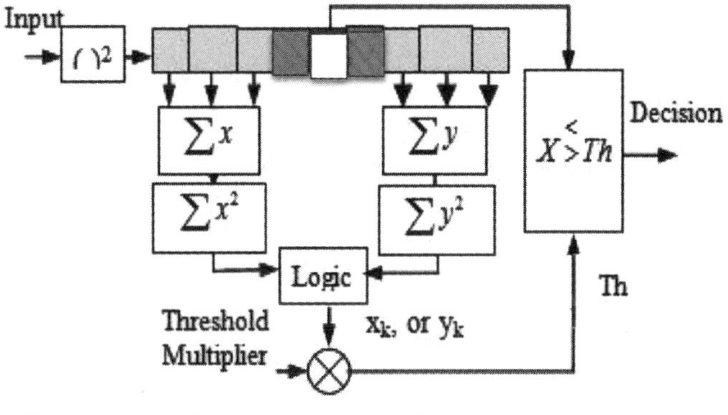

Figure 8.4 Operating principles behind the CFAR approach to processing.

detection method using a single sensor and CFAR is achieved by employing adaptive threshold techniques. Adaptive CFAR processing based on multiple distributed sensors and data fusion has been successfully employed using acoustic sensing. In these systems, M_{trig} ($<M$) distributed sensors (nearest neighboring nodes) provide observations to a group-averaging process operating within a homogeneous Gaussian background noise.

A prime example of employing the CFAR is provided through examination of automobile lane-departure camera systems. CFAR is used with multiple sensors to suppress background variations and multicamera data is accumulated, properly image-stitched, and input to a Hough-transform processor to extract key edges used to signal a driver of the lane departure.

8.5.2 Tracking Function

Tracking involves establishing a time-series of target positions. The objective of tracking is to present a trajectory of a target; that is, positions over time, along with estimates of direction and velocity. Having predicted trajectories allows a motefield to judiciously activate quiescent areas of sensor nodes to be prepared to gather observtion data. During tracking, numerous deleterious events occur that may disrupt the tracking process, including target drop-out (interruptions in observations), multiple-target crossings (confusion of which target is which), and complete lost of target (e.g., terrain hiding, excessive background noise, and loss of target signal).

Numerous algorithms have been developed to deal with various problems arising from the need to handle missing observations, erroneous data reception, corrupted processing, or maneuvering targets. Successful tracking requires the WSN system to estimate an initial point of target entry and to provide a trajectory with sufficient accuracy to the system (and MOC) within acceptable detection latency. Implicit in this requirement is the need for target localization. Tracking performance requirements dictate that tracking accuracy, the maximum difference between a target's actual and estimated position, be both bounded and specified (within practical limits). Excessive divergence of predicted versus active track trajectories usually ends with loss of target tracks due to one of the aforementioned difficulties.

8.5.2.1 The Alpha-Beta Tracker

Tracking maneuvering targets remains difficult and has been addressed through numerous algorithms, such as the classical alpha-beta (α–β) tracker. The α–β tracker is a model representation that performs well for simplistic scenarios where trajectories are dependent on position (x) and its first derivative, velocity

(v). Assuming that velocity remains constant over the sampling interval (Δt), updates to the position can be estimated at the next sampling time as

$$\hat{x}_k = \hat{x}_{k-1} + \hat{v}_{k-1}\Delta t \quad (8.2)$$

The hat variable represents estimation of a value derived from measurements of position and use of the velocity assumption. If the system gains information associated with how velocity deviates from the assumption of being constant velocity, (8.2) can accommodate this through modification of how $\hat{v}_k$ is updated. With these assumptions, we have, for velocity,

$$\hat{v}_k = \hat{v}_{k-1} \quad (8.3)$$

For position, the prediction of an updated value ($\hat{x}_k$) expects the previous estimates of position to incur errors due to noisy observations. The prediction error ($\hat{r}_k$) is called the residual estimation error, or deviation, and follows statistical filtering interpretation. Estimation of the prediction error is evaluated as the difference between observed position, x_k, and predicted position $\hat{x}_k$ as

$$\hat{r}_k = x_k - \hat{x}_k \quad (8.4)$$

The α–β tracker employs two constants (α, β) to modify both position and velocity estimates, respectively. Assuming a positive deviation, $\hat{r}_k > 0$, which indicates that the previous position estimate was low, a correction to predicted position and velocity values is made using the α, β constants and the predicted values for position and velocity modified as

$$\hat{x}_k \leftarrow \hat{x}_k + \alpha \hat{r}_k \quad (8.5)$$

$$\hat{v}_k \leftarrow \hat{v}_k + \beta \frac{\hat{\hat{r}}_k}{\Delta t} \quad (8.6)$$

Equations (8.5) and (8.6) are actually smoothed estimates of position and velocity. With a properly selected sampling period (i.e., fast enough), these corrections present small deviations of an estimate to the gradient direction. As corrections accumulate, the error in state vector estimates is reduced. For convergence and stability, the values of the alpha and beta multipliers should be positive and small and follow numerical bounds as [9]

$$0 < \alpha < 1;\ 0 < \beta < 2 \quad (8.7)$$

$$0 < 4 - 2\alpha - \beta \quad (8.8)$$

Noise is suppressed with $0 < \beta < 1$; otherwise noise entering the system is amplified. Values of alpha and beta are adjusted empirically. For example, larger values of alpha and beta produce a faster response when tracking transient shifts, whereas small alpha and beta gains reduce the level of noise introduced into state estimates. When balanced between accurate tracking and noise reduction, the α–β tracking algorithm is effective, and state estimates (8.1) converge to improved accuracy compared to direct measurements.

8.5.2.2 The Kalman Filter-Tracker

Improvements to the α–β tracker exist, such as Kalman filters, extended Kalman filters (EKFs), fuzzy logic modified α–β filters, and particle filters [10–12]. Kalman filtering estimates the values of state variables and corrects them in a manner similar to an alpha-beta filter but in a dynamic sense derived from modeling state inputs and process noise sources. The prominent differences between Kalman filters and alpha-beta filters are described as follows:

- Kalman filters use a detailed dynamic system model that extends beyond a simple two-state (position, velocity) model.
- Kalman filters operate using multiple observed variables to correct state variable estimates, which may or may not be direct measurements of individual system states.
- Covariance noise models for state characteristics and observations are embedded into Kalman filters, which automatically update coefficients compared to using fixed gain constants.

A Kalman filter designed to track a moving object using a constant-velocity target dynamics model with constant process noise covariance and measurement covariance converges to the alpha-beta filter algorithm. However, state gain values are computed recursively with each time prediction step using assumed process and measurement error statistics. In contrast, alpha-beta gain values are computed ad hoc. Accurate tracking of the target position using Kalman filters operate well to reduce residential errors but require relatively complex computations and memory capacity [13–15]. This computation load may not be suitable (or viable) for WSN sensor nodes. Another issue in using Kalman filters is verification of the complex modeling structure and use of numerous system parameters employed. This requires a complex, sophisticated diagnostic evaluation of the entire system.

8.5.2.3 The α-β-γ Tracker

An approach to improved tracking being employed with simpler complexity over Kalman filtering and suitable for applications as WSN-based processing is a return to alpha-beta tracker but with modification three gain parameters, the α-β-γ filter. This algorithm builds on the α–β tracker but assumes constant acceleration, versus constant velocity. System states are updated as per (8.5) and (8.6) but with a third parameter (γ) for acceleration. This results in updating the state vector via,

$$\begin{aligned} \hat{x}_k &\leftarrow \hat{x}_k + a\hat{r}_k \\ \hat{v}_k &\leftarrow \hat{v}_k + \beta \frac{\hat{\hat{r}}_k}{\Delta t} \\ \hat{a}_k &\leftarrow \hat{a}_k + \gamma \frac{\hat{\hat{r}}_k}{\Delta t^2}; \end{aligned} \tag{8.9}$$

Similar to the treatment of predicting velocity in the alpha-beta tracker, the update to the α-β-γ filter acceleration estimate uses the previous output for acceleration,

$$\hat{a}_k \leftarrow \hat{a}_{k-1} \tag{8.10}$$

8.5.2.4 Tracking Complexities and Approaches: Markov Chains and the Munkres Algorithm

Tracking complications arise when a sensor system must track an object against converging objects' trajectories or when tasked with tracking multiple targets simultaneously. Traditionally, tracking performance analysis is gauged based on terms of the probability of establishing a track and density of false alarms as considered using various tracking algorithms such as the Markov chain model and Munkres assignment [16]. Use of the Markov chain model describes the behavior of tracks in traditional track-oriented tracking algorithms based on simple time-history position (Markov chain) modeling. With Munkres, associations are established, sample-by-sample, using positional logic, such as the nearest-neighbor approach, to form bipartite graphs so that each object has an association with one another, measurement-to-measurement. Overall tracking performance for a sensor system is measured using two key characteristics (TPMs): (1) the probability of establishing a track from target returns and (2) computation of the density of false tracks produced, per unit surveillance volume per unit time.

The first performance measure is the probability of establishing a track using observations of a target. With multiple-hypothesis tracking algorithms, data-to-data association hypotheses are maintained as a set of consistent tracks. To establish tracks we must use reasonable logic operating on the hypotheses extracted using formulated tracks. In practice, tracking functions depend on complicated mechanisms that include the initiation of Gaussian distributions derived using a scarcity of data points, development of reliable data association logic, and elimination of possible target tracks to arrive at the best estimate target trajectory. Estimation of the density of false tracks produced by the same tracking logic is the other performance measure to be quantified. These two quantities, the probability of establishing a track and the false track density, are referred to as system operating characteristics, paralleling the frequently used receiver operating characteristics (RoC) of binary classification systems.

8.5.3 Discrimination/Classification Functions

Subsequent to detection, the processes of discrimination and classification begin. Semantics needs to be considered regarding these specialized functions because in practice, the objectives for these functions overlap [17, 18] and both functions require the use of statistical inference, as presented in Chapter 5. To differentiate between these two activities, we will adopt a formal meaning for discrimination and classification. The objective behind discrimination is to separate targets and objects into distinct groups based on direct measurements. The discrimination function is used to describe data statistics and strives to identify the best parameters (and values) useful in associating targets to designated groups. The discrimination function depends upon observable characteristics associated with a particular object that can associate that object to one of a set of predefined population groups (e.g., vehicle, personnel). How well the discrimination function performs depends highly on the criteria, in discriminate space (K), that separates population groups. The characteristic parameters used to statistically determine the individual populations must be selective enough to provide reasonable confidence in class assignment for an object through use of decision theory (Chapter 5). Errors will occur and can be correlated using the separation factor (K) and observables available to discern among competing class populations.

To aid this process, multiple discriminates are sought and applied to potential targets. Figure 8.5 depicts a multivariate function used to classify target groups and implements a comparison of observables to see if the target correlates with a particular group. How well targets are classified depends upon the separation of classification evaluation measures, K, between target and nontarget groups. If K represents the separation of varying classification groups, then error in classification begins to occur as discriminate separating groups begin to overlap the statistical populations gleaned from observed data for each target type.

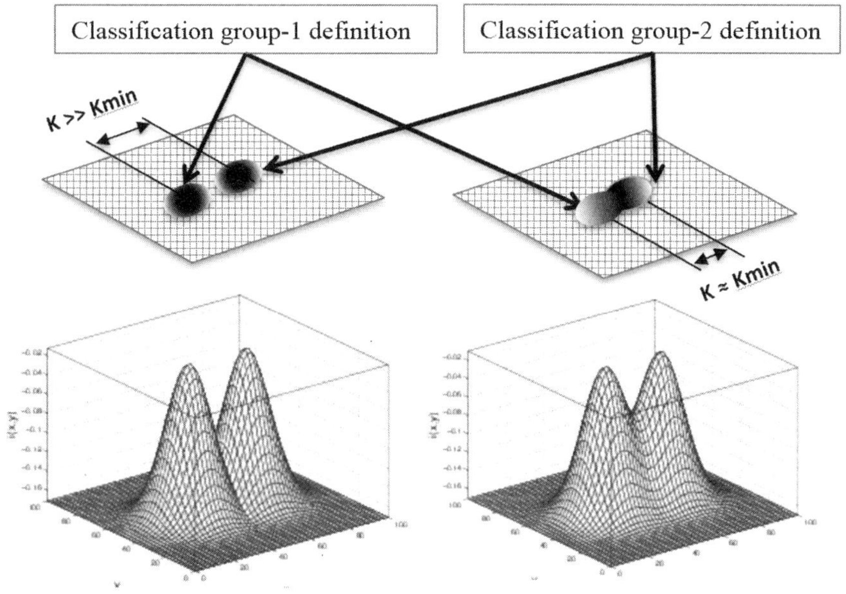

(a) (b)

Figure 8.5 Discrimination (*K*-factor) classification group separation : (a) Group-1 definition with discriminate sufficiently identified by being greater than *Kmin*, (b) discriminant space insufficient leading to Group-2 definition.

The objective associated with classification is to allocate newly viewed targets and objects to predefined group populations. Once an object has been observed, the appropriate discriminants place that object into one of multiple population classes. With classification, further refinement based on increasingly detailed discrimination is performed to arrive at specific identification of the target. Classification performance is evaluated through the use of costs associated with misclassification, which depends upon the severity of mission failure associated with erroneous classification. For correct classification, the cost is defined as zero. Costs of misclassification are not necessarily equal, meaning that misclassification of a group 1 object as a group 2 object may not be equivalent to misclassifying a group 2 object as a group 1 object [19–21].

8.5.4 Identification

The final stage, identification, associates a detected and classified target to a unique object. Identification is used to discern if the vehicle is one considered to be among a very specific target list, or not. Identification for T-ISR WSN is designed to designate a target/object class object as a type of vehicle (tracked, wheeled, diesel, gas, etc.) and whether the personnel are armed or unarmed. Sensor modalities aid with the identification function; however, information gleaned from tracking estimates of velocity and position aid in providing discrimination separation factors (K) useful to perform discrimination and classification of objects into target classes quickly. A large motefield can differentiate against background noise and objects using discrimination and classification rules associated with activation of a cluster of nodes versus activation of a larger area of nodes in a field. For example, thunderstorms would likely activate all acoustic monitoring in a motefield more or less simultaneously, whereas a diesel vehicle driving through the field would activate a group of nodes along its trajectory.

Figure 8.6 illustrates the use of grouping sensor nodes at a peer level (cluster) where the network implements aggregation and correlation of information as data travels from node-to-AP. The cluster supports handling assigned detection areas, where simple sensors with sensing field or regard (FOR) can be much less then 360 degrees. The clusters also work to support WSN network reliability as the sensor and RF range capabilities are deployed to provide overlap, providing redundancy of both. The cluster head handles the level of processing.

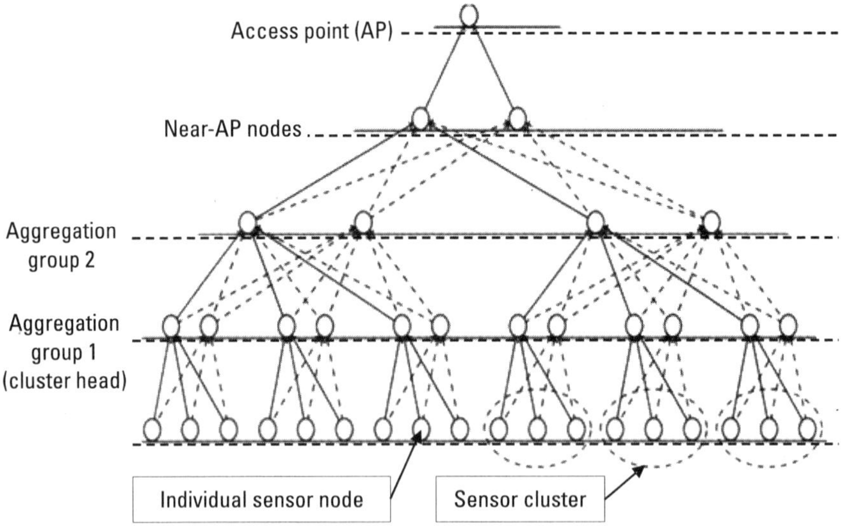

Figure 8.6 Multiple node and level data processing.

8.6 Power Management

One of the more critical supportive functions, power management, is addressed through a combination of middleware and hardware design. Of the severe failure modes for WSN, energy exhaustion ranks as the top cause of reliability issues, node loss, and network disruption. It is through the availability of energy, either via battery-based systems, solar, or energy harvesting that allows a WSN net to operate. As a result, power management must be considered in all aspects of the WSN system design: node capability, deployment pattern, and perhaps most importantly, adroit middleware design.

Of the subsystems that constitute a WSN sensor node, the radio transceiver consumes the majority of available energy regardless of whether the node is in an active or inactive (e.g., sleep) mode. During active operation, the averaged power drain for a RF transceiver is 400–500% that of the microprocessor or sensor subsystems [22]. During quiescent modes, the RF subsystem still consumes the larger portion of energy, followed by the microcontroller system as the sensors are powered off to achieve sleep mode.

There is a notable exception to this ranking of power use. When the implemented sensor modality is based on an active technology (e.g., microlaser and/or RF radar), power use shifts from the RF transceiver being the largest power drain to the sensor being such, but only when sensing is active. Active sensing nodes must therefore be handled accordingly, through use of heterogeneous motefields (a combination of active and passive sensors), specialized middleware, node redundancy, and augmented energy generation and storage (e.g., energy harvesting per observation period). Recall that the use of active sensors is typically a result of very particular T-ISR missions.

Considering that the RF transceiver consumes the preponderance of available power, it becomes obvious why the bulk of intensive research and papers on energy conservation have concentrated on energy-aware communication protocols [23–25]. Every aspect of the communication system must be explored and refined as much as possible to conserve energy while simultaneously providing efficient, robust, and timely delivery of sensor data. The RF transceiver hardware has been optimized repeatedly over the past decades, and chipsets and system-on-chip (Chapter 6) implementations have achieved tremendous reduction in energy use during transmission and sleep modes [26]. Significant improvements to power management lie with the control of the communications protocol, which is implemented through MAC middleware (Chapter 4). MAC middleware is responsible for monitoring, orchestrating, and controlling node access to the communication (RF) medium. To increase efficiency of WSN communication, and thereby, maintain control over power consumption, appropriate MAC designs must strive to minimize interference and packet collisions. Customization and refinement to MAC protocols have been considered by packet radio systems by optimizing channel access, packet length, modulation schemes, channel signaling and retransmission methods, and transmitter power.

8.6.1 MAC Consideration

For WSN, the emphasis of the MAC is to avoid packet collisions, minimize unintentional message reception, and reduce overhead transmission (e.g., clear-to-send, ready-to-send), and idle listening. WSN average traffic volume is relatively low. However, during triggering events, higher than average message rates occur, which is reflected as decreasing interarrival times between data packets. Additionally, a smaller number of nodes carries increased message traffic as the data streams propagate toward the access point (AP). As discussed in Chapter 6, network links are rather unpredictable and are best represented by probabilistic modeling. Solutions to lost links are manifold in the WSN design and use solutions such as: self-healing, retransmission, and/or rerouting.

Well-established MAC implementations can be categorized as either reservation-based (e.g., TDMA) or contention-based (e.g., CSMA) protocols. Unfortunately, neither TDMA nor CSMA are appropriate for WSN systems as the focus for WSN is on low-power operation first and foremost. Numerous low-power (LP) MAC protocols for WSN have been developed, extensively reviewed, and well presented [27–30]. A successful MAC protocol developed and applied to various WSN systems specifically designed to support T-ISR activities is known as Berkeley-MAC (B-MAC).

Although numerous LP-MAC protocols exist, we will focus on B-MAC given its background in use for T-ISR demonstrations and evaluations. To achieve low-power operation, B-MAC employs an adaptive preamble-sampling scheme to reduce duty cycle and minimize idle listening. B-MAC is lightweight and reconfigurable. WSN protocols may be implemented efficiently using B-MAC primitives. For example, sensor MAC (S-MAC) and timeout MAC (T-MAC) may be implemented as services that use B-MAC as the link protocol. S-MAC and MAC perform synchronization, organization, fragmentation, which work toward mitigating issues such as hidden terminal problem. B-MAC was designed to meet the requirements of WSN applications through incorporation of specific MAC characteristics conducive to WSN systems including the following:

- Low-power operation;
- Effective collision avoidance;
- Simple implementation, resulting in minimal code and associated memory;
- Efficient channel use at low and high data rates;
- Reconfigurable via network protocols;
- Tolerant to dynamic network configurations;
- Scalable to very large node count.

B-MAC builds on the foundation defined by CSMA/CA, employing clear channel assessment (CCA), which determines whether the channel is clear and includes carrier sensing and energy detection. The carrier sense (CS) mechanism consists of both physical CS and virtual CS. Physical CS is provided by the PHY layer and is a direct measurement of the received signal strength of a valid packet (symbol). If signal strength is above a certain predetermined (standard) threshold level, the medium is considered busy. Virtual CS is provided by the MAC and referred to as the network allocation vector (NAV), which operates as an indicator for the node as to when the medium is available. NAV is updated each time a valid frame is received that is not addressed to the receiving node, provided that the duration value is greater then the current NAV value. Through evaluation of NAV, a node can avoid transmitting even when the physical CS falsely indicates an idle medium.

Using a CSMA/CA-based protocol, a sensor node performs CCA before transmitting a frame to avoid causing a collision if the channel is busy. Recall that CCA determination is implemented through thresholding, and the channel is considered clear only if the measured signal is below the noise floor. Unfortunately, in a RF-noisy environment, thresholding may generate a large number of false positives because of the significant variance in channel energy. False positives lower the effective channel bandwidth and, thus, should be reduced. Although CCA is implemented at the PHY layer, it directly influences the MAC relating to power management and link throughput. In addition, CSMA/CA is unreliable due to the hidden node problem. Collisions can occur because CCA at a particular node preparing to transmit does not detect data messages directed at the intended receiving node; this third node is hidden (not detected) by the to-transmit node [31]. The hidden node issue has been addressed through the use of ready-to-send (RTS) and clear-to-send (CTS) short messages between linked nodes. However, RTS and CTS add to message traffic and do not convey data; in effect, these messages consume energy without accomplishing the main objective, the delivery of valid target data.

B-MAC protocol directly includes a technique based on outlier detection to improve the quality of CCA. With B-MAC CCA, a node searches for outliers in the received signal for channel energy significantly below the noise floor. If a node detects an outlier during channel sampling, it declares the channel clear because a valid signal would present outliers significantly below the noise floor only with low probability. If a node does not detect such outliers within five samples, CCA declares the channel to be busy. The outlier detection technique depends upon the accuracy of estimating the noise floor. B-MAC uses automatic gain control for estimating the noise floor to adapt to ambient noise changes. Each node takes signal strength samples at times when the channel is assumed to be clear, such as immediately after transmitting a frame. From these values, each node calculates an average value and uses it as a simple lowpass filter for the noise floor estimate. The resultant performance using outlier detection successfully outperforms CS thresholding [32].

Apart from collision avoidance and good channel utilization, accurate CCA provides additional power saving benefits. CCA allows a node to listen to preamble traffic while waiting to receive data frame to determine whether the channel is still busy. If the node detects that the channel has returned to idle before it receives the data, the node stops listening and goes back to sleep mode. By avoiding this reception, an accurate CCA improves preamble-sampling performance. Other aspects that directly influence power consumption include deployment pattern (redundancy in nodes, and links), node operation, and use of duty-cycling operations.

8.6.2 Low-Power Microcontroller Solutions

Low-power microcontrollers provide node designers with multiple power states that consume varying power levels, multiple wakeup latencies, and varying levels of peripheral support. LP-microcontroller systems are operated to remain in the lowest possible power state that can satisfy application requirements. Accurately determining this state requires knowing a great deal about the power state of associated subsystems memory, sensor(s), and ADC components. State transitions are correlated to microcontrollers receiving an interrupt as the power mode moves from a low power state to an active state until a RTOS scheduler identifies an empty task queue, at which time, the microcontroller is returned to a low power state. RTOSs, such as the venerable TinyOS 2.x, use multiple (for TinyOS 2.x, three) mechanisms to decide what low power state to employ with a microcontroller: status and control registers, a dirty bit, and a power state override. A dirty bit or modified bit is a bit that is associated with a block of computer memory and indicates whether or not the corresponding block of memory has been modified. The dirty bit is set when the processor writes to (modifies) this memory. The bit indicates that its associated block of memory has been modified and has not been saved to storage yet. The TinyOS scheduler puts a processor into a sleep state when the task queue is empty [31].

WSN node devices have been designed using off-the-shelf low-power microcontroller components and systems, which includes preconditioning and ADC circuits. Newer microcontroller architectures continue to be developed based on hardware specialization and power gating, resulting in additional power savings of two orders of magnitude [33]. An approach to decreasing microcontroller power use is focused at the front-end level where the power consumption is static [34]. Microcontroller design for low-power applications has been developed through analog CMOS VLSI processing units operating in the sub-threshold regime design technique [35].

8.6.3 Power Source: Battery Source

WSN design may be required to employ a stable rail voltage by using a voltage regulator or converter. The use of a DC-DC converter can allow a system to utilize more capacity of a battery if the rail voltage has a tight specification of the range of operating voltage. However, there is a power efficiency (e_v) associated with any DC-DC regulator that behaves as a function of source (battery) voltage (V_b) divided by regulated voltage (V_r) as

$$e_v \propto \frac{V_r}{V_b} \tag{8.11}$$

DC-DC regulation operates as designed as long as the headroom specification for the converter is observed. Voltage regulators operate correctly until the low dropout voltage level is reached. At this point, the regulated voltage output will begin to vary with the source voltage, and the constant rail level no longer exists. A consideration exists because of loss of power (L_p) by a regulator. This loss is evaluated using the product of the voltage difference between the input and output voltage levels with the draw current (i_d), as

$$L_p = \left(V_r - V_b\right) i_d \tag{8.12}$$

To minimize the overall power used by a sensor node, the battery pack can be connected to the power rail directly. However, by not regulating the power rails, an uncertainty is created, as the system voltage is not stringently defined. To ensure operation, the node designer should employ components that are rated for the maximum and minimum allowable battery voltage. Additionally, brownout circuitry will be required to reset the microcontroller system when voltage does drop below rated operating levels.

8.6.4 Power Source: Energy-Harvesting

Harvesting electrical power from environmental energy sources is an attractive and increasingly feasible option for WSN sensor nodes. However, designing highly efficient microscale energy-harvesting systems requires an in-depth understanding of various design considerations and trade-offs. Most of the challenge to energy harvesting arises from the form-factor constraint that constrains energy transducers to cubic-millimeter volumes. As a result, the maximum power output of these miniaturized transducers is extremely small, often only a few milliwatts. Therefore, a harvesting subsystem should be carefully designed to extract as much power as possible from the transducer and transfer it to the electronic system with minimal loss, which requires extremely energy efficient design [36].

Intensive research conducted throughout recent decades has evaluated harvesting energy from various natural and man-made energy sources to a scale useable by WSN. Numerous successes have occurred and continue to make inroads into augmenting sensor node power along with use of increasing energy-dense battery packs. Energy-harvesting mechanisms for WSN have considered four major sources of natural and man-made power sources: radiant energy (includes both solar and naturally-occurring and manmade electromagnetic fields), mechanical energy (wind-driven, water-driven, vibration using piezoelectric materials), chemical energy exchange (e.g., direct methanol fuel cell [37]), and thermal energy (geothermal differential extraction). Each approach has potential and each brings design changes that must be considered and evaluated. Descriptions of the approaches to WSN using thermal, radiant, and mechanical energy have been well-documented; see [38] for a comprehensive overview.

8.7 Reliability

WSNs consist of hundreds or thousands of sensor nodes with limited processing, storage, and battery capabilities. There are several strategies to reduce the power consumption of WSN nodes (by increasing the network lifetime) and increase the reliability of the network (by improving WSN QoS). However, there is an inherent conflict between power consumption and reliability: An increase in reliability usually leads to an increase in power consumption. In pursuit of reliability, routing algorithms may forward the same packet though different paths (multipath strategy), significantly increasing WSN power consumption. To ascertain the capability of a WSN system using a multipath strategy, evaluation modeling must consider the battery level as a key factor [39].

8.7.1 Reliable Transport Design

Techniques used by WSN systems to achieve reliability include data packet retransmission, network redundancy, and data error correction coding (ECC), where no reverse channels are required for retransmission. Most of the existing research focuses on traditional retransmission-based reliability, where reliable transmission of data packets is ensured in terms of recovering lost packets by retransmitting them. This might result in additional transmission overhead that not only expends energy resources but also contributes to network congestion and consequently impacts data transmission reliability. ECC contributes to link congestion, and data messages with ECC contain redundant information to allow for detection and correction of transmission errors, but the entire message (and the required message needed to signal error detection) would not have to be retransmitted.

Retransmission and redundancy can be achieved on a hop-by-hop basis, or through a process that operates, end-to-end. Hop-by-hop algorithms employ intermediate nodes to realize retransmission or to provide link redundancy. With end-to-end algorithms, retransmission or redundancy is conducted using only source and destination nodes. Algorithms that use a combination of these methods have also been evaluated based upon packet or event-level reliability. Packet reliability requires all packets from relevant sensor nodes to reach the access point, whereas event reliability ensures that the access point acquires sufficient information to discern that a particular trigger event has occurred. Several reliability schemes based on a hybrid combination of retransmission and redundancy techniques exist and have been studied in WSN reliability–focused literature [40].

8.7.2 Reliable Code Propagation

Another facet of reliability is associated with the middleware programming of nodes in a system (one or multiple motefields). For proper operation, a WSN system must be capable of asserting that all nodes are running the intended middleware version and that a robust process exists to provide over-the-air (OTA) reprogramming to allow for modifications, corrections, or updates to the middleware code residing on each node. WSN reliability must be able to reach nodes that are not within radio range of the access point (middleware update source), requiring updates through multihop propagation.

Inserting middleware updates must also consider methodologies to recover node operation in the event that errors are introduced during the middleware upload. Recoverability indicates that a sensor node is provided with the means to upgrade the middleware version being propagated and implemented, or is provided a fallback capability to return to a previous version and re-attempt the upload, assuming that the OTA propagation of the upload is repeated and/or reverberated throughout the system for a finite time window. The goal is to provide autonomous logic that prevents permanently locking out a node from reaching the middleware (or otherwise to consider providing a self-initiated total power-down capability). A number of recovery algorithms have been proposed for providing over-the-air or multihop wireless reprogramming functionality including in-network programming (XNP), trickle, multihop over-the-air programming (MOAP), and deluge.

All of these algorithms make assumptions about operation, including a core assumption that the recovery algorithm is intact and operating. Unfortunately, low-cost sensor nodes have been shown to operate in unscheduled, unplanned ways, precluding this core assumption. Consequently, any recoverability capability cannot be guaranteed unless a means of detecting a faulty upload is implemented, and the node is provided access to the network to repeat the attempt to upload.

8.7.2.1 Processor Protected Modes and Interrupt Timers

With traditional operating system implementations, a timer enables the operating system to maintain control of the processor. Prior to turning control of a processor to applications running in user mode, an operating system sets a timer to regain control of the processor; when the timer interrupt occurs, control is returned to the operating system. Instructions that modify timer operation are privileged, restricting execution of such instructions to occur only while in protected mode by the operating system. Unfortunately, many 8-bit microcontrollers do not provide a true protected mode of operation, leading to occurrences where applications assume control over node hardware and timers, and activate interrupts, which leaves the operating system with no mechanism to preempt a misbehaving application.

Hijacking an operating system can occur either accidentally or intentionally (a security concern). With past WSN middleware, the mode of compiling wirelessly reprogrammable (OTA) applications under RTOS as TinyOS was to create a monolithic code image that included the RTOS, a wireless reprogramming component, and the user application, which are expected to coexist in cooperative harmony. This presents a vulnerability that results in permanently disabling wireless reprogramming.

One example is pushing an upload with an application that does not include a wireless reprogramming module. Doing so eliminates the ability of a WSN system to perform OTA reprogramming. Existence of an accidentally inserted atomic block that never exits can also create vulnerability. Even with a robust RTOS that employs the use of watchdog timing, errant code with even a slight variation of an atomic block that does not exit globally may disable interrupts and loops endlessly, clearing the watchdog timer. With interrupts disabled and the watchdog timer periodically cleared, no mechanism exists for a RTOS to regain control. Even if interrupt vectors are within a protected code segment, common instructions exist so their presence in code is not unusual, especially with interrupt handlers. Automatic detection of rogue code is nontrivial. Given that manual reprogramming of all WSN nodes is impractical, there have been approaches to work this issue, one being the grenade timer used on sensor nodes (XSM) associated with DARPA's NEST program, discussed in Section 8.7.2.2.

8.7.2.2 Grenade Timer Approach

A grenade timer is like a watchdog timer that cannot be reset; once started, the timer cannot be stopped, only sped up [41]. Properties of a grenade timer are in order. Any software module may fire a microprocessor grenade timer at any time. The amount of time a grenade timer counts down to triggering a hard interrupt can be adjusted within bounds by a microprocessor (e.g., bootloader) until activated. Once the grenade timer is started, it cannot be stopped, and the countdown value cannot be changed (i.e., a one-shot then latch-out). As long as the grenade timer has not been activated, the real-time clock used to implement the grenade timer remains accessible to either the processor bootloader or application.

The grenade timer approach protects against an application not relinquishing control, but not against integrity or confidentiality violations. If an application can access and overwrite critical RTOS code, memory protection can be defeated. Placing the RTOS in ROM makes it safe from modifications; but the RTOS working area remains subject to corruption. With node designs that support dynamic updates, RTOS modules may be relegated to flash memory or RAM. An approach to solving this vulnerability uses the grenade timer circuit to intercept the processor address bus. The grenade timer, when activated, masks the upper bit address bus (one bit is sufficient in principle if one wishes to save on pin count) and replaces it with a fixed-page address defined by the RTOS prior to activating the timer. This prevents an application from reading or writing memory or I/O addresses outside the authorized memory page [42].

8.7.2.3 Node Image Integrity

If a node is detected to be operating erratically, or not at all through status checking and other diagnostic tools, and has been determined that the code has been corrupted through one of numerous causes, what can be done to regain control and load the correct code to a sensor node? Use of a grenade timer guarantees that the bootloader eventually regains processor control. By properly setting the processor fusemap, key RTOS components (bootloader, interrupt vectors, and interrupt handlers) are protected from an application. To recover the network interoperability, a network bootloader is devised to operate alongside a grenade timer. The network bootloader is implemented with set fuses so that it can be erased and only altered through manual reprogramming. An effective bootloader consists of a minimal network stack, a network-reprogramming module, grenade timer drivers, and supportive code. The network bootloader provides the logic that drives the capability of a WSN system to realize functions associated with the verification code or repair of code integrity.

Code verification works with a unique identifier that indicates the code image version, which is stored in nonvolatile memory protected from application code (e.g., flash memory). To further verify that the current image version is correct, the network bootloader interrogates neighboring nodes for their version value by broadcasting a request. The received messages provide version numbers, and values that can be compared to values for the requesting node. If values are identical, the code image is considered correct. If not, the larger version (assuming that incremental version numbers are used) is requested from the corresponding node and used to update the node image. To avoid iterative looping of uploads occurring, a network management function at the group level observes and tracks the status of uploading processes and intervenes if multiple transfers of code image continue beyond a predetermined count value. This prevents runaway messaging where erroneous version number values indicate newer code images, and the network repeatedly transfers code images back and forth, which has the potential to exhaust all energy stores available—or at least deplete significant energy in affected node areas. Erroneous code versions can also be handled through verification of a new application image and its version number using a message authentication code.

8.8 Security

WSN systems constitute an information source that is a mission-critical system component, and when applied to T-ISR missions, security and integrity of the system and associated data is paramount. The threats to T-ISR WSN systems are numerous, and varied. Security is a major issue that must be considered to allow successful application of WSN technology for T-ISR applications. Middleware has a relatively long history of addressing resource allocations, efficient communication, and management of WSN networks. When considering security measure for WSN, vulnerabilities include absence of tamper-resistance, exposure of wireless messaging using an open media, and the possibility of adversaries injecting misinformation into the network through malicious nodes or compromised nodes. Security aspects associated with WSN system services, resources, and data remain a challenge.

In consideration of security, it is necessary to consider several major categories: message security, node security, system security, and covert operation. Message security refers to the safety of the message transmission among sensor nodes. Node security addresses countermeasures to physical nodes. System security has to do with the network and overall system and addresses such security aspects as denial of service (DoS), malicious nodes, a compromise of the access point (AP), and/or an exfiltration process. Due to the minimal volume and low-cost requirements of sensor nodes, it has been difficult to address security measures from a hardware standpoint. Problems with public key (PK) algorithms and other security protocol measures have been experienced due to energy limitations and scarce memory and computational capacity to accommodate advanced encryption schemes. For successful security measures to be implemented, mechanisms must be considered and implemented during initial designs and analyses associated with WSN modeling, hardware, and protocols, with resource trade-offs between security lever and computational capacity.

8.8.1 Cryptographic Key Management

For those working in cryptography, the mantra to security design is that the entire mechanism of cryptography, except for the value of the cryptography key, is to be considered known to an adversary. In response, the security designer should not depend on algorithms remaining confidential, but rather rely on key security. Key management becomes the valued commodity and the basic component for successful cryptographic reliability.

An example of a key management approach investigated is one that is associated with groupings of nodes (clusters) and employs a distributed probabilistic clustering scheme. Nodes within a single hop are considered clusters and are regarded as a network. The WSN system is considered a collection of numerous, single-hop clusters, with a cluster key for in-cluster message authorization and a neighboring cluster key to enable cluster-cluster communications. With an effective localized key predistribution algorithm, each node needs to send only one cluster key announcement message to form a cluster key network.

8.8.2 Cryptocoprocessor

With advances continuing in CMOS and VLSI technologies, low power, low-cost cryptocoprocessors exist. Microprocessors with this level of dedicated processing capability built on security measures are experiencing increased security capabilities using increasingly complex algorithms [43]. Finally, regarding covert operation, if the existence of the WSN system is not known to adversaries, then a level of confidence exists that security of the system can be relied upon. This is the purview of LPD and LPI.

8.8.3 LPI and LPD

Due to the broadcast nature of radio wave propagation, adversarial eavesdroppers for interception purposes can readily intercept data packets being transmitted. LPI communication systems have been made available through the use of spread spectrum techniques. Spread spectrum refers to well-developed algorithms to provide secure communications by spreading the transmission signal over a large frequency band. The idea behind spread spectrum is to use more bandwidth than the original message while maintaining the same signal power. A spread spectrum signal does not have a clearly distinguishable peak in the spectrum, which renders the transmitted signal more difficult to distinguish from noise (LPD) and therefore more difficult to jam or intercept (LPI) [44].

8.8.4 Command Authenticity

Authentication protocols exist to ensure the authenticity of each node present in a particular WSN system to authorize or prevent access to the network and data messages. The usage of the authentication protocol helps to mitigate network attacks and security vulnerabilities by providing verified credentials to nodes to allow any access to the system. Authentication protocols protect the sensitive sensor data from compromise or attack by adversaries [45].

References

[1] Romer, K., O. Kasten, and F. Mattern, "Middleware Challenges for Wireless Sensor Networks," ACM Sigmobile Mobile Computing and Communications Review, 2002.

[2] "Chapter 3: The Extreme Scale Mote (XSM)," http://www2.ece.ohio-state.edu/~bibyk /ee582/XscaleMote.pdf.

[3] Ahmed, S., "Middleware for Sensor Networks," https://www.slideserve.com/erling /middleware-for-sensor-networks-powerpoint-ppt-presentation.

[4] Khalid, Z., N. Fisal, and M. Rozaini, "A Survey of Middleware for Sensor and Network Virtualization," *Sensors*, 2014.

[5] Anderson, S., "Target Classification, Recognition and Identification with HF Radar," *RTO SET Symposium on Target Identification and Recognition Using RF Systems*, 2004.

[6] PixInsight, "Image Integration," https://pixinsight.com/doc/tools/ImageIntegration /ImageIntegration.html.

[7] Chang, K., S. Mori, and C. Chong "Performance Evaluation of a Multiple-Hypothesis MultiTarget Tracking Algorithm," *29th IEEE Conference on Decision and Control*, 1990.

[8] Uner, M., and P. Varshnev, "Distributed CFAR Detection in Homogeneous and Nonhomogeneous Backgrounds," *IEEE Transactions on Aerospace and Electronic Systems*,1996.

[9] Asquith, C., "Weight Selection in First Order Linear Filters," U.S. Army Missile Command, Report no. RG-TR-69-12, 1969.

[10] Gelb, A., *Applied Optimal Estimation*, The MIT Press, 1974.

[11] Ligorio, G., and A. Sabatini, "Extended Kalman Filter-Based Methods for Pose Estimation Using Visual, Inertial and Magnetic Sensors: Comparative Analysis and Performance Evaluation," *Sensors*, 2013.

[12] Hue, C., J-P. Le Cadre, and P. Perez, "Tracking Multiple Objects with Particle Filtering," *IEEE Transactions on Aerospace and Electronic Systems*, 2002.

[13] Olfati-Saber, R. "Distributed Kalman Filtering for Sensor Networks," *Proceedings of the 46th IEEE Conference on Decision and Control*, 2007.

[14] Caceres, M., F. Sottile, and M. Spirito, "Adaptive Location Tracking by Kalman Filter in Wireless Sensor Networks," *IEEE International Conference on Wireless and Mobile Computing, Networking and Communications*, 2009.

[15] Medeiros, H., J. Park, and A. C. Kak, "Distributed Object Tracking Using a Cluster-Based Kalman Filter in Wireless Camera Networks," *IEEE Journal of Selected Topics in Signal Processing*, 2008.

[16] Munkres, J., "Algorithms for the Assignment and Transportation Problems," *J. Soc. Indust. Appl. Math*, 1957.

[17] https://maitra.public.iastate.edu/stat501/lectures/Classification-I.pdf

[18] Helwig, N., "Discrimination and Classification," University of Minnesota, course notes, 2017, http://users.stat.umn.edu/~helwig/notes/discla-Notes.pdf. 2017.

[19] Jeffrey, T. W., "Target Classification, Discrimination, and Identification," *Phased-Array Radar Design: Application of Radar Fundamentals*, IET Digital Library, 2009.

[20] Debes, C., et. al., "Target Discrimination and Classification in Through-the-Wall Radar Imaging," *IEEE Transactions on Signal Processing*, 2011.

[21] Ostwald, J., H. Schnitzler, and G. Schuller, "Target Discrimination and Target Classification in Echolocating Bats," in *Animal Sonar*, P. E. Nachtigall, and P.W. B. Moore (eds.), Springer, 1988.

[22] Bachir, A., M. Dohler, and T. Watteyne, "MAC Essentials for Wireless Sensor Networks," *IEEE Communications Surveys & Tutorials*, 2010.

[23] Malewski, M., D. Cowell, and S. Freear, "Review of Battery Powered Embedded Systems Design for Mission-Critical Low-Power Applications," *International Journal of Electronics*, 2018.

[24] "Chapter 2. Hardware: Sensor Mote Architecture and Design," http://feihu.eng.ua.edu/NSF_TUES/sensor_hardware.pdf.

[25] "Chapter 8: Power Management," https://www3.nd.edu/~cpoellab/teaching/cse40815/Chapter8.pdf.

[26] La Rosa, R., et.al., "Strategies and Techniques for Powering Wireless Sensor Nodes through Energy Harvesting and Wireless Power Transfer," *Sensors*, 2019.

[27] Aslam, S., F. Farooq, and S. Sarwar, "Power Consumption in Wireless Sensor Networks," *FIT'09*, 2009.

[28] Sha, M., G. Hackmann, and C. Lu, "Energy-efficient Low Power Listening for Wireless Sensor Networks in Noisy Environments," *2013 ACM/IEEE International Conference on Information Processing in Sensor Networks (IPSN)*, 2013.

[29] Agnelo Silva, A., M. Liu, and M. Moghaddam, "Power-Management Techniques for Wireless Sensor Networks and Similar Low-Power Communication Devices Based on Nonrechargeable Batteries," *Journal of Computer Networks and Communications*, 2012.

[30] Jain, R., "Energy Management in Ad Hoc Wireless Networks," https://www.cse.wustl.edu/~jain/cse574-06/.

[31] Snoeren, A., "Lecture 17: 802.11 Wireless Networking," https://cseweb.ucsd.edu/classes/wi13/cse222A-a/lectures/222A-wi13-l17.pdf.

[32] Szewczyk, R., P. Levis, and M. Turon, "Microcontroller Power Management," *Wired/Wireless Internet Communications: 7th International Conference, WWIC 2009*, 2009.

[33] Pasha, M., S. Derrien, and O. Sentieys, "A Complete Design-Flow for the Generation of Ultra Low-Power WSN Node Architectures Based on Micro-Tasking," *Design Automation Conference*, 2010.

[34] Antao, U., C. Dibazar, and T. Berger, "Low Power, Long Life Design for Smart Intelligence, Surveillance, and Reconnaissance (ISR) sensors," *2012 IEEE Conference on Technologies for Homeland Security (HST)*, 2012.

[35] Zhai, B., et al., "Energy-Efficient Subthreshold Processor Design," *IEEE Transactions on Very Large Scale Integration (VLSI) Systems*, 2009.

[36] Lu, C., V. Raghunathan, and K. Roy, "Micro-Scale Energy Harvesting: A System Design Perspective," *Proceedings of the Asia and South Pacific Design Automation Conference*, 2010.

[37] Knight, C., J. Davidson, and S. Behrens, "Energy Options for Wireless Sensor Nodes," *Sensors*, 2008.

[38] Satish, G., and P. Varma, "Energy Management System in Ad Hoc Wireless Networks," *International Journal of P2P Network Trends and Technology (IJPTT)*, 2013.

[39] Damaso, A., N. Rosa, and P. Maciel, "Reliability of Wireless Sensor Networks," *Sensors*, 2014.

[40] Mahmood, M., W. Seah, and I. Welch, "Reliability in Wireless Sensor Networks: A Survey and Challenges Ahead," *Computer Networks*, Vol. 79, 2015, pp. 166–187.

[41] Stajano, F., and R. Anderson, "The Grenade Timer: Fortifying the Watchdog Timer Against Malicious Mobile Code," *Proceedings of 7th International Workshop on Mobile Multimedia Communications (MoMuC 2000)*, 2000.

[42] Xie, L., et. al., "A Tamper-Resistance Key Pre-distribution Scheme for Wireless Sensor Networks," *Proceedings of the Fifth International Conference on Grid and Cooperative Computing Workshops (GCCW'06)*, 2006.

[43] Panic, G., O. Stecklina, and Z. Stamenkovic, "An Embedded Sensor Node Microcontroller with Crypto-Processors," *Sensors*, 2016.

[44] Bash, B, D. Goeckel, and S. Guha, "Hiding Information in Noise: Fundamental Limits of Covert Wireless Communication," *IEEE Communications Magazine*, 2015.

[45] Riaz, R., et. al., "SUBBASE: An Authentication Scheme for Wireless Sensor Networks Based on User Biometrics," *Wireless Communications and Mobile Computing*, 2019.

9

WSN Sensor Modalities

Throughout this text, the main objective for realization of wireless sensor networks is to deploy sensors where (and when!) observation data is needed. This chapter discusses sensor modalities associated with WSN, reviewing design and performance for sensor modalities that meet WSN constraints while providing the necessary sensitivity performance. Merriam-Webster defines "sensor" as, "a device that responds to a physical stimulus (such as heat, light, sound, pressure, magnetism, or a particular motion) and transmits a resulting impulse (as for measurement or operating a control)" [1]. Here, however, we use a more specific definition as we consider applications, sensor capabilities, and T-ISR mission requirements. For T-ISR, a sensor is a transducer that reacts, predictably, to an external physical force and/or field and converts an input force (field) to an appropriately defined electrical signal. With this working definition, we attach conditions to our selection and design of sensors. To ensure predictable response to input signals, consider two aspects: fundamental sensor capability and sensor calibration. With attention toward T-ISR missions, sensor nodes can function merely to interrogate if an object has been detected or can be used to conduct sophisticated levels of ISR such as providing unique identifying characteristics belonging to the detected objects. The correlation of the intended purposing of a sensor node and system expectations defines sensor performance specifications as well as the rigor for calibration of the sensor. The use of an "appropriately defined" signal in our working definition is to indicate that the sensor, alone, does not necessarily provide a readily available signal for T-ISR system end users. Instead, sensors involve electrical circuits and processing algorithms that extract, process, and format measurement data.

The modalities discussed in this chapter follow the hierarchy of that presented in Figure 9.1, where the gray boxes indicate the prominent modalities developed for WSN sensor nodes. Optical, RF-based, and acoustic sensors are categorized as active if the sensor emits interrogation signals (e.g., radar), whereas passive sensors depend upon the target emitting a signal that is sensed by the detector. For optical systems, this separates laser radars from imagers (cameras). Regarding RF sensors, radars provide an active role whereas passive RF receivers can be used to intercept emissions from targets. Other passive modalities have been employed for decades with unattended ground sensors in T-ISR applications. (Consider, for example, the discussion of UGSs in Section 1.2.) These sensors include those based on various physical stimuli such as acoustic, seismic, magnetic, and chemical-biological (chem-bio) agents.

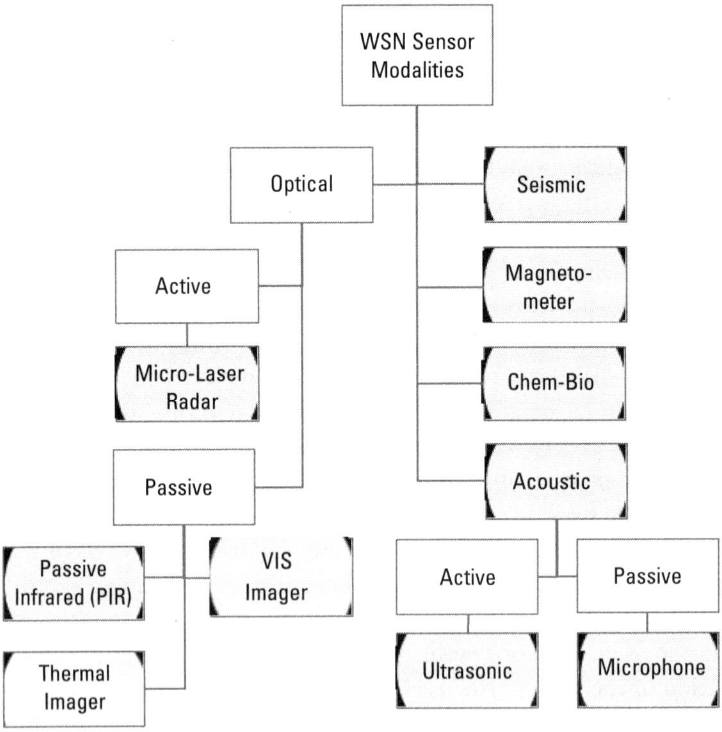

Figure 9.1 WSN sensor modality hierarchy.

Given the restriction of resources by WSN nodes, adapting scaled versions of larger sensor devices is severely limited. However, WSN systems possess a distinct advantage over sophisticated sensors based upon the large number of disbursed nodes. This affords an advantage to sensor node designs compared to more complex, but standalone sensors. Most notably, coverage required by each WSN node is a small fraction of that associated with a standalone sophisticated sensor. Unlike sophisticated sensors, WSN nodes, given their large numbers, have no need to address large fields of view or sensitivity areas, a valuable byproduct of having several nodes providing various point-of-view perspectives of targets.

9.1 Sensor Operational Considerations

There are operational concerns that impact all sensor modalities, including terrain, weather, time-of-day (ToD), diurnal effects, background interference, and target characteristics (including density and number). In evaluating sensor performance and comparing a particular sensor to alternative modalities, fundamental aspects of sensor characteristics should be thoroughly understood. In response to unknown scenarios and targets, sensor nodes typically support two or more sensor modalities to aid with target detection as well as reduction of false alarms. In addition to sensor considerations, correlating sensor performance to the wireless (RF) transceiver capability is conducted as these two functions compete for the same limited resources. WSN designers seek to maximize the range of communication (R_{RF}) as well as that for the sensors (R_{sen}). To ensure reliable network integrity and robustness through highly reliable wireless connectivity, particular node designs have strived for RF range capable of multiples of sensor range. As node density is increased, the RF-to-sensor range ratio, $\mathcal{R}$, approaches unity (1:1). However, designing for a range ratio depends on both mission and node operational characteristics. Missions that employ active or acoustic/seismic sensors can easily result in range ratio values approaching 1:100. Considering sensor and RF systems suitable for inclusion in WSN designs, we can bound range ratio values to

$$0.1 \le \mathcal{R} = \frac{R_{RF}}{R_{\mathrm{sen}}} \le 10 \tag{9.1}$$

The bounds in (9.1) are based on design goals and performance capabilities experienced with actual T-ISR motefields. First, sensors and RF systems are considered of use when their effective ranges are 10m or more. Sensor ranges shorter than 10m place a burden on the number of nodes necessary to cover a significant AOI (e.g., 1×1 km^2). Transceivers matched to typical T-ISR mission lifetimes and available node power, where nodes are placed at ground level ($1/r^4$ attenuation, Figure 6.3), are considered well-balanced when an average RF range of approximately 100m is achieved. Some sensors are able to operate up to (and beyond) the 1,000-m range, but most sensors for WSN typically operated with the 10–30-m range.

Figure 9.2 presents an example of WSN nodes that span the range ratio values of (9.1) within a single deployed system. A notional mission, protection of a long roadway, can be implemented using a combination of individually monitored areas with a linear sensor of a 1-km length. Associated with a mission as depicted by Figure 8.3, Figure 9.2 considers the use of a heterogeneous motefield consisting of: acoustic/seismic, microlaser radar (MLR) nodes, and PIR nodes. The acoustic/seismic sensors provide wide-FOV coverage (the grayed circular areas) and relatively long-range sensing but are unable to precisely locate a target unless robust triangulation is implemented. (Phase shift localization using correlation of distributed acoustic signals is overly complex for WSN-class sensors.) With MLR nodes, the interstitial area between the seismic/PIR-covered areas is monitored. As with all nodes by design, the RF range (R_{RF}) even for NLR mote nodes is insufficient to support a 1-km link;

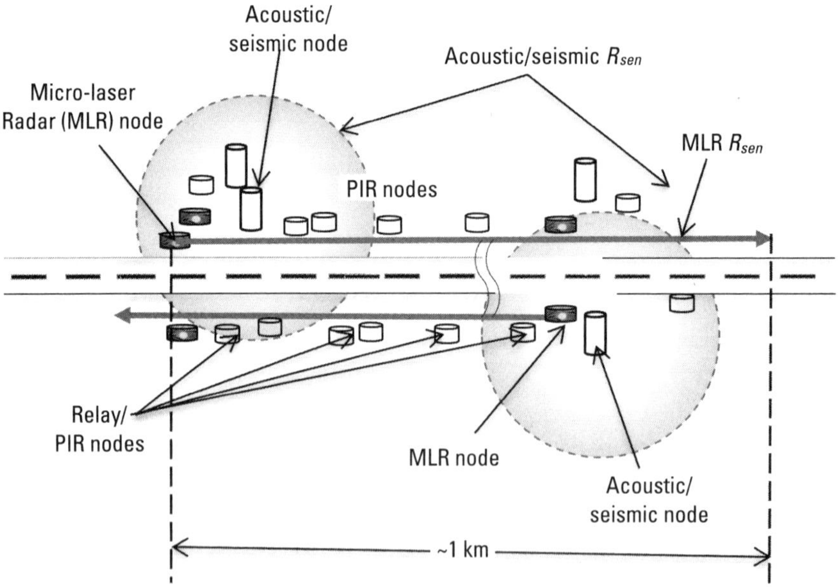

Figure 9.2 Heterogeneous WSN system design having extreme range-ratio variation.

however, such systems are supplemented, cost-effectively, using PIR nodes. These filler nodes provide a secondary layer of detection and tracking and provide a relay capability via multihop MAC (Chapter 4).

With regard to sensor performance requirements, a WSN designer must consider mission objectives as well as key sensor characteristics as detection range (R_{sen}), field-of-view (FOV), and resolution of the sensor signal for targets of interest while operating in the mission environment and background. Other aspects that drive sensor design include node resources (power, volume, mass), measurement accuracy, security, and sampling rate(s). For spherically propagating signal radiation, detection range directly correlates to received signal power (P_{rec}), source transmitted power (P_{scr}), and strongly on range (r) between sensor and the target as,

$$P_{\text{rec}} \propto \frac{P_{\text{src}}}{r^2} \tag{9.2}$$

Using (9.2), and introducing a term to capture noise power (P_{noise}), the SNR defines sensor range, using decibels units, which can be expressed as

$$SNR \doteq 10\log\left(P_{\text{src}}\right) - 10\log\left(P_{\text{noise}}\right) - 20\log(r) \tag{9.3}$$

Recall, from Chapter 8, the discussion of incoherent integration using multiple sensor samples and/or multiple sensor node data, combined, which results in an effective SNR gain of $N^{1/2}$. As a result, having N sensors (SNR_N) provides a SNR with a sampling gain compared to a single sensor node, SNR_1 of

$$SNR_N - SNR_1 = 10\log(N) \tag{9.4}$$

9.2 Passive Optical Sensor Modalities

Optical sensor designs deployed with sensor nodes operate in both the visible (380–800 nm) and infrared (IR) bands (1–14 um). For WSN, visible-band imaging has relied on intrinsic silicon (photodiodes) and CMOS detector array technologies. In the IR, WSN sensor designs used uncooled detectors due to SWAP2 constraints. Passive optical sensors' complexity spans from single-detector photodiodes (useful to indicate the presence or absence of light) to high-resolution video imagery using large detector arrays. Such elements can be worked with either a simple photodiode (simplistic motion detection) or require the capability (and support electronics) of an imager.

WSN sensors operate at the electromagnetic radiation (EMR) wavebands within the visible and infrared optical wavelength bands. Near infrared (NIR) is defined as 0.75–1.4 um, short-wave infrared (SWIR) as 1.4–3-um, mid-wave infrared (MWIR) as 3–8 um, and long-wave infrared (LWIR) as 8–15 um [2]. For use with IR sensors, Figure 9.3 presents spectral radiometric for a room-temperature (T = 300K) object, which indicates peak radiance for humans at 9.5 um. With IR, the sensor can operate day and night as the target emits energy in the IR bands as a graybody (i.e., blackbody with emission < 1) based on the object's temperature. For visible-band sensors, reflected light is required. Although visible sensors provide excellent resolutions, they are relegated to conditions containing illuminating sources.

9.2.1 PIR

A single optical detector can indicate motion through measurement of the variability of light falling on the photodiode surface. These simple photodiode detectors are extremely cost effective and require minimal power; however, they are prone to a high FAR. Any scene change within the single-detector FOV can trigger an event. With exterior deployments, even light variations from the movement of clouds can trigger a photodiode detector. Using a pyroelectric detector that operates in the 5–14-um range as a PIR mitigates FAR using segmented fields of view. Sensor response in this waveband is optimum

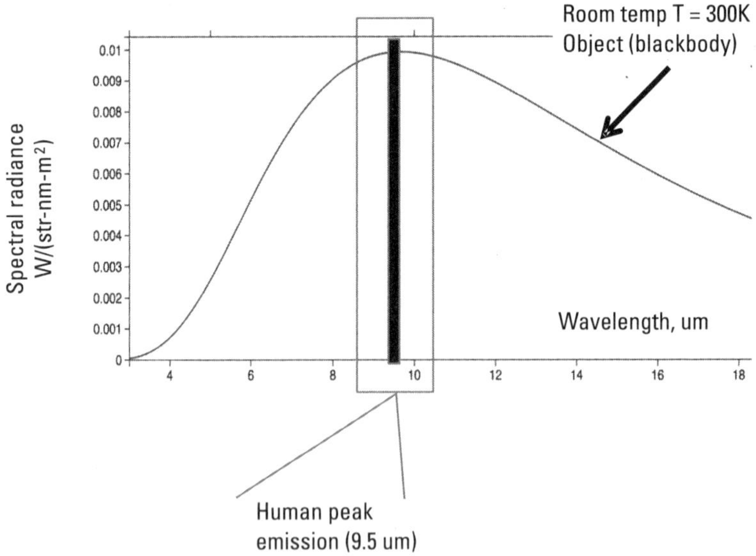

Figure 9.3 T-ISR target spectral radiance.

for T-ISR as it readily functions captures the 9.5-um-centered spectra emitted by 300-K objects (e.g., the human body). Figure 9.4 illustrates the principle behind a PIR sensor. Using Figure 9.4(a), an observation volume is divided into two segments, each assigned to one of two opposite-polarity series-connected pyroelectric detectors. Light is directed through an optical filter, which rejects

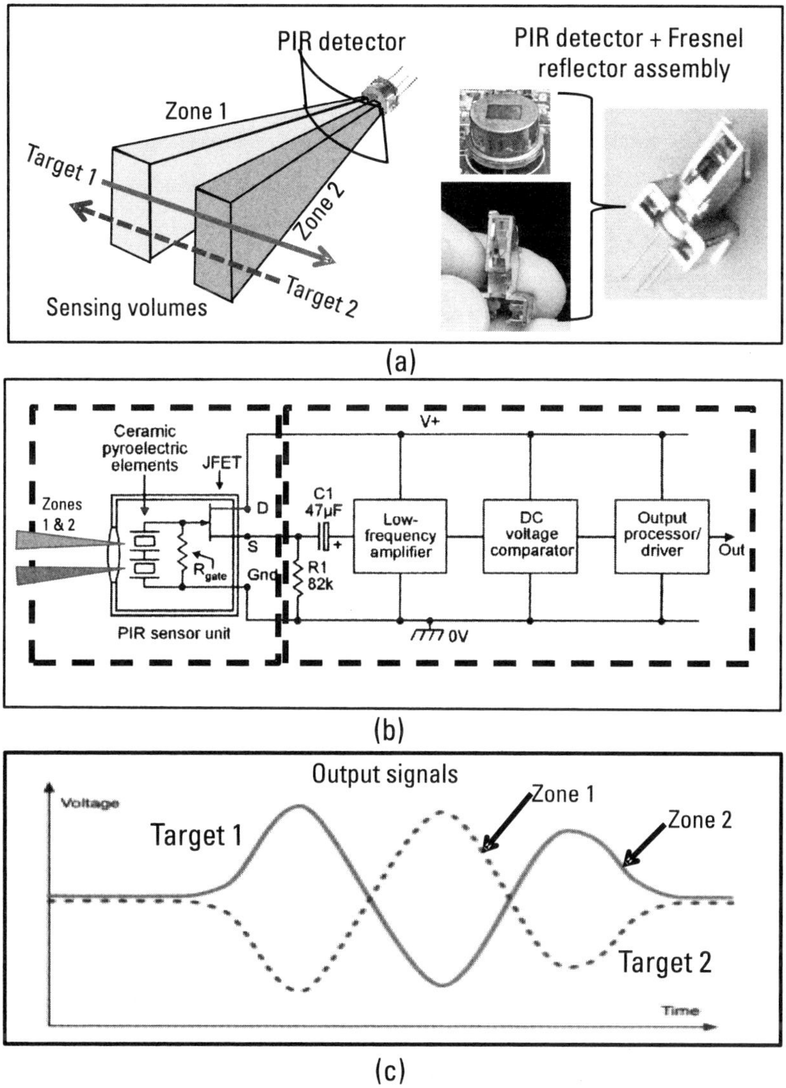

Figure 9.4 PIR sensor principle: (a) optical sensing, (b) PIR and signal conditioning block diagram, and (c) resultant output signal.

unwanted light as depicted in Figure 9.4(b). Figure 9.4(b) also depicts the circuit function of the detector and associated processing. With two pyroelectric detector elements series-wired in opposition, a polarity change occurs in the output signal for an object moving from one detector FOV to the other. The resulting voltages from the IR signals are combined and output-buffered via a JFET source-follower. The output from the detection section [Figure 9.4(a)] is fed into a signal-conditioning circuit to suppress the DC-component, filter out high frequencies from the signal, and provide amplification, voltage window comparison, and output to a microcontroller. Power consumption by a PIR detector and associated signal processing ranges from 50 to 65 mW. It is important to note that the final output voltage is proportional to the difference between output voltages of the two ceramic elements.

Figure 9.4(c) illustrates targets in the observed scene moving in opposite directions with respect to the sensing volume. As target velocity across the FOV increases, output signal frequency content also increases. Although the FAR associated with PIR detectors is not zero, it is reduced by virtue of PIR detectors operating in a complementary configuration. Complementary detector behavior also minimizes common-mode interference due to local electric field interactions and cancels simultaneous (i.e., background) thermal signals arriving at the PIR sensors. Range for a PIR sensor can be determined through calculation of the SNR for the sensor, which varies as R^{-2}. Low-cost PIR sensors can detect thermal differences (target-to-background) and realize reliable detection typically to 10m for human targets. Range is directly related to the net thermal difference between target and background, range to target, optical performance, and target orientation with respect to the sensor. To obtain an SNR equation for a PIR sensor, it is necessary to use radiometric terms and assumptions regarding infrared radiation propagation.

9.2.1.1 Developing the PIR Target Signal

As described, the resultant PIR signal from the presence of a target is based on the differential luminance signal produced between target and background within the FOV. This is determined through energy integration over the optical passband (determined by the PIR window or filter) of the spectral luminance (L_λ) derived from the spectral exitance, $M_\lambda(T, \lambda)$, which is radiant flux emitted by a surface per projected unit area [3, 4]. Expressing all terms as spectrally dependent variables, spectral distribution of luminance generating the signal can be expressed as [5],

$$L_\lambda(T,\lambda) = \frac{\epsilon_\lambda(\lambda) M_\lambda(T,,\lambda)}{\pi} \tau_a(\lambda) \tau_o(\lambda) \tag{9.5}$$

Using Planck's law for spectral exitance, $M_\lambda(T, \lambda)$ [6],

$$M_\lambda(T,\lambda) = \frac{2\pi hc^2}{\lambda^5} \frac{1}{e^{\left[\frac{hc}{\lambda kT}\right]} - 1} \quad \text{watts}/\left(\text{cm}^2\text{-um}\right) \tag{9.6}$$

Equation (9.6) depends on three physical constants, Planck's constant (h, $6.62607015 \times 10^{-34}$ Joule-sec), Boltzmann's constant (k, $1.38064852 \times 10^{-23}$ m^2kgs^{-2} K^{-1}), and the speed of light (c, in vacuum, 299792458 ms^{-1}). The units used address spectral response with wavelength as measured in microns. Calculating total target luminance (L_{tgt}) with (9.5) requires an estimate of the averaged temperature for the target ($\overline{T_{tgt}}$), and integration over the bandpass ($\lambda_2 - \lambda_1$) of the PIR filter,

$$L_{tgt}\left(\overline{T_{tgt}}\right) = \frac{\epsilon_{tgt}}{\pi} \int_{\lambda_1}^{\lambda_2} L_\lambda\left(\overline{T_{tgt}},\lambda\right) \, d\lambda \tag{9.7}$$

Similarly, to calculate the background luminance using an average temperature, $\overline{T_{bg}}$, results in,

$$L_{bg}\left(\overline{T_{bg}}\right) = \frac{\epsilon_{bg}}{\pi} \int_{\lambda_1}^{\lambda_2} L_\lambda\left(\overline{T_{bg}},\lambda\right) d\lambda \tag{9.8}$$

To estimate a signal output (S_{tgt}) occurring due to the presence of a target, we form the difference between the target and background luminance values, which must be integrated over the passband as

$$S_{tgt}\left(\overline{T_{tgt}},\overline{T_{bg}}\right) = K \int_{\lambda_1}^{\lambda_2} \left[L_\lambda\left(\overline{T_{tgt}},\lambda\right) - L_\lambda\left(\overline{T_{bg}},\lambda\right)\right] \, d\lambda \tag{9.9}$$

The coefficient, K, combines the constants involved, such as the pyroelectric responsivity and optical collection efficiency. Emissivity values can be assumed constant, which allows only <1% error [7].

9.2.1.2 Developing SNR for PIR

The noise term for PIR sensors is based on a combination of background noise, detector noise, and amplifier noise, with background noise being the dominant source. Placing a bound on the main noise component, the thermal fluctuation of the background, and using a minimum (T_{bg}^{MIN}) and maximum (T_{bg}^{MAX}) background temperature range with (9.9) results in a noise term of

$$S_{bg}\left(T_{bg}^{\mathrm{MIN}},T_{bg}^{\mathrm{MAX}}\right)=K\int_{\lambda_1}^{\lambda_2}\left[L_\lambda\left(T_{bg}^{\mathrm{MAX}},\lambda\right)-L_\lambda\left(T_{bg}^{\mathrm{MIN}},\lambda\right)\right]d\lambda \qquad (9.10)$$

To evaluate SNR, we use S_{tgt}/S_{bg}, the geometry and observable (projected) areas of the sensor aperture and target. For targets close to the sensor, the sensor FOV (Ω_{sens})is completely illuminated, and the target is considered an extended target; the entire receiver FOV is flooded with target energy. For targets at a distance such that the sensor FOV is only partially filled by the target area (Ω_{tgt}), the target is considered a resolved target. To determine the PIR sensor SNR, we make use of (9.9) and (9.10),

$$SNR=\frac{\Omega_{tgt}f^2}{R^2A_d}\frac{\int_{\lambda_1}^{\lambda_2}\left[L_\lambda\left(\overline{T_{tgt}},\lambda\right)-L_\lambda\left(\overline{T_{bg}},\lambda\right)\right]d\lambda}{\int_{\lambda_1}^{\lambda_2}\left[L_\lambda\left(T_{bg}^{\mathrm{MAX}},\lambda\right)-L_\lambda\left(T_{bg}^{\mathrm{MIN}},\lambda\right)\right]d\lambda} \qquad (9.11)$$

where the term, $\Omega_{tgt}f^2/R^2A_d$, represents the overlap of projected areas of the detector's field-of-view solid angle at the range (R) for the target. The solid angle, measured in units of steradians (sr), is used in radiometry to determine object energy emitted as well as energy received by a sensor using

$$\Omega=\frac{da}{R^2} \qquad (9.12)$$

Figure 9.5 depicts the solid angle definition per (9.12), which is the solid angle (Ω) projecting area (a) on a spherical surface at a distance (R) from a center of the sphere. This equates to the area of a hemisphere, $2\pi R^2$ for the solid angle of 2π sr (Figure 9.4). Two additional terms used by (9.1) are the focal length of the optical system (f) and detector area (instantaneous FOV, IFOV) of the detector, a_d. Range performance is obtained through determination of the integrated luminance values and a selection of a minimal SNR value for a level detection probability, which are determined as per Chapter 5.

In practice, PIR detectors require specific changes to the simple single-PIR detector optical setup and associated electronics used with typical applications of PIR detectors to meet detection performance requirements for T-ISR. Detection performance improvements required over typical PIR application include reliable detection ranges beyond 10m and detection of slow-moving targets. PIR efficiency when viewing slowly moving objects is low because temporal changes of thermal radiation from targets occur at the same rate as background thermal fluctuations. Moreover, to detect slowly moving targets

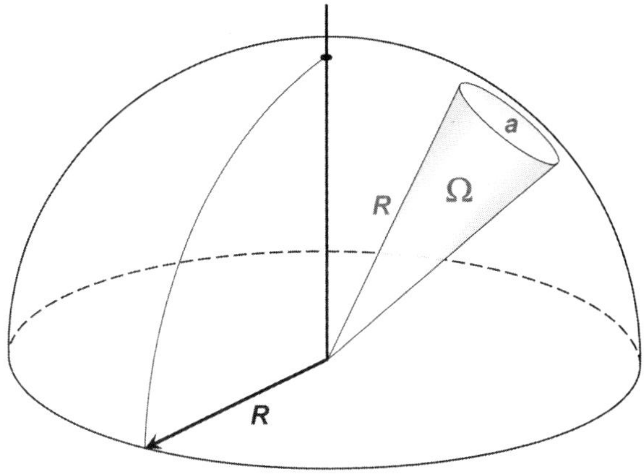

Figure 9.5 Defining solid angle for use with radiometric quantities.

(e.g., crawling people), the lower frequency limit of the transmission band for a PIR detector must approach zero. When this condition is fulfilled, an increase in low-frequency noise has been observed. To detect slow-moving targets, a number of PIR detectors (>2) oriented along the same LOS are employed to increase detection zones while also increasing SNR per detector due to each being assigned a smaller observation area. In addition to orienting the optical sensing volumes via multiple detectors is a voltage drift at the changes of ambient temperature. (A pyroelectric sensor is equipped with a field transistor playing a role of a voltage follower.) Thus, to have a sensor efficiently detecting slowly moving people, it is necessary to develop an algorithm eliminating changes of a signal from a sensor caused by background noise as well as the changes of a signal caused by temperature drift of a pyroelectric sensor [8].

9.2.2 Passive Imaging Sensors

Moving from single-detector optical sensors to an array of detectors results in passive imaging sensors (cameras) that, given proper range to target and pixel size, can spatially resolve objects within the field-of-view (FOV) to perform target recognition and identification. Additionally, forming a time series of 2D images (frames) introduces processes capable of increasing the SNR through background suppression and time-integration of images. Using an image sequence allows for frame-to-frame correlation, which is highly useful in establishing and tracking individual moving targets (e.g., Munkres' approach to object tracking discussed in Chapter 8).

9.2.2.1 Evaluating Imager Performance

Imager performance evaluation, for both visible and thermal cameras, is expressed through the measurement and evaluation of several characteristics, notably scene contrast, luminance (image brightness), scene resolution, data digitalization noise (binning), induced image aberrations, and noise level. Each of these measurable characteristics is directly considered in determining an imaging sensor correctly representing an observed scene. In discussing imagers, visible and infrared imagers can be evaluated using similar key performance parameters, such as resolution, optical system response, contrast, and SNR. However, there are differences as thermal cameras operate using emitted radiance from objects whereas visible cameras depend upon radiance from reflected light off of objects. Imagers are evaluated using measurements of resolution, which is a function of pixel count (detector array size), optical light gathering capability, and optical system quality. Resolution measurements are obtained through comparison of image data against well-defined input test scenes. Resolution represents the ability of an optical system to respond properly to the spatial frequency represented by the input radiance, which is evaluated using modulation transfer function (MTF) measurements.

For infrared (IR) imaging sensors, discussions of performance involve a term denoted as minimum resolvable temperature difference (MRTD), which is determnned through a primary operational evaluation. MRDT is a function of spatial resolution and is defined as the SNR required to resolve standardized (test) bar targets. IR imagers are evaluated as with visible imagers by using measurements of resolution, which is a function of pixel count (detector array size), optical light gathering capability, and optical system quality. As with imagers, resolution measurements are obtained through comparison of image data against well-defined input test scenes. Another widely accepted metric used for describing the resolution performance of infrared systems is noise equivalent temperature difference (NETD). NETD is defined as sensor sensitivity to input temperature change (ΔT) to produce an output signal (ΔV_s) equal to the rms noise level. How MRTD is directly related to this noise quantity and how NETD is evaluated are both described in Section 9.2.2.6. For visible optical sensors, point spread function (PSF) is used to evaluate the ability of the system to resolve two point objects to evaluate resolution. With visible imagers, resolving power can be evaluated through two approaches: (1) resolving power of refraction based on Airy patterns at the focal plane or (2) determining the optical system modulation transfer function (MTF). MTF, which represents the ability of the optical system to accurately pass a scene's spatial frequency content, is considered an optical system transfer function.

9.2.2.2 PSF

The impulse response for an optical system is its response to a point source, denoted as the PSF. Assuming an imaging system as a linear shift-invariant (LSI), the output is identical regardless of which spatial point is involved. With ideal optics, each image point is a perfectly scaled replica of a corresponding scene point. Unfortunately, the input radiance is collected by a finite aperture, which introduces inevitable diffraction at the edges resulting in slight blurring of the scaled image. Instead of having sharp, point-to-point mapping between scene and image, each scene point undergoes diffraction in the image. For a point source (e.g., star), diffraction by the optics aperture results in concentric rings of bright (white) and dark (black) rings, a pattern known as the Airy disk. Considering a circular aperture for the imaging sensor, the radiance entering the sensor can be modeled as a three-dimensional function with circular-symmetric *x-y* cross-section and intensity on the *z*-axis, per Figure 9.6(a). With *r* defined as radius, the function used to describe the assumed circular symmetric radiance is the circular function, circ(*r*), which has the following properties,

$$z = \text{circ}(r) = \begin{Bmatrix} 1 & r < 1 \\ 0.5 & r = 1 \\ 0 & \text{otherwise} \end{Bmatrix} \tag{9.13}$$

What we seek is the expression for the distribution of incident radiance, assumed as incoherent light at the image plane of the optical system. Expressing radiance at the circular aperture as a function of radius for an aperture having diameter, *D*, we use circ(*r*/*D*). From Fourier optics, we employ the Fraunhofer (far-field) diffraction integral that operates mathematically, as the Fourier-Bessel transform (referred to as Hankel transform) of the aperture distribution. Using a normalized radius, r', (9.14) results in the square of the Fourier transform of the exit pupil (at the detector) of the Bessel function of the first kind, $J_1(x)$, as presented by (9.15) [9, 10]. Selecting wavelength (λ) and focal length (f) for the optical system, (9.15) becomes the Hankel transform of (9.13) as

$$r' = \frac{\pi D}{\lambda f} r \tag{9.14}$$

$$PSF(r') = \left[2 \frac{J_1(r')}{r'} \right]^2 \tag{9.15}$$

Figure 9.6(a, b) presents circularly incident radiance distribution, circ(r) and its Hankel transform (Airy pattern), respectively. Figure 9.6(c) is a cross-section profile through the optical center of an Airy pattern. To determine resolution capability, a separation distance is considered. For Rayleigh criterion, well-known in establishing resolution for optical systems, the minimum resolved separation occurs at the first zero of the Airy pattern, (r_A):

$$r_A = \frac{1.22\lambda f}{D} \tag{9.16}$$

Figure 9.7(a) presents the Rayleigh resolution criterion. Figure 9.7(b, c) illustrate two alternate criteria, the Abbe and Sparrow criteria. The Abbe criterion defines resolution where the separation distance occurs where the PSF profiles cross at the 50% maximum signal strength (luminance) value. The

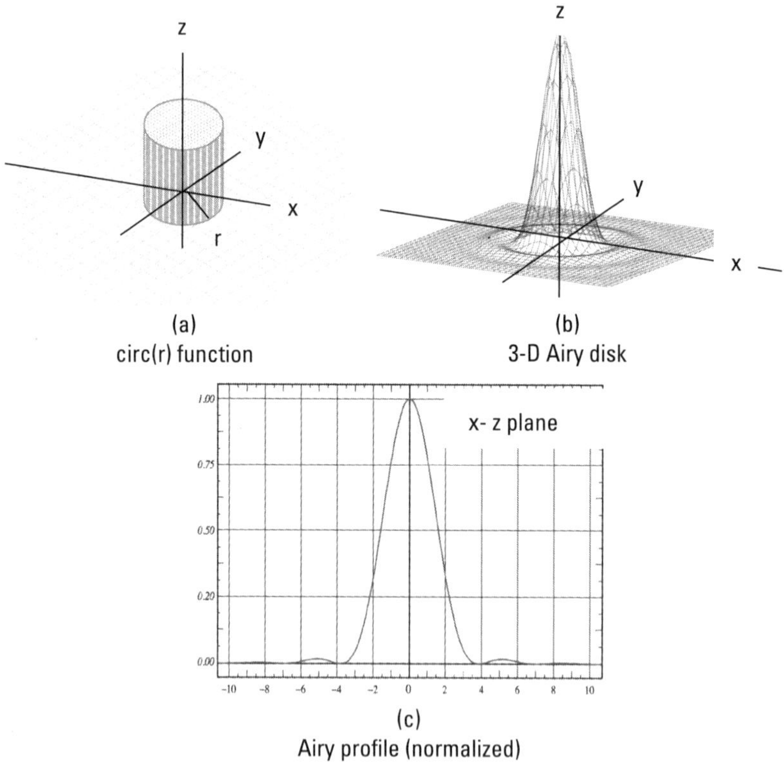

Figure 9.6 (a) Aperture function, (b) two-dimensional Airy pattern (PSF), and (c) PSF profile.

Sparrow criterion occurs when the PSF distributions superimpose without a perceptible ripple in the summed signal level. Figure 9.7 illustrates the defined resolution distances for each criterion [11].

9.2.2.3 Imager Resolution

The approach to determining resolution assumes a different sense with the advent of digital imagers and cameras. Over recent decades, most discussions of camera resolution are defined using imager pixel count (i.e., $M \times N$ array size). This would suffice if it were not for our optics being limited in collecting incoming light via a finite aperture, which we know produces a diffraction pattern at the focal plane. PSF, defined via Equation (9.15), describes this diffracted distribution of light at the focal plane due to the entrance aperture. Referred to as diffraction-limited limit, the distribution of incoming light onto the focal plane is the theoretical best focusing an optical system can attain. Unfortunately, there are design and implementation detractors from ideal diffraction-limited optical systems due to detector array configuration (pixel size and pixel number), spacing between pixels, digitalization noise, and the existence of optical aberrations.

The onset of digitalization noise arises from the limited spatial capability of a focal plane. Light falling on a pixel (detector) is binned to form an image. Aberrations result from the difficulty of correctly relaying incoming light onto the focal plane due to optical surface imperfections, chromatic dispersion, and errors in optical paths. With the aberrations, the central lobe of the diffraction pattern of Figure 9.6 broadens accordingly, reducing the central peak magnitude. This resultant image is distorted and/or blurred. Using the

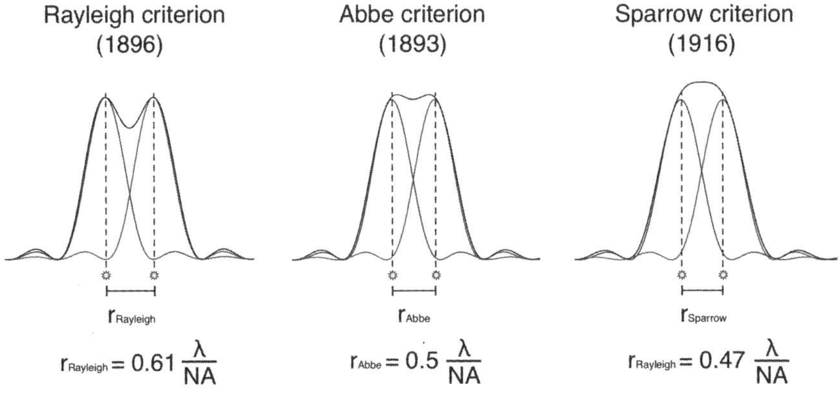

Figure 9.7 Optical imager resolution using various criteria: (a) Rayleigh, (b) Abbe, and (c) Sparrow.

ratio of the peak PSF height for an aberrated imager to a diffraction-limited peak PSF value is defined as the Strehl ratio [12]. Finally, additional noise can be introduced by having light outside the optical FOV directed to the focal plane (e.g., stray light).

9.2.2.4 Imager Contrast

We indirectly addressed contrast for imagers using (9.9), where we related radiant intensity changes based upon the presence (or absence) of a target. Contrast, C, can be expressed as a normalized ratio of signal output for the target of interest (S_{tgt}) versus background (S_{bg}) as

$$C = \frac{S_{tgt} - S_{bg}}{S_{tgt} + S_{bg}} \tag{9.17}$$

Contrast is directly linked to modulation transfer function (MTF), but how is contrast linked to resolution? Decades-old resolution measurement, which requires distinguishable line pairs per millimeter (lp/mm), corresponds to spatial frequencies where contrast (MTF) is between 5% and 2% (0.05–0.02). Perceived image sharpness (as distinguished from traditional lp/mm resolution) is closely related to the spatial frequency where MTF is 50% (0.5), where contrast has dropped by half. Regarding resolution, consider the Rayleigh criterion. At the Rayleigh limit, image intensity drops to 73.5% of the peak intensity of the two point sources. Using an MTF equation [(9.19), developed in Section 9.2.2.5], the associated cutoff frequency (9.20) evaluates the MTF value to 8.94%—resulting in $C = 0.09$ (or, MTF = 0.09) and is implied in the definition of the Rayleigh diffraction limit. Recall, however, that this is for one point on the MTF curve at one particular wavelength (evaluated using the helium d-line, 587.56 nm) [13].

9.2.2.5 MTF

Thermal and visible imagers both use MTF to indicate how accurately the spatial frequency content of an observed scene is relayed by the optical system to the image. MTF gauges sharpness, contrast, and clarity of the original scene (input) as captured by its image; it is equivalent as evaluating contrast (C) at a given spatial frequency relative to low frequencies. MTF, and through direct association contrast, are evaluated using spatially defined target charts that provide line pairs (analogous to cycles/sec, hertz, to electronic systems). Calibration charts present lines that vary with decreasing separation between the line pairs (line pairs/mm, lp/mm) to evaluate frequency performance, and is presented as intensity amplitude versus spatial frequency (cycles/pixel, cycles/

mm, or line widths/picture height). Figure 9.8(a) illustrates the use of a MTF chart (lp/mm) as an input to an optical system, with the resulting optical system capability indicative of a deteriorated representation of the original scene (degraded system response). Figure 9.8(b) presents an increasing spatial frequency MTF test chart. Figure 9.8(c) presents a graphical description of the MTF input scene as a sinusoidal having a DC-offset. The optical system

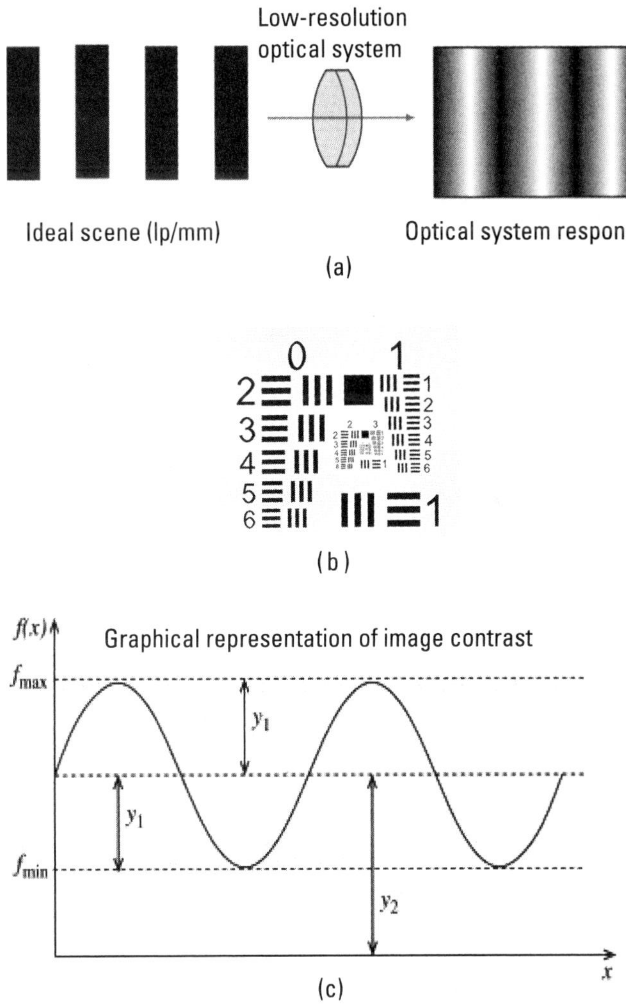

Figure 9.8 (a) MTF target chart, (b) graphical representation of image contrast, and (c) shows the image contrast over distance (*x*) with a variation of *y*1 against an offset background of *y*2.

response represents the radiance transmitted and provides the contrast modulation, which is nonnegative. Evaluating contrast, C, using terms, y_1 (modulation amplitude) and y_2 (DC offset), defined by Figure 9.8(c),

$$C = \frac{y_1}{y_2}: \text{ where } y_1 > 0, y_1 < y_2, \text{ and } 0 < C < 1 \tag{9.18}$$

The variables, y_1 and y_2 presented in Figure 9.8(c), are the maximum and minimum of the optical response function, $y = f(x)$. Using Figure 9.8(c), contrast can be expressed similarly as in (9.9) and verifies that expressed by (9.18).

$$C = \frac{\left(f_{\max} - f_{\min}\right)/2}{\left(f_{\max} + f_{\min}\right)/2} = \frac{\left(y_2 + y_1\right) - \left(y_2 - y_1\right)}{\left(y_2 + y_1\right) + \left(y_2 - y_1\right)} = \frac{y_1}{y_2} \tag{9.19}$$

The response function, $f(x)$, is a measurable characteristic that reflects output-to-input contrast, such as transmittance. To arrive at the optical system response, we form the ratio of output-to-input contrast values, $H(x)$, as,

$$H(x) = \frac{C_{\text{OUTPUT}}(x)}{C_{\text{INPUT}}(x)} \tag{9.20}$$

Equation (9.20) provides the result for the modulation transfer function at a single spatial frequency. Figure 9.9(a) illustrates an MTF curve based on the detector pixel size and pitch. Figure 9.9(b) illustrates various MTF components of the imaging system, including the circular-lens MTF. The calculated MTF is the resultant MTF, which is the product of all MTF components

To obtain the lens system response, the MTF[f], autocorrelation of the exit pupil (aperture, diameter = $2w$) is required. For the diffraction-limited case, autocorrelation is used to generate an expression for MTF for a circular aperture. Using the approach of two displaced circles [14], as shown in Figure 9.10(a), the area of overlap (shaded) is equal to four times the area, B, in Figure 9.10(b). Using the following definitions: z_i = distance from exit pupil to image plane, f_x = spatial frequency in the X-direction (Figure 9.10), and λ = wavelength, we determine the area of the circular sector, which is $A + B$ using the geometry presented by Figure 9.10(b). Equations (9.21) to (9.24) use the approach of areas to arrive at a MTF with spatial frequency being the root-sum-square (RSS) of the x and y frequency components, as

$$Area\left[A + B\right] = \left[\frac{\theta}{2}\right]\left(\pi w^2\right) = \left[\frac{\arccos\left(\lambda Z_i\, f_x/2w\right)}{2\pi}\right]\left(\pi w^2\right) \tag{9.21}$$

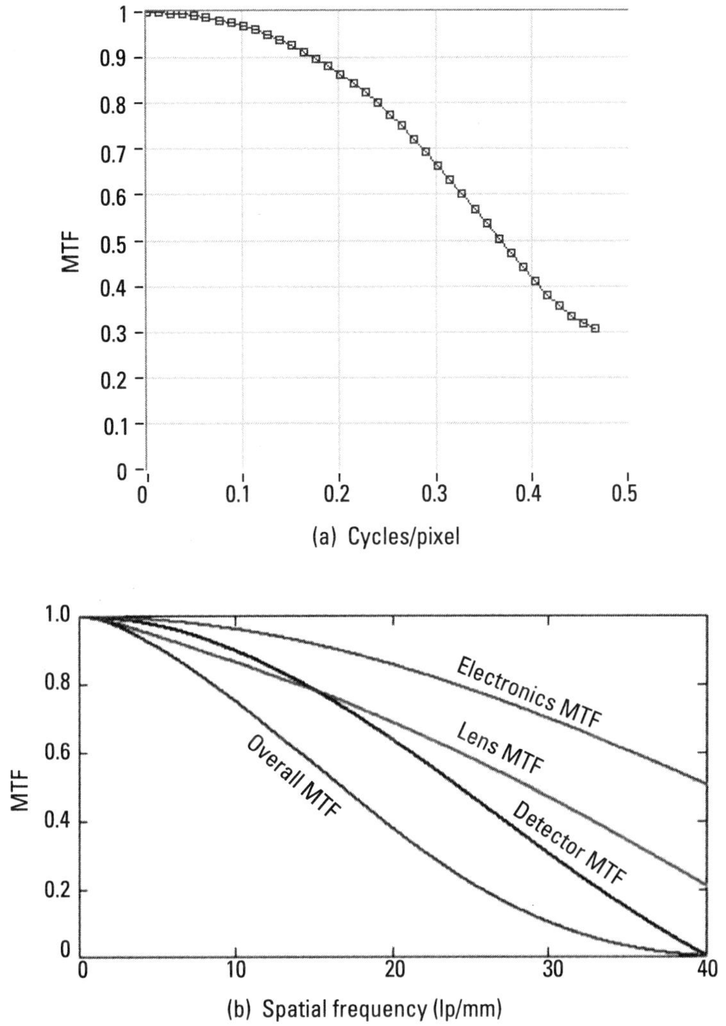

Figure 9.9 Modulation transfer function: (a) typical roll-off with spatial frequency, and (b) optical components to overall MTF.

$$Area[A] = \frac{1}{2}\left(\frac{\lambda Z_i f_x}{2w}\right)\sqrt{w^2 - \left(\frac{\lambda Z_i f_x}{2w}\right)^2} \tag{9.22}$$

$$MTF[fx,0] = \frac{4\left[Area[A+B] - Area[A]\right]}{\pi w^2} \tag{9.23}$$

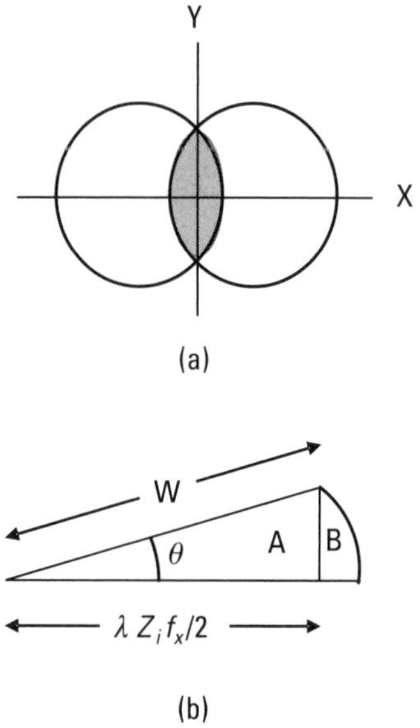

Figure 9.10 Deriving the MTF expression for a circular aperture.

Using f as the radial frequency, RSS $[f_X, f_Y]$, and f_0 as the cutoff frequency,

$$MTF[f] = \left\{ \begin{array}{ll} \frac{2}{\pi}\left[\arccos\left[\frac{f}{2f_0}\right] - \frac{f}{2f_0}\sqrt{1-\left(\frac{f}{2f_0}\right)^2}\right] & f \le 2f_0 \\ 0 & \text{otherwise} \end{array} \right\} \tag{9.24}$$

The spatial cutoff frequency presents a precise method to quantify the smallest object that is resolvable by an optical system. Using the optical system $f\#$ value, defined as focal length/aperture diameter, results in an optical system cutoff frequency of

$$f_0 = \frac{w}{\lambda Z_i} = \frac{1}{\lambda f\#} \tag{9.25}$$

The effect of diffraction is determinable through the convolution of an ideal image with the Fraunhofer diffraction pattern of the lens pupil. Of note, and evidenced by the MTF curves as shown in Figure 9.9, optical systems operate as lowpass filters to the scene being viewed. As a result, there is a level of blurring and a decrease in contrast as a result of the optics [14].

9.2.2.6 Thermal Imager: Minimum Resolvable Temperature Difference

Discrepancies exist regarding the level of resolution required by thermal cameras to detect, recognize, or identify objects. Performance criteria and resolution requirements have both similarities and differences for thermal compared to visible imagers. In defining thermal imager capability, a widely accepted standard is to employ the Johnson criteria of detection, recognition, and identification (DRI). In evaluating IR cameras for use on ISR missions, Johnson established performance criteria through the use of effective resolution of varying objects to achieve DRI objectives [15, 16]. For clarify, DRI is defined as follows:

- Detection addresses the imager ability to distinguish an object from the background.
- Recognition enables the ability to classify the object class (e.g., vehicle, human, or animal).
- Identification provides an in-depth assessment of an observed object to provide unique details (e.g., an individual person, type of vehicle) [17].

The 1958-era Johnson criterion defines DRI performance based on the ability of an observer to achieve a successful DRI assessment with a minimum of 50% of the sample images. Minimum Resolvable Temperature Differnce (MRTD) provides a connection between Johnson's concept of resolvable bars across the target critical dimension and thermal imager system performance. MRTD conducted in a laboratory environment results in accuracy and repeatability, which provides a successful approach in evaluating performance in an operational environment. MRTD can be determined through the application of modeling codes (e.g., NVTherm) as described in open literature [18, 19] beginning with the simplified expression using MTF, SNR, and noise equivalent differential temperature (NETD) as,

$$MRTD(f) = \frac{SNR\ NETD}{MTF(f)} \tag{9.26}$$

Equation (9.26) is optimistic at low spatial frequencies and pessimistic at higher spatial frequencies. Although limited in scope and fidelity, for decades the application of Johnson metrics to assess thermal imager capabilities in an operational scenario has met with success in establishing bounds of performance. Thermal imager technology development provides high-quality thermal imagers that can deliver a higher level of target acquisition performance but that cannot override the basic limitations related to geometrically range limitations and atmospheric transmission losses [20]. As used in (9.26), NETD provides a fundamental measure of MWIR/LWIR thermal camera capability. The NETD of a thermal imager indicates sensitivity of an infrared sensor to an input temperature change (ΔT). With output signal (ΔV_s) equal to the rms noise level (Nrm_s), we define $SNR = 1$ [21], and can express NETD as

$$NETD = \Delta T \text{ when } SNR(\Delta T) = 1 \tag{9.27}$$

While normally thought of as a system parameter, detector NETD and system NETD are the same except for system losses (due to conservation of radiance). NETD can be further defined using thermal and signal changes as a function of photon flux (Q),

$$NETD = N_{rms} \frac{\partial T / \partial Q}{\partial V_s / \partial Q} = N_{rms} \frac{\Delta T}{\Delta V_s} \tag{9.28}$$

Equation (9.28) reverts to (9.27) when SNR(ΔT) = 1 by substituting the following identity

$$SNR = \frac{\Delta V_S}{N_{rms}} \tag{9.29}$$

Additional information exists, including discussions on background-limited performance (BLIP) and spatial noise considerations using NETD in the open literature [15, 20, 22–24].

9.2.3 Thermal Imaging for WSN

For T-ISR, electromagnetic radiation (EMR) wavebands of interest and of most use for IR imagers are optical frequencies within the MWIR and LWIR bands (Figure 9.3).

To achieve maximum sensitivity providing single-photon response, thermal imager detectors are cryogenically cooled to avoid signal loss due to detector self-radiation and typically operate at −200°C (−328°F). Cryo-cooling and the cost of the high-sensitivity detector array are not conducive to low-cost (and low-power) WSN nodes, but fortunately, recent research into solar array technology has resulted in acceptable uncooled detector performance through use of amorphous silicone (a-Si) microbolometer detector arrays. Modern micromachining technology has enabled relatively low-cost production of thermal detector arrays of microscopic elements (pixels) that have response times similar to those of photon detectors. Uncooled microbolometer imagers have demonstrated operation with sensitivities to 75 mK and capable of detection distances of the 150–500-m range, which easily satisfies range objectives associated with WSN sensors. The a-Si detector arrays not only eliminate the need for a cryogenic cooler and produce sufficient sensitivity in the 7–14-um band, but also meet with cost and power constraints associated with WSN node designs [25].

With emphasis on WSN-class thermal sensing (range < 100m to objects), there are numerous thermal imaging devices that provide design options for low-power, low-cost thermal imagers and have been developed from the small 4×4 detector arrays [26] based on MEMS-ASIC technologies to 384×288 and 640×480 arrays at 9-Hz frame rates [27, 28]. Studies have indicated that microbolometers are able to provide high-fidelity measurements of a type traditionally associated only with cooled-IR imagers. One study in particular obtained commercially available vanadium-oxide microbolometers that have been compared to a scientific-grade cooled indium antimonide imager under similar conditions. Although NETD measurements for microbolometers are approximately two to six times inferior to cooled imagers, direct comparison indicates that microbolometers consistently outperform the costlier and more complex InSb imager [29]. Microbolometers are continually undergoing improvements; the ULIS ATTO640 presents a 60-Hz VGA format (640×480) with a 12-um pitch, stressing low SWAP [30].

9.2.4 Visible Imaging (Camera) for WSN

Early in WSN deployments, WSN nodes were used to support the operation of individual T-ISR UGS cameras. These cameras (e.g., REMBASS, Chapter 1) were relatively power-hungry and required portable battery packs (bags) that provided significant energy, such as multiple BA55690U batteries. These batteries, which weigh 2.25 lb each, were successfully used in groups of three or more to provide 14/28VDC. WSN nodes were used to augment a deployed sensor field to activate UGS sophisticated sensor units (SSUs) to preserve battery life (each 5,590 provides 64,000 mA-hr, or 2.5A continuous) over extended periods [31]. SSU devices were located separate from the WSN motefield and operated in standby mode (or off) using WSN nodes to provide constant monitoring. When a target was detected and classified as a target of interest by the WSN system, an alert message was autonomously generated and forwarded to the appropriate SSU to fully sequence to an operational mode. This helped to reserve power for the cameras by offloading the monitoring function to low-power WSN nodes and avoided unnecessary SSU operation and power use triggered by a false alarm. Although appropriate at the time, SSU camera systems combined with 5,590-battery weight and size are prohibitive for use directly by WSN nodes.

With the rapid commercial development of cellphones and associated cameras, low-cost and low-power camera systems became readily available, and viable to WNS node designers. Cellphone advancements, which also strived for low-cost and low power consumption, began with the first truly (fully integrated hardware) cellphone camera arriving in 2000, specifically the Sharp J-Phone (J-SH04), which provided a resolution of 0.11M pixels with 256 colors [32]. With modules such as the 2005 VGA Mobile Camera Module VS6524, 2M pixels at VGA frames became viable node subsystems with power requirements of 3VDC supply at 30mA maximum (streaming) and 10 uA quiescent. The dimensions were also conducive to insertion into nodes for imaging, being 7.0 × 7.0 × 4.5 mm. Former (and current) WSN node designs, employed the venerable VGA (640x480) video/photo frame-size CMOS-based camera associated with the cell phone industry. Figure 9.11

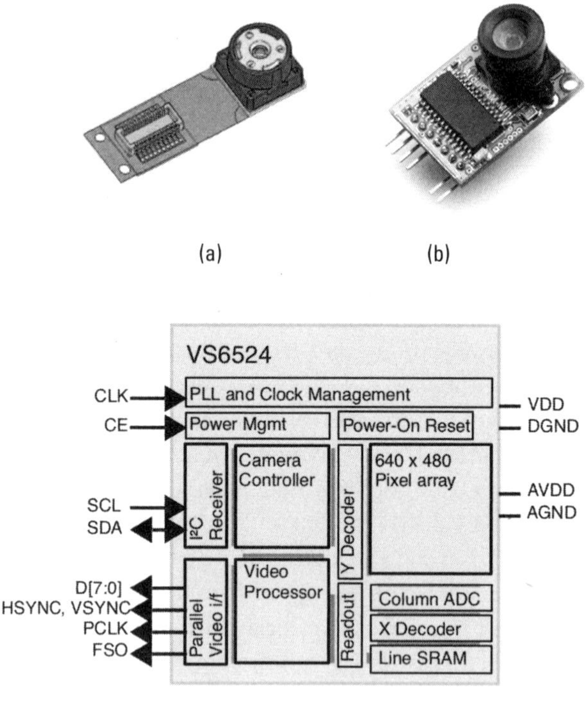

(a) (b) (c)

Figure 9.11 CMOS camera examples: (a) VS6524 camera, (b) OV2640/OV5642, and (c) VS6524 electronics.

presents examples of camera modules scaled for WSN node design. Figure 9.11(a, b) illustrate the physical packages for the VGA VS6524 [33, 34] and OV2640/OV5642 cameras, respectively. The 2- and 5-M pixel OV2640/OV5642 incorporate high-definition serial peripheral (SPI, 4-wire) and inter-integrated circuit (I2C, 2-wire) interfaces, which reduced the complexity of the camera control interface and made it possible to add multiple cameras to a single microcontroller [35]. Figure 9.11(c) provides the attendant VS6524 support electronics [34].

As with imaging sensors (UGS) in the 20th century, WSN-based CMOS cameras can be programmed for various operational modes to conserve power and to restrict data flow congestion. T-ISR imagers can be operated to provide individual images, image bursts, or variable frame rates of video based on system resource management procedures upon command. Design and manufacturing improvements to low-power camera capabilities have accelerated with cellphone resolutions attaining 108M pixels (e.g., Qualcomm Snapdragon 865 chipset within Xiamoi Mi CC9 pro phone), and with near-term expected designs to surpass 200M pixels (e.g., the Xiamoi Mi10 device) [36].

9.3 Active Optical Sensor: MLR

Unlike passive systems, active systems (laser-based) require significantly more energy to operate continuously and thereby challenge use by WSN sensor node designers. However, there are T-ISR applications that require long-range (>>10m) nodes that are not required to operate at a high duty cycle. Examples, as depicted in Figure 9.2, include protection of large linear objects or access to terrain of interest not conducive to direct physical emplacement of nodes (waterways, valleys, and gullies). This highlights another advantage of WSN motefields over sensor nodes, which are heavily linked to the nearest neighboring nodes and are dependent on the ensemble versus the individual node to provide the required data. A joint DoD effort, working with the Defense Intelligence Agency (DIA) and the DoD's USSOCOM, supported the merging of a low-power, low-cost laser radar with an existing WSN node capability, resulting in a MLR mote (MLRmote). Figure 9.12 shows the MLRmote, which was demonstrated for various T-ISR missions including force protection, asset protection, and border support. The node design was built using a TI MS430F1611 microcontroller, CC2420 2.4GHz RF transceiver and supported a SIRFIII GPS chipset to conduct localization. The sensor package included an infrared laser rangefinder and a PIR to aid in duty-cycling the MLRmote. The MLR mote is made up of the following components:

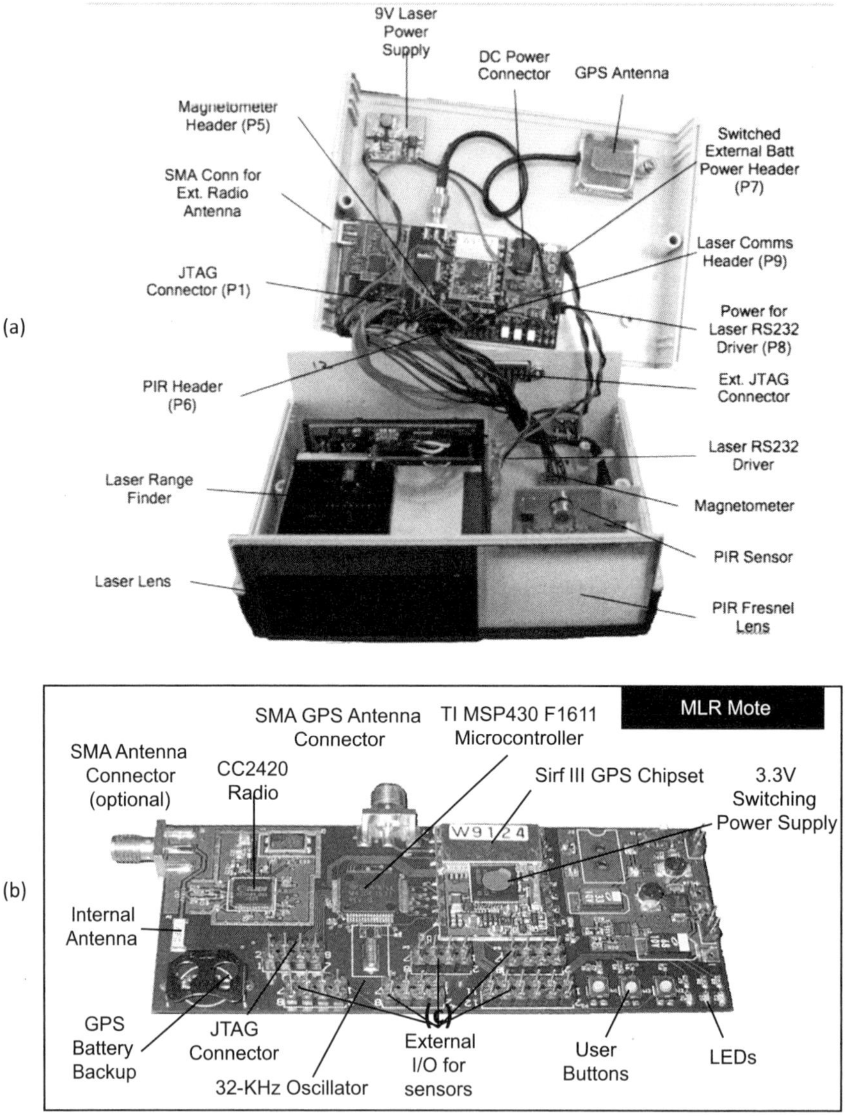

Figure 9.12 MLRmote : (a) the assembled MLRmote, and (b) closeup of the MLRmote logic board.

- Custom development board using a TI MSP430F1611 processor;
- Chipcon CC2420 2.4-GHz radio;
- 3.3-V National Semiconductor LM2621 switching power supply;
- ET301 GlobalSAT Sirf III GPS, with powered external antenna;
- Perkin Elmer PIR sensor with MSP430F2013;

- PNI two-axis Magnetometer;
- Opti-Logic RS100 RS232 IR Class-I (905 nm, eye safe) Laser Rangefinder;
- 9-V National Semiconductor LM2731 power supply for laser;
- 3.7-V, 6,600-mAh lithium-ion rechargeable battery module.

Although a LOS sensor, which requires clear viewing to operate at its 100-m range, the MLRmote was also tested at ground level in dense grass and foliage environments and observed to operate up to 10m. Range was strongly dependent on azimuth, indicating highly variable pathways, and reflectivity through the dense vegetation was experienced. Laser communications for WSN applications have also been addressed and face similar challenges as the MLRmote [37].

If SWAP2 is a guarded WSN commodity, why consider a laser radar system? For simple applications, it's the extension in range for target detection and ranging (leading to target localization). WSN laser-based motes (e.g., the MLRmote) cannot be expected to operate continuously; otherwise, the node's expected lifetime would be measured in hours versus months. However, if augmented with extremely low-powered nodes, as in Figure 9.2, an augmented WSN system can operate for prolonged periods of time. With a 6,600-mA-h power source, the MLRmote duration depends heavily on the duty cycle (how often the IR laser is energized and the firing pulse repetition rate (PRF), which can be set, typically to 2 Hz. With 60-sec bursts at PRF = 2 Hz, the system is estimated to operate for 60 days. Obviously, from this relatively large power by the laser per ranging operation (400 mA at 3.3-VDC), the duty cycle and PRF must be judiciously controlled to reach T-ISR mission duration. To support duration, suppression of false alarms coming from the supportive (passive sensing) motefield is paramount. There are mitigating implementation and technology paths to making adaptation of laser-based motes to T-ISR more palatable. The need for a >100-m sensor range can be reduced drastically to the 10–100-m range and still provide a critical detection and location function not capable of by passive nodes. The goal in introducing the use of low-cost direct-detection laser rangefinders is to provide WSN designers with a solution to meeting mission requirements outside of the capability of typical WSN sensor modalities, without resorting to the use of complex (large and costly) sophisticated sensor units (SSUs).

To provide a sense of range capability for the direct-detection laser radars, range equations for both unresolved and resolved targets are developed and presented. The unresolved arrangement addressees the physical location of the sensor and the target so that the laser floods the target. The backscattered signal is then that portion of transmitted power intercepted and directed back to the laser radar receiver. The expression for the power received back at the sensor (P_r) can be developed by referring to numerous laser radar (and radar) handbooks [38–41]. For unresolved target results, received power directed back at the sensor is,

$$P_r = \left[\frac{P_T}{4\pi R^2}\right] G_a \left[\frac{\sigma}{4\pi R^2}\right] A_e \tau_e \tag{9.30}$$

Equation (9.30), received power, is the product of groups of parameters: isotopic laser radiated power, transmission antenna gain, far-field target cross-section, and reradiation, receiver antenna gain, and end-to-end transmissivity. The various terms used are defined as transmitted optical power (P_T), optical antenna transmit gain (G_a), cross section of the target (σ), range from laser-to-target (R), optical receive antenna gain (A_e), and effective transmission end-to-end (τ_e). Solving for the estimate of maximum range results in

$$R_{\text{MAX}} = \left[\frac{P_T G_a A_e \tau_e}{(4\pi)^2 P_r}\right]^{1/4} \tag{9.31}$$

For resolved targets, where the transmitted beam falls on the target, the laser range equation (LRE) is expressed as,

$$P_r = P_T \frac{\rho_t}{R^2} A_e \tau_e \tag{9.32}$$

Using (9.32) to determine maximum range results in,

$$R_{\text{MAX}} = \sqrt{P_T \frac{\rho_t}{P_r} A_e \tau_e} \tag{9.33}$$

With (9.32), P_r can be defined in terms of total noise equivalent power (NEP_T) and SNR, or

$$P_r = \left(NEP_T\right)\left(SNR\right) \tag{9.34}$$

Total NEP addresses numerous source of noise, through summation of contributing noise sources: Poisson behavior of photons (Q), local oscillator noise, background and scattered radiation, dark current within the detector(s), thermal current, amplifier noise (F), $1/f$ noise (flicker, not optical frequency), and ADC quantization noise. Table 9.1 presents a list of noise sources, expressions for mean-square noise currents, and parameter definitions used.

9.4 Seismic Sensors

Human activities on the ground generate characteristic vibrations from points of contact as seismic waves. Seismic vibrations are primarily transmitted as Rayleigh waves that disseminate along the surface with the remaining energy propagating as bulk waves perpendicularly to the Rayleigh waves. The use of seismic-based UGSs to address ISR objectives over previous decades has proven extremely useful because of the transmission direction and preservation of frequency content. From data extracted from seismic UGSs, frequency-dependent attenuation and environmental ground characteristics have been modeled. The ground environment strongly influences the ability to conduct footstep and vehicle detection, such as decreased effective range during daylight hours due to heightened noise levels, and seismic signaling velocity varies depending on the vibration frequency [42].

Low-powered geophone sensors have been developed that are less sensitive to Doppler effects of the environmental variations than acoustic sensors. Two types of coil-based geophone sensors are used: single-axis and three-axis geophones, with improved bearing estimation afforded with the single-axis geophones, which strive for maximum sensitivity along one axis only. Further improvement for WSN-based geophone applications arrive via MEMS technology via use of ultra-low noise MEMS accelerometers. An advantage

Table 9.1
Laser Noise Source Expressions of Mean-Square Noise Currents

Noise sources	Mean-square currents	Parameters
Signal photons	$2qP_sR_\lambda B$	P_s—sig power
Background photons	$2qP_bR_\lambda B$	P_b—background power
Dark current (total)	$2qi_DB$	i_D—mean dark current
Thermal noise	$4\,kT\,B/R_L$	T—device temperature
		R_L—load resistance
Amplifier noise	$4k(F-1)\,T\,B/R_L$	F—noise figure
$1/f^n$ noise	detector specific	f—modulation frequency
(flicker, contact states		n—appropriate empirical factor (high pass filter)

of MEMS accelerometers is the capability to detect seismic signatures of a variety of targets in frequency bands where coil-based geophones are not able to function [43]. Extracting key characteristics for both seismic and acoustic sensors resides in the post-acquisition signal processing to accurately resolve and extract signal frequency content. Fast Fourier spectral analysis has been applied and produces satisfactory results [44]. However, applying FFT and windowed-FFT algorithm to data always implies assumptions concerning the underlying data, as sampled times series represent an infinite signal and frequency content within the sampled data window is stationary. An approach used with laser vibrometry data has been demonstrated to not only provide improved detection by reducing the requisite demodulation CNR threshold but to improve the resolution and accuracy of spectral estimation of the baseband signal through the reduction of cross-frequency products, typical of Fourier processing [45]. In addition to maximizing the extracted signals from seismic and acoustic sensors, signal analysis is critical to reduce FAR to discriminate mobile objects (humans, animals, and vehicles) from single vibration events (e.g., thunder, tree falls, and rockslides). Use of a neurobiology-motivated algorithm to detect approaching vehicles and to identify vehicles has been considered [46].

9.5 Acoustic Sensors

There are many different types of microphones and acoustic transducers available that can be used as acoustic sensors, including dynamic, electret, condenser, ribbon, and piezoelectric varieties. Electret microphones are sensitive, resilient, and inexpensive while maintaining a small size. The electret design is a derivative of the using the condenser microphone, where capacitance creates a signal voltage proportional to the sound waves generated. In electrets, bias voltage is inherently built in, which reduces the amount of power needed for operation.

Electret devices using MEMs are commercially available and offered in small surface-mount packaging. Electret microphones can be assembled and soldered directly onto a printed circuit board (PCB) using standard automated manufacturing processes, which results in low cost and increased reliability. For acoustic signals, signal data is more meaningful and compact in the frequency domain than time domain; therefore, sensor processors compute a fast Fourier transform (FFT) on collected time domain signals for spectral analysis. If sensor power efficiency were not a high priority, the processor would continuously sample audio data and compute FFTs. However, sampling data and computing FFTs requires running the ADCs and clocking a processor at a higher rate than that needed for quiescent operation. These processes translate into higher current draw and, thus, greater power consumption. Limiting sensor power consumption is a high priority, requiring another solution. Under one approach, the sensor can be held in a power-saving state until background audio levels exceed a preset threshold. Once triggered, the sensor can sequence to a collection state where during that time frame, ADCs and processor operate at the higher rate [47].

9.6 Magnetometers

Magnetic sensors detect targets containing ferromagnetic materials, which distort the Earth's magnetic field. There are numerous approaches to designing a magnetometer, as there are multiple effects that can be exploited to sense and measure magnetic fields. Galvanomagnetic effects occur when a material carrying an electrical current is exposed to a magnetic field. Low-cost and simple magnetometer design employs the Hall effect associated with the generation of electric potential perpendicular to electric current flowing along conductive material when immersed in an external magnetic field at right angles to the current direction. Further refinement uses anisotropic magnetoresistance (AMR) sensors, which respond to parallel fields and sense both magnetic poles, making AMR more sensitive than Hall-effect sensors unless additional circuitry is provided. Modulation of the resistance by a magnetic field occurs with magnetoresistance, which is exploited by devices such as the Honeywell HMC-1002 magnetometer sensor [48], where magnetoresistive elements are configured in as a Wheatstone bridge [49] to produce a differential voltage output (Figure 9.13). Recent activity in reducing magnetometer size, weight, power, and cost, while increasing sensitivity and resolution, has embraced MEMS technology [50].

Magnetic permeability (μ), defined as the ratio of magnetic flux density (B) to magnetic field strength H), can increase or decrease the resultant magnetic field inside a material compared with the external magnetizing field M. Magnetic flux density B is a measure of the actual magnetic field within a material considered as a concentration of magnetic field lines, or flux, per unit cross-sectional area. Magnetic field strength H is a measure of the magnetizing field produced by electric current flow in a coil of wire. In vacuum, magnetic flux density is equal to magnetizing field because there is no matter to modify the field. (In CGS units, permeability is dimensionless and has a value of 1.) With MKS and SI units, B and H have different dimensions, and

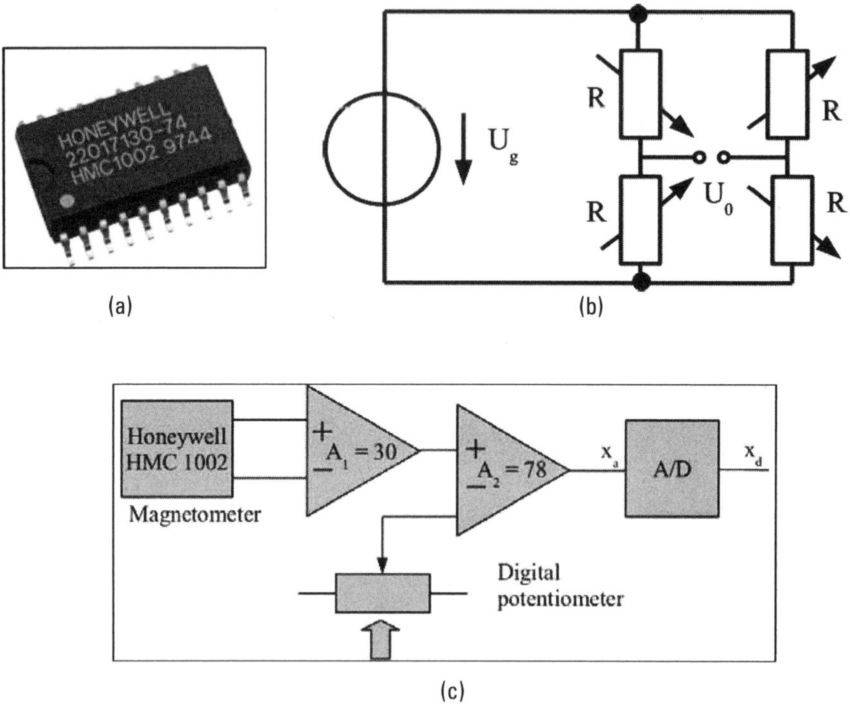

Figure 9.13 Wheatstone bridge arrangement for magneto-resistive magnetometer : (a) an example of an AMR 2-axis magnetometer chip (Honeywell HMC1002) [48], (b) Wheatstone bridge configuration [49], and (c) block diagram of a HMC 1002-based magnetometer [49].

free-space permeability (μ_0) is defined as $4\pi \times 10^{-7}$ weber per ampere-meter. In these systems, permeability ($\mu = B/H$), is called the absolute permeability μ of the medium. The relative permeability μr is then defined as the ratio μ / μ_0. In statistical signal processing, a sensor model of a time-invariant system can be expressed using state vectors, where Y_k is a measurement, X_k is the state of the system, and n_k is measurement noise at time instant k Ts (Ts being the sample period). This is represented as,

$$Y_k = h(X_k) + n_k \tag{9.35}$$

An acting force on an electron, F, is produced in the presence of an electromagnetic field (electric and magnetic field, E and B, respectively) on a carrier charge (q) creating a current (based on carrier velocity, v). With objects that can be magnetized, the magnetic induction vector induces a magnetic field, which can be measured by a magnetometer.

$$F = q\left(E + v x B\right) \tag{9.36}$$

$$B = \mu\mu_0 H \tag{9.37}$$

Considering the target to act as a magnetic dipole, the target field can be modeled as a magnetic dipole field. An expression for this field is derived using Maxwell equations and is expressed through a nonlinear model [51],

$$h\left(x_k\right) = B_0 + \frac{\mu_0}{4\pi}\frac{3\left(r_k \cdot m_k\right)r_k - ||r_k||^2\, m_k}{||r_k||^5} = B_0 + J^m\left(r_k\right)m_k \tag{9.38}$$

$$J^m\left(r_k\right) = \frac{\mu_0}{4\pi\, ||r_k||^5}\left(3r_k r_k^T - ||r_k||^2\, I_3\right) \tag{9.39}$$

In (9.38), B_0 is a bias, r_k is the position of the target relative to the sensor, m_k is the magnetic dipole moment of the target, and n_k is zero-mean white Gaussian noise. The Jacobian of r_k, $J^m(r_k)$, is used in (9.38) [52], and the resulting sensor model has the state of the system being expressed as

$$X_k = \left[B_0^T\; r_k^T\; m_k^T\right]^T \tag{9.40}$$

The dipole model is an approximation of the induced magnetic field that considers the target to be a point source. This approximation is valid if the target is far from the sensor compared to its size.

The bias, B_0, is assumed constant given reasonable thermal variations and is equal to the Earth magnetic field. In practice, the bias includes magnetic distortion induced by other metallic objects in the environment, also assumed to be constant. With the magnetic field of an object expressed as a multipole series expansion, and knowing that individual magnetic poles do not exist, the lowest-order dipole decays as $1/r3$, with higher-order multiples decaying with correspondingly higher powers of the distance. For ranges approaching the target dimension, the dipole moment dominates the signal, resulting in the problem that locating and characterizing a target becomes equivalent to locating a magnetic dipole and measuring its moment vector. Total-field magnetometers produce only a single datum and cannot solve the unconstrained localization problem. A single-axis vector component magnetometer loses the total-field sensor's motion-noise immunity, without providing any advantage in localization ability [53].

9.7 Chemical-Biological Sensors

Chemical-biological sensors are analytical transducer devices that require application of a sensing material. With changes in the input associated with the arrival of chemical or biological particles, the sensing characteristics respond by altering characteristics and creating a readable form of energy. The resultant signals are processed to discern the presence and concentration of chm-bio species in the sampled volume (usually, ambient environment. The energy-transduction principles employed for chemical-biological sensing involve radiant, electrical, mechanical, and thermal types of energy. Of note, chem-bio sensors have relatively low response selectivity to analytes of interest when presented within complex samples that contain high levels of interferences. Also, chem-bio sensors suffer from short-term stability. Various toxins, hazardous materials, explosives, taggants, oxidizers, organic compounds, bacteria, and viral particulates have all been addressed by sensor designs [54].

In the context of a WSN system, chem-bio equipped nodes can provide a highly detailed evolution of subject agents through a time series of detections and concentrations relayed across a deployed WSN network. To achieve a comprehensive capability for WSN, chem-bio sensor response must perform the following additional functions:

1. Identification and stabilization of sensing material;
2. Matching sensing material to appropriately sensitive transducer;
3. Development of algorithms compatible with WSN processing capability that allow for discriminate (FAR rejection).

References

[1] "Sensor," in Merriam-Webster.com, retrieved April 12, 2020, from https://www.merriam-webster.com/dictionary/sensor.

[2] Kopackova, V., and L. Koucka, "Integration of Absorption Feature Information from Visible to Longwave Infrared Spectral Ranges for Mineral Mapping," *Remote Sensing*, 2017.

[3] Zalewski, E., "Chapter 24: Radiometry and Photometry," in *Handbook of Optics, Volume II: Design, Fabrication and Testing, Sources and Detectors*, Third Edition, McGraw-Hill, 2009.

[4] Norkus, V., et al. "Performance Improvements for Pyroelectric Infrared Detectors," in *Proc. SPIE 6206, Infrared Technology and Applications XXXII*, 62062X, May 18, 2006, https://doi.org/10.1117/12.664389, 2006.

[5] Kastek, M., et al., "Long-Range PIR Detector Used for Detection of Crawling People," *Proc. SPIE 7113, Electro-Optical and Infrared Systems: Technology and Applications*, 2008.

[6] Korites, B., Korites, B., *Python Graphics,* Berkeley, CA: Apress, 2018, pp. 355–357.

[7] Madura H., T. Piatkowski, and E. Powiada, "Multispectral Precise Pyrometer for Measurement of Seawater Surface Temperature," *Infrared Physics & Technology*, 2004.

[8] Whatmore, R., and R. Watton "Chapter 5: Pyroelectric Materials and Devices," in *Infrared Detectors and Emitters: Materials and Devices*, Chapman and Hall, 2001.

[9] Goodman, J., *Introduction to Fourier Optics* (Second edition), McGraw-Hill, 1996.

[10] Sheppard, C., Diffraction Optics, in *Handbook of Biomedical Optics*, D. A. Boas, C. Pitris, and N. Ramanujam (eds.), CRC Press, 2011.

[11] Weixing, L., "Single Molecule Cryo-Fluorescence Microscopy," Ph.D. dissertation, Georg-August-Universitat, 2016.

[12] Nugent, P.,et al., "Measuring the Modulation Transfer Function of an Imaging Spectrometer With Rooflines of Opportunity," *Optical Engineering*, 2010.

[13] Nakamura, J., *Image Sensors and Signal Processing for Digital Still Cameras*, CRC Press, 2017.

[14] Vollmerhausen

[15] Barela, J., et al., "Determining the Range Parameters of Observation Thermal Cameras on the Basis of Laboratory Measurements," in *Proc. of SPIE Electro-Optical and Infrared Systems: Technology and Applications X*, 2013.

[16] Tracy, A., et al., "History and Evolution of the Johnson Criteria," SANDIA REPORT SAND2015-6368,

[17] Johnson. J., "Analysis of Image Forming Systems," *Selected Papers on Infrared Design. Parts I and II*, Vol. 5, 1985.

[18] Moyer, S., "Modeling Challenges of Advanced Thermal Imagers," Ph.D. dissertation, Georgia Institute of Technology, August 2006.

[19] Maurer, T., et al., "2002 NVTherm Improvements," *Proceedings of SPIE Infrared and Passive Millimeter-wave Imaging Systems: Design, Analysis, Modeling, and Testing*, 2002.

[20] Peric, D., et al., "Thermal Imager Range: Predictions, Expectations, and Reality," *Sensors*, 2019.

[21] Chrzanowski, K., "Noise Equivalent Temperature Difference of Infrared Systems Under Field Conditions," *Optica Applicata*, Vol. X, 1993.

[22] Rogalski, A., "Infrared Detectors: An Overview," *Infrared Physics & Technology*, 2002.

[23] Andersson, J., L. Lundqvist, and J. Borglind, "Dark Current Characteristics and Background-Limited (BLIP) performance of AlGaAs/GaAs Quantum Well Detectors," *Proceedings Optoelectronic Integrated Circuit Materials, Physics, and Devices*, Volume 2,397, 1995.

[24] Joseph Kostrzewa, J., et al., "TOD Versus MRT When Evaluating Thermal Imagers That Exhibit Dynamic Performance," *Proc. SPIE 5076, Infrared Imaging Systems: Design, Analysis, Modeling, and Testing XIV*, 2003.

[25] Tissot, J., et al., "Uncooled Microbolometer Detector: Recent Developments at ULIS," *Opt-Electronics Review*, 2006.

[26] "High Sensitivity Enables Detection of Stationary Human Presence," OMRON D6T MEMS Thermal Sensors, OMRON datasheet, https://omronfs.omron.com/en_US/ecb/products/pdf/en-d6t.pdf.

[27] "Seek Thermal Introduces Two New Series of Low Cost, High-Resolution OEM Thermal Cameras," *Seek Thermal*, December 10, 2019.

[28] "Micro Core Ultra-Compact, Low Cost, High-Performance Thermal Imaging Core with 200 × 150 Sensor Resolution," *Seek Thermal*, 2020.

[29] Rajic, N., and N. Street, "A Performance Comparison Between Cooled and Uncooled Infrared Detectors for Thermoelastic Stress Analysis," *Quantitative Infrared Thermography Journal*, 2014.

[30] "ULIS Releases ATT0640™, World's Smallest 60 Hz VGA/12 Micron Thermal Image Sensor," Andrew Lloyds & Associates, 2019.

[31] "Li-SO2 Primary Battery System BA 5590 B/U One Battery for Various Military Applications," *Saft*, December 2005.

[32] "Sharp J-SH04: World's First Ever Phone with Integrated Camera," Gadgetizor.com, 2010.

[33] "STMicroelectronics Expands Phone Camera Family with Highly Integrated 2-Megapixel Module," EmbeddedTechnology.com, 2007.

[34] "CMOS OV5642 Camera Module, 1/4-Inch 5-Megapixel Module Datasheet," OmniVision, 2013.

[35] Jackson, L., "ArduCAM Mini Released," ArduCam, 2015

[36] Dolcourt, J., "Qualcomm: Your Next Phone Could Have an Enormous 200-Megapixel Camera," *C|Net*, 2019.

[37] Ghosh, A., et al., "Free-Space Optics based Sensor Network Design Using Angle-Diversity Photodiode Arrays," *Free-Space Laser Communications X*, 2010.

[38] Eaves, J., *Principles of Modern Radar*, Van Nostrand Reinhold, 1987.

[39] Skolnik, M., *Introduction to Radar Systems*, McGraw-Hill, 1980.

[40] Minkoff, J., *Signals, Noise, & Active Sensors*, John Wiley & Sons, 1992.

[41] Jelalian, A., *Laser Radar Systems*, Norwood, MA: Artech House, 1992.

[42] Koc, G., and K. Yegin, "Hardware Design of Seismic Sensors in Wireless Sensor Network," *International Journal of Distributed Sensor Networks*, 2013.

[43] Beresík, R., J. Puttera, and F. Nebus, "Seismic Sensor System for Security Applications Based on MEMS Accelerometer," *IEEE 2014 International Conference on Applied Electronics*, Pilsen, Czech Republic, 2014.

[44] Sharma, N., et al., "Detection of Various Vehicles Using Wireless Seismic Sensor Network," *IEEE 2012 International Conference on Advances in Mobile Network, Communication and Its Applications*, 2012.

[45] Cole, T., and A. S. El-Dinary, "Estimation of Target Spectra from Laser Radar Backscatter Using Time-Frequency Distributions," *SPIE Proceedings Applied Laser Radar Technology*, 1993.

[46] Debaser, A., et al., "Recognition of Acoustic and Vibration Threats for Security Breach Detection, Close Proximity Danger Identification, and Perimeter Protection," *IEEE Conference on Technologies for Homeland Security*, 2011.

[47] Shimazua, R., et al., "MicroSensors Systems: Detection of a Dismounted Threat," *Proceedings of SPIE Unmanned/Unattended Sensors and Sensor Networks*, 2004.

[48] [48] Honeywell Device, HMC1001/1002/1021/1022 Technical Specifications, https://aerospace.honeywell.com/content/dam/aero/en-us/documents/learn/products/sensors/datasheet/N61-2056-000-000_MagneticSensors_HMC-ds.pdf.

[49] Koszteczky, B., and G. Simon, "Magnetic-Based Vehicle Detection with Sensor Networks," *2013 IEEE International Instrumentation and Measurement Technology Conference (I2MTC)*, 2013.

[50] Herrera-May, A., et al., "Resonant Magnetic Field Sensors Based on MEMS Technology," *Sensors*, 2009.

[51] Wahlstrom, N., and F. Gustafsson, "Magnetometer Modeling and Validation for Tracking Metallic Targets," *IEEE Transactions on Signal Processing*, 2014.

[52] https://en.wikipedia.org/wiki/Jacobian_matrix_and_determinant.

[53] Czipott, P., et al., "Magnetic Detection and Tracking of Military Vehicles," Technical Report, AD-A409217, NASA/STI 2002.

[54] Potyrailo, R., N. Nagraj, and C. Surman, "Wireless Sensors and Sensor Networks For Homeland Security Applications," *Trends Analyt. Chem.*, 2012.

10

WSN System Deployment and Integration

Deployment of a WSN system involves a number of considerations, from preplanning based on mission objectives and associated operating terrain to post-deployment integration with supportive infrastructure to remotely access the system by end users. Deployment planning is a multistep process of answering the why, where, and how questions regarding an assigned T-ISR mission. Example considerations include:

- Is the deployment area outdoors, across ravines, waterways, into tunnels, within buildings, or does it consist of combinations of the aforementioned?
- Is the system to be deployed to operate as a covert, or overt, asset?
- Can the physical emplacement of sensor nodes employ a variety of deployment mechanisms, or is it limited to a few (or perhaps one) specialized deployment approach due to accessibility or security considerations?
- Once deployed, are there multiple exfiltration points and capabilities?
- Are updates to mission, or operational parameters to be expected?
- How much autonomy can be embedded into the nodes versus access point relay?

- What services exist to seamlessly bring the deployed WSN system online within a context of existing T-ISR systems?
- How do we ensure fast and reliable linkage with remote end users?
- Are WSN output products properly formatted to present information to remote end users using expected T-ISR standards for displaying and presentation?

So far, we have discussed the rationale and objectives associated with T-ISR systems using WSN. In addition, we have reviewed the functionality and implementation of WSN systems with a focus on subsystem designs and sensor node implementations. Moreover, this book has presented relevant theoretical considerations and models, along with practical implementation considerations. Finally, we have discussed subsystems and technologies involved with WSN-based systems such as middleware functionality and design, enabling technologies, MAC design, localization approaches, and node-based sensor modalities. With this knowledge of approaches to evaluating system elements and an understanding of overall system functionality, we now address deployment and insertion of WSN nodes into mission environments. WSN-based systems are a capable tool for T-ISR missions, and as with the preceding UGS-based systems [1], successful deployment and integration into an overall T-ISR/ISR infrastructure requires consideration and planning.

10.1 Deployment Considerations

As with any ISR tasking, preplanning is paramount to achieving a successful mission. The military speaks of this as joint intelligence preparation of the operational environment (JIPOE). JIPOE is the analytical DoD process used by joint intelligence organizations to produce intelligence estimates and other intelligence products in support of the joint force commander's decision-making process. It is a continuous process that includes defining the operational environment, describing the impact of the operational environment, evaluating an adversary, and determining possible courses of action of both blue and red forces [2]. Regarding deployment of WSN-based systems, the preplanning stage must address critical aspects of the mission assignment, including remote sensing requirements, proximity to human activities, deployment terrain, and AOI weather/climate. Once sensor nodes are deployed, individual nodes are expected to autonomously sequence through initialization, wirelessly connect with other nodes to form an ad hoc network, and commence sensor operation to provide persistent surveillance.

With UGS, there was typically a system operator charged after deployment with the verification that the system was properly deployed, initialized, and operational. This requirement is avoided with WSN systems as sensor nodes are designed to be preprogrammed and capable of establishing and maintaining a cohesive network without intervention. This places an emphasis on the checkout stage for WSN components and subsystems prior to deployment. Upon deployment, verification of the final network formation, operation, and linkage to exfiltration relays occurs remotely. An exception may be if deployment is accomplished via hand emplacement. Here, an on-site operator, as with many UGSs, is in proximity of the deployed system. However, loitering is not always possible, and with increasingly regularity, motes are required to be deployed by aerial or vehicle means.

The basis of operation depends of a WSN system depends upon mission objectives. A WSN-based sensor field can be categorized as a proactive, or reactive, T-ISR system. Sensor nodes in a proactive system sample and forward data periodically to an exfiltration relay, whereas a reactive system transmits data back to the user only when a predetermined event triggers the system. Reactive WSNs are particularly suitable for event-driven applications as intrusion detection can take full advantage of the low cost and flexible design and easy-to-deploy features of WSNs [3].

10.1.1 Mission Objectives

WSN systems, which are part of the IoT and IoBT, are used in peacetime to support numerous civil applications (e.g., manufacturing control, property monitoring, disaster search and rescue, forest fire monitoring, and customs and border patrol). Although mission objectives differ greatly between civil and military systems, system capabilities and performance characteristics are surprisingly similar. The application defines how the WSN system should be designed, deployed, and operated. Defensive applications (e.g., perimeter surveillance) indicate that the user knows the areas to be monitored and what target types are to be expected. Knowing an AOI provides an advantage to the WSN end user, especially if provided with full control and access to the surveillance areas. Points of penetration are better understood for defensive missions as vulnerabilities become self-evident given familiarity (e.g., bridges, gates, fences, and ravines). Even the issue of being overt, versus covert, comes into play. A defensive application may prefer having the sensor system be self-evident to fend off would-be intruders. Finally, with defensive scenarios, the logistics of getting the WSN system to the AOI(s) and deploying the system is easier and usually under the users' control.

Offensive applications deviate drastically from a defensive installation. In offensive applications, the mission typically requires the system to be inserted, initialized, and operated in a covert manner. This implies a difficult deployment process followed by surreptitious operation. Sensor nodes must be located within proximity of assigned AOIs supportive of both sensor and RF-link range capabilities. Counter to a defensive application, offensive system requirements typically conceal the sensors and exfiltration relay(s) from the opposition. Several logistical and deployment issues have to be considered and answered. Dispersion and location of the sensor nodes must ensure reliable sensor and RF range values, adequately observe AOIs, and maintain sufficient operational covertness. Numerous approaches and tools exist to attempt to optimize deployment circumstances; however, regardless of the fidelity, results obtained from these preplanning tools are only estimates. To improve the probability of a successful deployment, it may be necessary to garner in-depth information regarding topology, soil characteristics, surrounding conditions, and the opposition's capability (e.g., on-foot personnel, or vehicular presence, and number of expected targets to consider) [4]. As with any RF-based technology, understanding the physical and electrical properties (e.g., soil moisture) of the operating environment differentiates success from failure.

10.1.2 Proximity to Human Activities

In setting up a T-ISR system, proximity to human activities directly influences the approach to disbursing sensors, deployment configuration, and operation of the WSN system. It is not just the possibility that a passerby could observe sensor emplacement but the fact that human activity leads to RF interference, numerous false alarms due to innocuous incursions (civilians and domesticated animals), and /or intentional tampering or theft. Covert operations cannot afford any possibility of discovery. Deployment may not only be required to occur rapidly, it may have to occur during inopportune times (e.g., after nightfall or during inclement weather). As for electronic interference, WSN motes are designed to operate using adaptive, spread-spectrum waveforms, and if designed appropriately, should be capable of operating satisfactorily with harsh RF interference. Regarding false alarms due to highly traveled pathways and associated activities, astute use of sensor modalities, overall motefield extent, and orientation can be employed to provide higher levels of target discrimination. Finally, addressing tampering or theft, covert applications must employ surreptitious WSN elements along with countermeasures against tampering or exploitation of the nodes.

10.1.3 Terrain Considerations

The terrain in which a system is to operate presents questions regarding sensor and RF LOS, surface and soil composition, vegetation, and accessibility. These factors determine sensor detection range, viable emplacement methods, and RF propagation behavior. For example, with hard compacted soil, seismic sensors experience relatively good detection ranges. Ambient interference such as vehicular activity, surf, volcanic activity, Earth tremors, or even running water degrades the quality of seismic sensor performance. Similarly, emissions from power lines and other electronic sources can disrupt magnetic sensors. As for LOS, seismic and acoustic sensors can handle loss of LOS, but they do require reduced ambient seismic/acoustic activity and precautions such as avoiding placement of acoustic sensors near waterfalls. For increased sensor range, or operation across ravines, waterways, or such disjointed terrain, the use of optical sensors may be appropriate.

Although vegetation provides cover and concealment for sensors and relays, the presence of vegetation makes antenna placement difficult by interfering with the RF communications link. One of the hard lessons learned early with a 2.4-GHz RF-linked WSN is that vegetation, and in particular, conifers, present a difficulty not anticipated: the water content and needle size of conifers resonates at 2.4 GHz and readily attenuates that frequency. RF attenuation in the ISM band by vegetation becomes significant with increasing frequency. For RF frequencies (3.2–3.9 GHz), mean attenuation values have been measured as 21±1.6 dB for cedar trees (with an average canopy thickness of 6.5m) [5]. As demonstrated by Figure 10.1, ISM-band RF attenuation is significant in the presence of vegetation.

With the presence of terrain features such as ravines, slopes, waterways, and other topographical forms, the type and capabilities of sensor modalities must be considered. When viewing an AOI [or perhaps more accurately, a volume of interest (VOI)], the altitude aspect of node localization is emphasized. (Recall Figure 3.1.) As the AOI (VOI) geologic features become complex, both the deployment approach and selection of sensor modality become critical. Certainly in either a defensive or offensive stance, effective system performance must consider that manmade threats would view an area and respond to the same features—and, perhaps, see opportunities for their use (e.g., ravines for troop movement).

10.1.4 Weather/Climate

T-ISR systems operate in areas where weather conditions can change drastically and quickly, causing system performance degradation, which is most notable

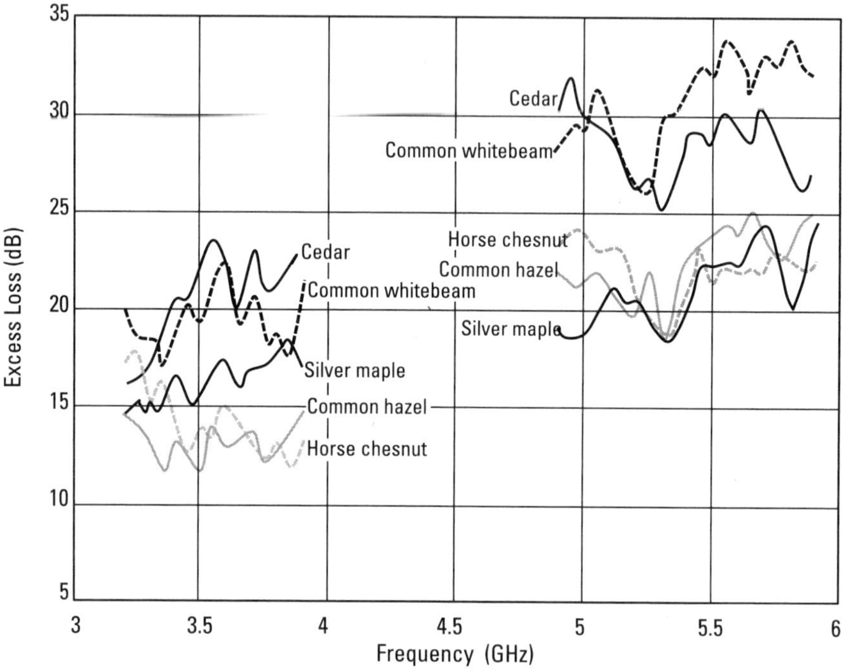

Figure 10.1 ISM-band RF attenuation per vegetation types [5].

for sensor and RF link reliability. Regarding sensor impact, various sensor equations (Chapter 8) incorporate coefficients and parameters that associate atmospheric conditions to the probability of detection (and false alarm). For RF link reliability, weather-associated parameters are considered in evaluating their impact upon received signal strength, which can be characterized by RSSI. Effects of temperature and humidity on radio signal strength in outdoor wireless sensor networks have been empirically characterized.

Of particular note is the negative effect of increasing temperature on ISM RF signal strength [6]. Rain and fog have been studied in the evaluation of WSN RF link reliability. However, at 2.4 GHz (and lower), the impact does not appear to be as severe as thermal effects [7]. The more significant impact on RSSI has been repeatedly correlated to thermal changes compared to other weather effects, assuming adequate protection of the nodes from being inundated by water or buried in snow. The thermal dependency is considered a result of low-cost componentry in the sensor node design, and in particular, transceiver performance. To help mitigate the deleterious thermal effects, node designs have adopted increased use of frequency diversity.

10.2 Deployment Planning Approach and Tools

Conventional approaches to deployment using geometric or random distributions have both met with success. However, to assess a WSN-based system as successful is ultimately determined through reliable and persistent coverage and network connectivity provided by the selected deployment scheme. As with any wireless communication architecture, gauging how well a network operates is associated with a well-known cellular term, QoS [8]. With packet-switched telecommunication networks, QoS refers to traffic prioritization and resource reservation control mechanisms rather than the achieved service quality. With sufficient QoS, a WSN network provides the ability to handle various priorities assigned to multiple applications, users, or data flows and/or to guarantee a level of performance to a data flow.

Different mission types and deployment environments require uniquely tailored optimization goals, and although the focus is on AOI (or, VOI) coverage, it is persistent connectivity and energy efficiency that enable a WSN system to fulfill T-ISR mission requirements [9]. With the maturity of GISs, numerous deployment models have been developed, evaluated, and categorized [10]. This includes mathematical codes that have been developed to express the probability of complete coverage even as environmental characteristics vary [11, 12], with several codes becoming WSN mission-planning tools, including Geographic Area Limitation Environment (GALE) [13], Sensor Deployment Planning Tool (SDPT) [14], and ISR Synchronization Tool (IST) [2]. To ensure delivery of a robust WSN configuration requires that the associated physical network topology provide alternative pathways (see Chapter 6). Approaches to addressing fault-tolerant capability include overloading deployment with an abundance of nodes to provide significant overlap of node-to-node RF range. This might guarantee a robust WSN, but at the expense of excessive nodes being deployed. To minimize node count yet ensure that each sensor node deployment is availed to *k*-node disjoint links, centralized preplanning algorithms, such as GRASP-ARP [15], have been developed and evaluated.

10.3 Deployment Configuration (AOI Coverage)

Astute energy management is critical to providing sufficient power to conduct persistent operation. With operating lifetime meeting the mission requirement, the next critical elements required for a T-ISR system are adequate sensor capability and reliable wireless connectivity. To provide the required coverage, the number of nodes used for deployment must consider both sensor range (R_{sen})

and RF link range (R_{rf}), as discussed in the development of (9.1) range ratio (R_{rf}/R_{sen}). Sensor coverage refers to obtaining ample range with an associated measurement at the resolution necessary for accurate target detection and identification. RF coverage refers to sufficient capability to successfully establish a RF link and to handle network issues such as congestion, link failure (e.g., weather-induced effects), and/or node failure. As a result, the deployment scheme we seek should minimize the number of nodes required to meet mission requirements while striving to keep a control of cost, complexity, and maintenance. (Additionally, with minimal nodes to be deployed, an offensive system has a better probability of maintaining a covert profile.)

Geometric patterns (e.g., hexagonal arrays and lattices) have been considered, where the system operates based on closely spaced distances described by assuming fixed ranges (R_{sen}, R_{rf}), such as disk-sensing models or dual triangular pattern deployment, which employs data fusion [2, 16]. Figure 10.2 presents

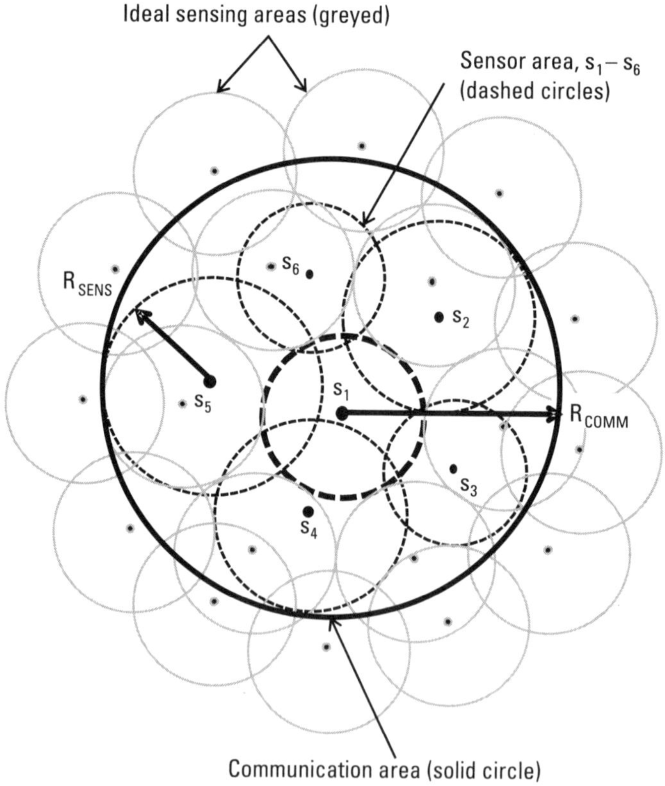

Figure 10.2 Disk-sensing models of sensor-node deployment.

node-deployment patterns considered by disk-sensing models. Unfortunately, it is seldom that actual terrain and vegetation, or even the mission requirements and constraints, allow for such regularly spaced deployment. Further, in covert operations, the last characteristic one wishes to express is any sign of a periodic structure or manmade element in the environment, to avoid arousing the suspicion of the offense.

More appropriate alternative approaches use statistically (randomized) distributed sensor deployment. One such approach is through application of negatively skewed lognormal distribution, which presents the logarithm of the node distribution along the *X*- and *Y*-axis as a Gaussian-distributed variable. The resultant deployment places the highest density of sensor nodes at the periphery of the operating area (AOI), as depicted in Figure 10.3. The negative skew pushes the preponderance of the nodes onto the periphery with a sparse density of nodes near the center, ostensibly where an exfiltration relay is located. (This is similar to an approach considered that employs a reverse-Gaussian

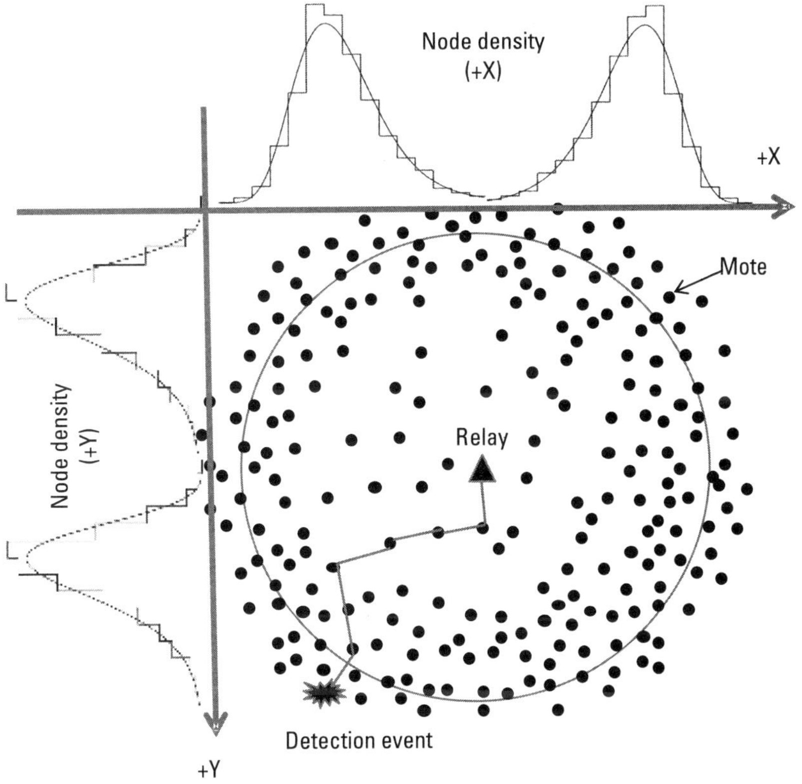

Figure 10.3 Negatively skewed lognormal-distributed WSN deployment.

[17].) With periphery-emphasized deployment, nodes can cover a large area yet maintain a reasonable (and, one hopes, undetectable) distance between nodes. The density at the periphery also ensures adequate AOI coverage at a maximum range from the vertex (useful for a defensive mode system).

Development of application-specific wireless monitoring systems can benefit from deployment patterns that emphasize particular system attributes required by the T-ISR mission [18]. The overall function of a WSN system is to provide a critical process such as false alarm filtering, event detection, priority alerts, interval observation, and scheduled interleaving (i.e., cycling through the node operational modes of sleeping, sensing, reception, and transmission) [19]. Regardless of the approach and modeling employed by the WSN designer, rarely do targets of interest or associated AOIs provide idealistic behaviors that correlate nicely with deployment models. Deployment pattern models provide a reasonable approach to the deployment process for the system designer; however, inevitable tailoring is the norm, and the final deployment configuration requires a combination of a priori planning with a last-minute injection of mission realism.

10.4 Deployment Mechanisms

The physical aspects of deployment must align to mission support constraints as much as the sensor selection and dispersal layout. For numerous deployments, tried-and-true hand emplacement has been used, but conducting deployment using this emplacement approach may place those conducting deployment in harm's way, especially for T-ISR offense missions. Even for defensive scenarios, terrain and/or vegetation may prove difficult for defenders to emplace sensor nodes as required. Regarding hand emplacement, node design and system implementation must consider equipment weight, initialization sequence, ease of emplacement (including the manipulation of antenna and/or seismic spike appendages), and ergonomic aspects. Teams conducting offensive mission emplacement shun node-based initialization cues that are verified through use of acoustic or lighting (LED) signals. Furthermore, the system designer must account for the number of sensor nodes required for a deployment, which may exceed payload limits for on-foot personnel. An example is incorporating a node power switch that is indented in the node packaging. Complex physical access creates difficulty for personnel wearing gloves (e.g., to flip a recessed switch, or push a recessed button). Removing gloves to complete mote emplacement becomes overly difficult as motefield sizes approach large numbers. In cold climes or similar, longer loitering time spent to conduct deployment impedes and endangers personnel.

An approach apart from hand emplacement has been successfully used for numerous UGS and ISR sensors: aerial drop from rotary and fixed-wing aircraft (and recently, UAVs). The pivotal use of aerial drop of sensors occurred with sonobuoys (code named High Tea, World War II, 1942) [20] and with ground-based UGSs during the Vietnam War under the auspices of the Muscle Shoals program (recall Operation White Igloo in Figure 1.2). Numerous missions dropped seismic sensors along various combatant trails (e.g., the Ho Chi Minh trail/Truong Son Road) [21]. A similar approach has been adapted to WSN for both military and civilian uses, including SAR demonstrations [22]. Techniques have also been evaluated using a maple leaf/butterfly packaging design, as used by the nefarious PFM-1 anti-personnel mine deployments. The PFM-1 is a small explosive device designed for deployment through a myriad of mechanisms: aerial vehicles, artillery, and specially designed mortar systems [23]. Adaptation of the versatility of PFM-1 deployment approaches to suitably designed nodes has been and remains under consideration. This aerial dispersal approach would lead to a cost- and operationally effective means to delivering a large number of motes over a large area via UAV-based dispensers. Demonstrations and systems have been designed for such deployments, including aerial dispersion of motes using a predetermined grid pattern using a rotary UAV platform [24].

10.5 WSN System Integration

Chapter 6 presents a description of the exfiltration relay [access point (AP)] hardware, which provides the extraction of sensor and housekeeping data from a WSN motefield and injects operational commands and updates to the field. As Figures 6.13 and 6.14 illustrate, both aerial and hand-emplacement relay variants exist. The relays typically are set up to communicate on multiple RF bands. The various bands are used to support direct communication with the WSN system (i.e., ISM RF bands, including 833-MHz, 2.4-GHz, and 5.2-GHz), to provide long-range VHF (e.g., SEIWG-005A) to interconnect with sophisticated sensors (SSUs), and to allow for connectivity to long-haul networking via SATCOM (or cellular) capability to bridge the WSN system to a worldwide communication infrastructure. In addition to ground-based sensor nodes, MANETs are also employed using ground and/or aerial mobile platforms. WSN systems (T-ISR and other applications) benefit from what is considered an overlay wherein MANET nodes collaboratively join a fixed-node WSN system to augment sensing coverage, RF linkage capability, or both. Overlays using MANET-based nodes can be used to dynamically differentiate and expedite critical datagram traffic (sensor data and commands) using low-latency MANET paths by integrating with the latest emergent standards and specifications for WSN data collection [25].

As the basic nature of T-ISR missions spans across numerous operating requirements, system integration has embraced other design aspects, including adaptability of the transport protocol (Chapter 4). Development of various transport protocols continues to push network efficiency. For example, protocols that employ pump slowly, fetch quickly (PSFQ), have been defined, simulated, and tested. Early WSN networks tend to be application-specific and hard-wired to realize specific tasks efficiently at low cost. With T-ISR systems, repurposing a deployed system often occurs, challenging the original design. Protocols based on PSFQ support simple, yet robust and scalable performance, which is adaptable to meet the needs of different reliable data applications [26].

10.5.1 Open Geospatial Consortium

The Open Geospatial Consortium (OGC) is an international voluntary consensus standards organization that was originated in 1994. With more than 500 commercial, governmental, nonprofit and research organizations worldwide, the OGC collaborates in the development and implementation of open standards for geospatial content and services, sensor web, and IoT [27]. There are numerous, incompatible standards in the geographic information technology area that exacerbates sharing of computerized geographical data (geodata) among GISs. The objective behind OGC standards is to normalize encoding of information in software systems (i.e., data format standards and data transfer standards), to name features and feature relationships (i.e., data dictionaries), and to present cognitive framework for descriptions of data sets (metadata) [28]. OGC standards are technical documents that present the details of these standard addressing interfaces and encodings. OGC maintains open access to these documents to enable GIS developers to develop open interfaces and encodings regarding GIS products and services [29].

10.5.1.1 Sensor Web Enablement

The term "sensor web" describes a wireless sensor network architecture where individual sensing elements interconnect to form a network using multihop communication, and operate as a coordinated homogenous system. The use of this term describes a specific type of sensor network composed of an amorphous network of spatially distributed sensor platforms that wirelessly communicate with each other [30]. The sensor web concept has been largely influenced by wireless sensor networking (and subsequently, IoT and IoBT). Summarizing, the sensor web is an infrastructure that enables the collection, modeling, storage, sharing, processing, and visualization of sensor and associated sensor metadata data via the World Wide Web in a standardized way. Sensor web promotes extensive monitoring that provides timely and continuous observations and is considered the sensor resource as the internet is to general information sources—an infrastructure allowing end users to easily share and access resources via a well-defined method.

A key challenge in building a sensor web is how to automatically access and integrate various types of geolocated data obtained across a variety of sensor devices or through simulation models. Also, sensor resources have been developed using applications that integrate specific mechanisms rather than leveraging off a well-defined and established integration layer. Integrating diverse sensors into a universal observation system has and continues to address incompatible services and encodings (even within the same T-ISR system). This issue remains the driving force for the OGC and the basis behind the concept of sensor web enablement (SWE) [31]. SWE is a suite of standards developed and maintained by OGC to break down stovepipe GIS systems and to leverage open-source services readily available across the Web. SWE assumes a service-oriented architecture (SOA) approach in providing OGC standards to sensor and sensor system developers to promote discovery and accessibility of remote sensors, transducers, and/or associated sensor data repositories using the Web. The goal for SWE is to design net-centric systems that are sensor-system agnostic where virtually any sensor or model system can be supported [32]. Why involve SWE with WSN, and in particular, a WSN-based T-ISR system design and operation? Through the use of SWE standards, T-ISR systems are inbubed with increased capabilities, such as to

- Rapidly discover sensors and sensor data (secure or public) based on sensor location, sensor data, data reliability and quality, and an ability to task sensors to perform observations required.
- Obtain sensor information via standard encoding understandable by system user software without a priori knowledge;
- Readily access sensor observations in a common manner in a form specific to mission needs.
- Enable subscriptions in order to receive alerts from sensor systems that detect and observe a predefined event.

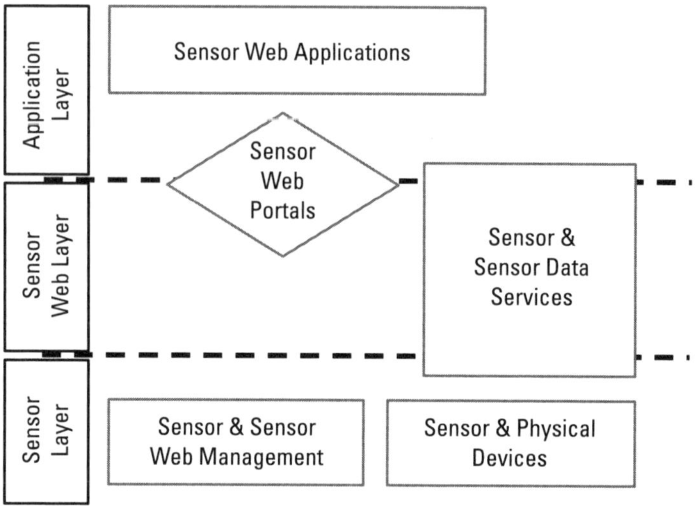

Figure 10.4 SWE environment architecture.

Figure 10.4 presents an overview of the SWE environment tiered architecture, which spans from sensor devices (sensor layer) to the application layer via the sensor web layer (supporting services and portal interfaces) [33].

OGC SWE standards aid interoperability to enable and sustain a sensor web. A prime example that collaborated with WSN demonstration and test programs for T-ISR, is Northrop Grumman's PULSENet [34]. PULSENet provides a standards-based framework for autonomous discovery, access, use and control of heterogeneous sensors, associated metadata, and observation data [35]. A combination of open-source SWE code, COTS products, and custom development were employed to construct PULSENet, which was designed to accommodate legacy and non-SWE standards sensors into SWE-based architectures. In real time, WSN sensor data has been successfully accessed using PULSENet to provide alerts and pointing instruction for high-resolution cameras collocated near WSN motefields. An important aspect is the implementation of OGC SWEs by sensor manufacturers via related standards (e.g., IEEE 1451 and CCSI), and OGC SWE interfaces. Services are loosely coupled, meaning that the service interface is independent of the underlying implementation, and interacts over the network using a protocol such as REST or Simple Object Access Protocol (SOAP) [36]. Developers or system integrators can converge one or more services into an application without necessarily knowing how each was implemented.

10.5.1.2 SOA

As an IoBT system, WSN leverages the internet (e.g., DoDIN) infrastructure to disseminate vast amounts of ISR data worldwide. With increasing complexity, higher-level abstractions in support of continued development of T-ISR applications are required. An approach to wielding high complexity codes is through SOA, which emerged in the early part of this century as an evolution of distributed computing. Before SOA, services were understood as the end result of the application development process. With the advent of SOA, the application itself is composed of services that can be delivered individually or combined as components in a larger, composite service. In particular, SOA is a software design tool that provides services through the use of application components via a network protocol. Having a SOA tool enables middleware developers to combine large functionality modules to form applications rapidly by integrating reusable application components. As SOA promotes loose coupling among services, functions are separated into distinct units (services). Different services can be used in conjunction with specific code modules provide the functionality of a large software application. TinySOA [37] is an example of a SOA designed for WSN, which opens access to functional blocks for WSN middleware developers. Access is performed from middleware applications under development using a simple service-oriented API and programming language of choice.

Linking and fusing platform routing information, sensor exploitation results, and database setup and distribution (e.g. GISs) results in improvement to sharing situation awareness and to increase mission effectiveness. Within the information fusion community, research efforts appeal to open standard approaches to aid convergence of heterogeneous network components into an accessible framework [38]. WSN, via DoDIN, gathers vast amounts of ISR data worldwide by having large-scale processing applications available through the application of SOA. TinySOA [37] allows middleware developers to access WSN functional blocks via the DARPA NEST repository via a simple service-oriented API. Through the use of Smart Transducer Web Services, smart transducers cognizant of IEEE 1451 (discussed in the next section) can be efficiently and quickly integrated to OGC-SWE [39].

10.5.2 IEEE 1451: Smart Transducer Interface Standards

IEEE 1451 is a set of smart transducer interface standards developed by the IEEE Instrumentation and Measurement Society's Sensor Technology Technical Committee. These standards describe a set of open, common, network-independent communication interfaces for connecting transducers (sensors or actuators) to microprocessors, instrumentation systems, and control/field networks [40]. The goal of the IEEE 1451 standards is to allow the access of transducer data through a common set of interfaces whether the transducers are connected to systems or networks via a wired or wireless means. A key element of IEEE 1451 is the definition of Transducer Electronic Data Sheets (TEDS) for each transducer. TEDS is a memory device attached to the transducer that stores metadata available to remote queries. TEDS metadata consists of transducer identification, calibration, correction data, and manufacturer-related information. TEDS formats are defined with the IEEE 1451 set of smart transducer interface standards developed by the IEEE Instrumentation and Measurement Society's Sensor Technology Technical Committee. With IEEE 1451, a set of open, common, network-independent communication interfaces is described and established for connecting transducers to microprocessors, instrumentation systems, and control/field networks. Figure 10.5 illustrates the layered approach to the IEE 1451 standards.

TEDS can be implemented in two ways. TEDS can reside in embedded memory, typically an EEPROM, within the transducer itself and be connected to the measurement instrument or control system. Alternately, a virtual TEDS can exist as a data file accessible by the measurement instrument or control system. A virtual TEDS extends the standardized TEDS to legacy sensors and applications where embedded memory may not be available [41].

10.6 User Integration

With the successful deployment of sensor nodes, supportive relays, and sophisticated sensor units (SSUs), the T-ISR system is readied for use and control by end users. Effectively distributing data to these users requires WSN-based systems to operate seamlessly, using legacy ISR systems and associated standards, such as MILSTD-2525D [42] symbology. Inventing another new format or display is counterproductive given the longstanding regimen associated with ISR analyst interaction with electronic and remote sensing ISR systems for over 50 years.

10.6.1 Legacy Integration

The U.S. military employs numerous COP systems to associate sensor observations within a GIS-based overlay. These systems have evolved from several

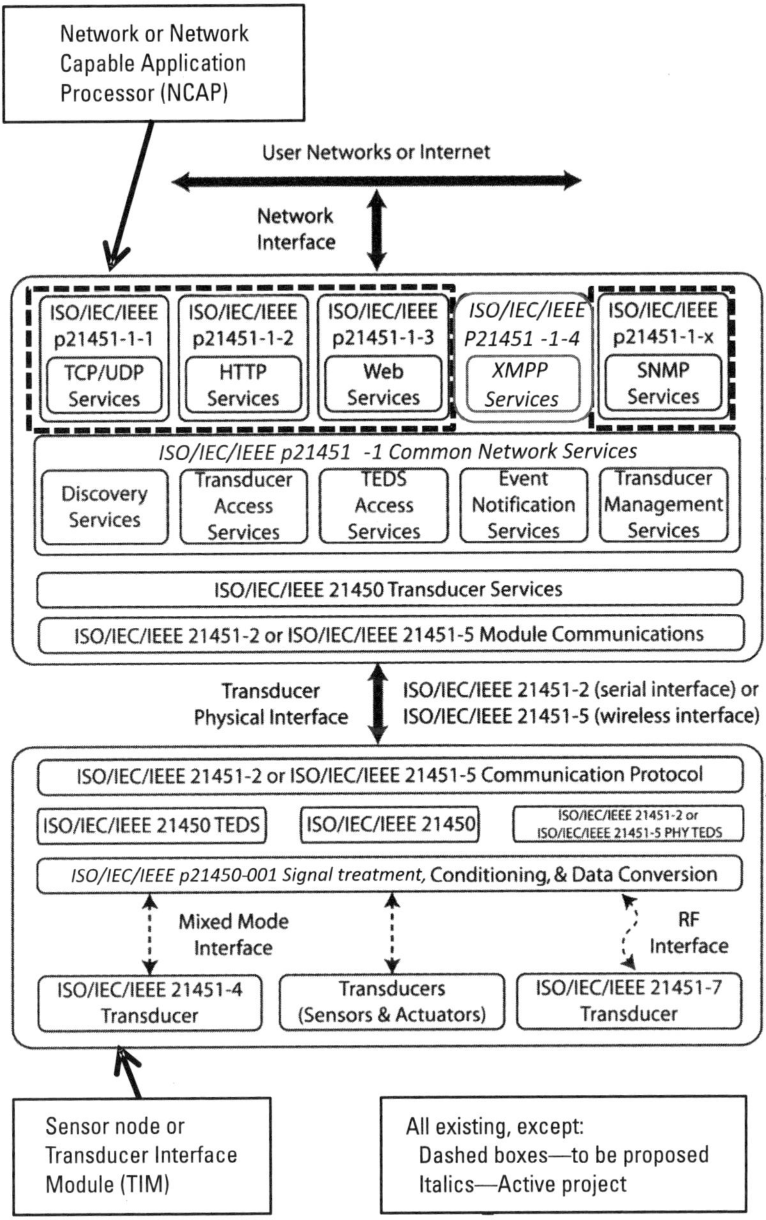

Figure 10.5 The IEEE 1451 model for sensor harmony standards.

years of operation and upgrades and have been augmented to handle new sensor platforms as they avail themselves, including WSN-based T-ISR systems. The code used to provide the necessary translation is referred to as injectors. The more common T-ISR displays used with WSN have been Northrop Grumman's Command and Control PC (C2PC) [43], Georgia Tech Research Institute's Falcon View [44], and MITRE's Cursor on Target (COT) [45]. There are innumerable derivatives to ISR displays, and depending upon the mission, variants to these displays have been developed to support ISR and INTEL analysts using T-ISR sensor data.

10.6.2 C2PC Common Operating Picture

The C2PC tactical display environment is Northrop Grumman's widely deployed PC-based product. C2PC facilitates creation and visualization of a common tactical picture (or COP). Using a familiar Microsoft Windows interface, C2PC provides a geographic map display and tactical picture with live tracking capability, integrated messaging services, and a range of planning and decision tools. C2PC also provides an open software foundation to build on or to integrate into other systems; it is easy to scale and provides rapid track promulgation and dissemination. C2PC's versatility is suitable for use in a wide range of operational environments from MOC to the battlefield. Associated integrated communications systems are efficient, allowing effective operation in low-bandwidth environments (e.g., Combat Net Radio and SATCOM) and are tolerant to unreliable connections. C2PC is comprised of three components: a global command and control system (GCCS) unified build (UB) host machine, the C2PC client, and the C2PC gateway. The UB host machine is the central source that feeds information to a C2PC gateway and associated C2PC clients. C2PC requires a UB host to receive automatic updates to information that is being tracked. The UB host provides overlays and operator notes (OPNOTES) that may be imported into C2PC. The UB host receives quick reports, OPNOTES, or overlays sent out by C2PC, and processes them appropriately, resulting in the Tactical Common Operational Picture.

The C2PC gateway component processes track information received from and sent to the UB host. The gateway must be up and running for the C2PC client to receive track updates. The workings of the gateway are transparent to the end user whereas the C2PC client component displays a map window and C2PC menu structure. Each C2PC workstation must run a C2PC client to run C2PC. Figure 10.6 shows the use of C2PC in the Simulation Experiment

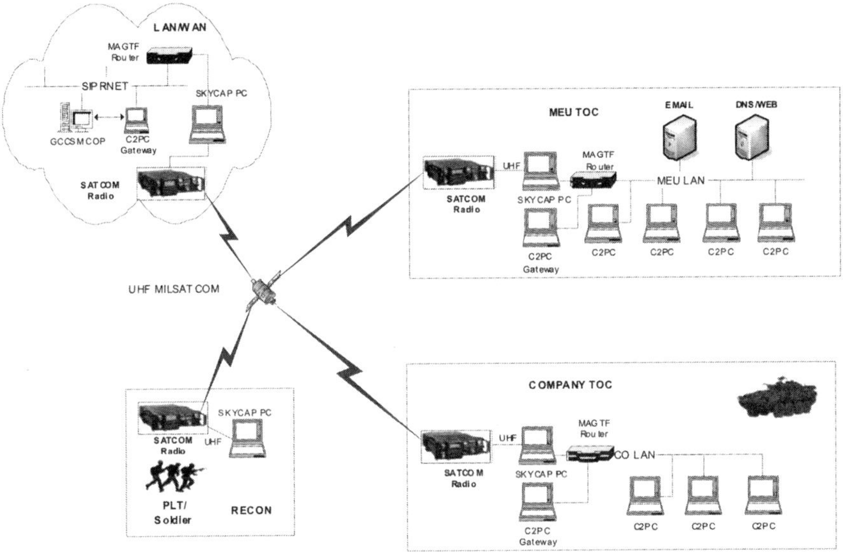

Figure 10.6 C2PC data flow.

(SIMEX) environment that includes the implementation of a Target Workflow System (TWS) client as a C2PC overlay [46]. TWS employs Atlas (a common application framework and tactical graphics rendering on various map views) for its GUI to display track icons and map data and Tactical Management System (TMS) API to acquire track data.

The C2PC display follows MILSTD-2525D symbology standards to present map overlays, friendly unit locations, sensor locations and metadata, projected plans of movement, and hostile unit locations. The versatility of C2PC is based on rapid information exchanges enabled by C2PC between staff sections, adjacent, subordinate and higher headquarters.

10.6.3 FalconView

FalconView is a Windows mapping system that displays various types of maps and geographically referenced overlays. FalconView, created by the Georgia Tech Research Institute, was initially developed for the Windows operating systems; however, versions for Linux and mobile operating systems have been developed. Many types of maps are supported, but the ones of interest to most users are aeronautical charts, satellite images and elevation maps.

FalconView supports a large number of overlay types that can be displayed over several map backgrounds with the typical overlay targeted toward military mission–planning users and oriented toward aviators and aviation support personnel. FalconView is an integral part of the Portable Flight Planning Software (PFPS) suite, which includes FalconView, Combat Flight Planning Software (CFPS), Combat Weapon Delivery Software (CWDS), Combat Air Drop Planning Software (CAPS), and several other software packages built by various software contractors [47]. Current work includes the development of FalconView as part of XPlan, the DoD's most recent mission planning system. The Joint Mission Planning System is also being added to FalconView as a plugin.

10.6.4 Cursor-on-Target

Cursor-on-target (COT) is a simple exchange standard used to share information about targets. COT is a loosely coupled design that has led to multiple implementations and is used to facilitate interoperability of several systems with fielded military software. COT was originally developed by MITRE in 2002 in support of the U.S. Air Force Electronic Systems Center (ESC) and was first demonstrated during a combined joint task force exercise in 2003, during which a Predator unmanned aircraft was able to operate in coordination with manned aircraft [40]. COT Event data model defines an XML data protocol for exchanging the time-sensitive position of moving objects among tactical COP systems. The COT data strategy is based on a terse XML schema and a set of extensions amenable to exchanging internet information via bandwidth-limited hardware. The state of the art for such communication exchanges involves the TCP/IP stack, optimized for sending small packets of information frequently. In contrast, systems that enjoy high-bandwidth capacity are operated economically by exchange of a rich amount of information versus frequent sparse information very often.

The Web Services protocol, and Web Services are more effective when systems exchange information for thousands of moving objects through a single Web Service call rather than relying on thousands of Web Service calls with individual moving object information. As a result, the COT data schema has been adapted to Web Service–based communication exchanges maintaining COT characteristics of terseness and extensibility [48].

References

[1] Coster, M., J. Chambers, and A. Brunck, "SCORPION II persistent surveillance system with universal gateway," *SPIE Proceedings Volume 7333, Unattended Ground, Sea, and Air Sensor Technologies and Applications XI*, 2009.

[2] "Commander's Handbook for Persistent Surveillance," Version 1.0, Joint Warfighting Center Joint Doctrine Support Division, 2011.

[3] Sharma, A., and P. Lakkadwala, "Performance Comparison of Reactive and Proactive Routing Protocols in Wireless Sensor Network," *IEEE Proceedings of 3rd International Conference on Reliability*, 2014.

[4] Remote Sensor Operations, USMC, MCRP 2-10A.5 (Formerly MCRP 2-24B), 2004.

[5] Adegoke, A., "Measurement of Propagation Loss in Trees at SHF Frequencies," Doctor of Philosophy Thesis, University of Leicester (Department of Engineering), 2014.

[6] Luomala, J., and I. Hakala, "Effects of Temperature and Humidity on Radio Signal Strength in Outdoor Wireless Sensor Networks," *IEEE Proceedings of the Federated Conference on Computer Science and Information Systems*, 2014.

[7] Wennerstrom, H., "Meteorological Impact and Transmission Errors in Outdoor Wireless Sensor Networks," Uppsula University (Department of Information Technology), 2013.

[8] Abdollahzadeh, S., and N. J. Navimipour, "Deployment Strategies in the Wireless Sensor Network: A Comprehensive Review," *Computer Communications*, 2016.

[9] Thakur, A., D. Prasad, and A. Verma, "Deployment Scheme in Wireless Sensor Network: A Review," *International Journal of Computer Applications*, 2017.

[10] Sharmaa, V., et al., "Deployment Schemes in Wireless Sensor Network to Achieve Blanket Coverage in Large-Scale Open Area: A Review," *Egyptian Informatics Journal*, 2016.

[11] McKitterick, J., "Sensor Deployment Planning for Unattended Ground Sensor Networks," *Proc. SPIE 5417, Unattended/Unmanned Ground, Ocean, and Air Sensor Technologies and Applications VI*, 2004.

[12] Boudriga, N., "On a Controlled Random Deployment WSN-Based Monitoring System Allowing Fault Detection and Replacement," *International Journal of Distributed Sensor Networks*, 2014.

[13] "Advanced Remote Ground Unattended Sensor (ARGUS)," GlobalSecurity.org, https://www.globalsecurity.org/intell/systems/arguss.htm.

[14] ENSCO, "Sensor Modeling & Mission Planning Tools," ENSCO, 2020.

[15] Sitanayah, L., K. N. Brown, and C. J. Sreenan, "Fault-Tolerant Relay Deployment for k Node-Disjoint Paths in Wireless Sensor Networks," 2011 IFIP Wireless Days (WD), 2011.

[16] Cheng, W., et al., "Regular Deployment of Wireless Sensors to Achieve Connectivity and Information Coverage," *Sensors*, 2016.

[17] Li, H., et al., "A Reverse Gaussian Deployment Strategy for Intrusion Detection in Wireless Sensor Networks," *2012 IEEE International Conference on Communications (ICC)*, 2012.

[18] Vales-Alonsoa, J., et al., "On the Optimal Random Deployment of Wireless Sensor Networks in Nonhomogeneous Scenarios," *Ad Hoc Networks*, 2013.

[19] Brusey, J., E. Gaura, and R. Hazelden, "WSN Deployments: Designing with Patterns," *IEEE Sensors*, 2011.

[20] Holler, R., "The Evolution of the Sonobuoy from World War II to the Cold War," *U.S. Navy Journal of Underwater Acoustics*, 2014.

[21] Correll, J., "Igloo White," *Air Force Magazine*, 2004.

[22] Bernard, M., et al., "Autonomous Transportation and Deployment with Aerial Robots for Search and Rescue Missions," *Journal of Field Robotics*, 2011.

[23] Smet, T., and A. Nikulin, "Catching 'Butterflies' in the Morning: A New Methodology for Rapid Detection of Aerially Deployed Plastic Land Mines from UAVs," *The Leading Edge*, 2018.

[24] Corke, P., et al., "Autonomous Deployment And Repair of a Sensor Network Using an Unmanned Aerial Vehicle," *Proceedings IEEE International Conference on Robotics and Automation*, 2004.

[25] Bellavista, P., et al., "Convergence of MANET and WSN in IoT Urban Scenarios," *IEEE Sensors Journal*, 2013.

[26] Wan, C., A. T. Campbell, and L. Krishnamurthy, "Pump-Slowly, Fetch-Quickly (PSFQ): A Reliable Transport Protocol for Sensor Networks," *IEEE Journal on Selected Areas in Communications*, 2005.

[27] "Open_Geospatial_Consortium," https://en.wikipedia.org/wiki/Open_Geospatial_Consortium.

[28] "OGC's Role in the Spatial Standards World: An Open GIS Consortium (OGC)," White Paper; https://www.ogc.org.

[29] "Open_Geospatial_Consortium Standards," https://www.ogc.org/standards.

[30] Delin, K., P. Shannon, and P. Jackson, "Sensor Web: A New Instrument Concept," *SPIE Proceedings Vol. 4284: Functional Integration of Opto-Electro-Mechanical Devices and Systems*, 2001.

[31] Rouached, M., S. Baccar, and M. Abid, "Restful Sensor Web Enablement Services for Wireless Sensor Networks," *2012 IEEE Eighth World Congress on Services*, 2012.

[32] Botts, M., et al., "OGC Sensor Web Enablement: Overview and High Level Architecture," *GeoSensor Networks*, 2006.

[33] Broring, A., et al., "New Generation Sensor Web Enablement," *Sensors*, 2011.

[34] Thompson, S., J. Kastanowski, and S. Fairgrieve, "PULSENet," *IEEE Computer Society Proceedings MILCOM 2006*, 2006.

[35] Fairgrieve, S., J. A. Makuch, and S. R. Falke, "PULSENet™: An Implementation of Sensor Web Standards," *IEEE 2009 International Symposium on Collaborative Technologies and Systems*, 2009.

[36] Halili, F., and E. Ramadani, "Web Services: A Comparison of Soap and Rest Services," *Modern Applied Science*, 2018.

[37] Avilés-López, E., and J. A. García-Macías, "TinySOA: A Service-Oriented Architecture for Wireless Sensor Networks," *Service-Oriented Computing and Applications*, 2009.

[38] Chen, G., et al., "Services-Oriented Architecture (SOA)-Based Persistent ISR Simulation System," *Proc. SPIE 7694, Ground/Air MultiSensor Interoperability, Integration, and Networking for Persistent ISR*, 2010.

[39] Song, E., and K. Lee, "Integration of IEEE 1451 Smart Transducers and OGC-SWE Using STWS," *2009 IEEE Sensors Applications Symposium*, 2009.

[40] "Universal Core (UCORE)," https://en.wikipedia.org/wiki/Universal_Core.

[41] "IEEE Standard for a Smart Transducer Interface for Sensors and Actuators Wireless Communication Protocols and Transducer Electronic Data Sheet (TEDS) Formats," in IEEE Std 1451.5-2007, 2007.

[42] "Department of Defense Interface Standard Joint Military Symbology, MIL-STD-2525D," MIL-STD-2525C, 2014.

[43] "Northrop Grumman Brochure on C2PC," https://www.militarysystems-tech.com/sites/militarysystems/files/supplier_docs//NG%20MS%20C2PC-Tactical%20DS.pdf.

[44] FalconView," https://en.wikipedia.org/wiki/FalconView.

[45] Kristan, M., et al., "Cursor-on-Target Message Router User's Guide," MP090284 MITRE PRODUCT, 2009.

[46] Parks, E., Integrating the Target Workflow System (TWS) with the Command, Contract No.: DAAB07-98-C-C201, Dept. No.: W407 Project No.: V17A, 2003.

[47] National Geospatial-Intelligence Agency (NGA), FalconView, https://www.nga.mil/ProductsServices/Pages/-FalconView.aspx.

[48] Konstantopoulos, D., and J. Johnston, J., "2006 CCRTS: The State of the Art and the State of the Practice," Data Schemas for Net-Centric Situational Awareness, MITRE report, 2006.

11

WSN Application to T-ISR

This book began by addressing the history, objectives, and definitions associated with remote sensing as it pertains to ISR. ISR systems, such as Chain Home and the Vietnam-era unattended ground systems (UGSs) were reviewed, as these systems are vanguards for the approach and development of T-ISR remote sensors. Each previous system provided lessons regarding the difficulties associated with remote sensing, with several of these difficulties persisting regardless of underlying technology development. With adaptation of WSN systems to address T-ISR missions, ongoing challenges such as data volume, radio connectivity, and sensor reliability join the new issues stemming from low cost, size, and power, which necessitate compromises to resilient design, qualification testing, and system complexity. In addition, with a WSN-based system, disbursed nodes operate at or near ground level, severely limiting reliable wireless and sensor performance range. To address these issues, a series of programs sponsored by DARPA conducted significant research, development, and testing that has drastically accelerated WSN maturity to respond to functions associated with T-ISR.

In discussing WSN, the book deconstructs WSN to the elemental subsystems and related technologies (see Figure 11.1) and presents the theory and application of each WSN topic along with design examples. This chapter reassembles these elements and discusses how WSN systems opperate based on testing and demonstrations within realistic T-ISR scenarios. In addition,

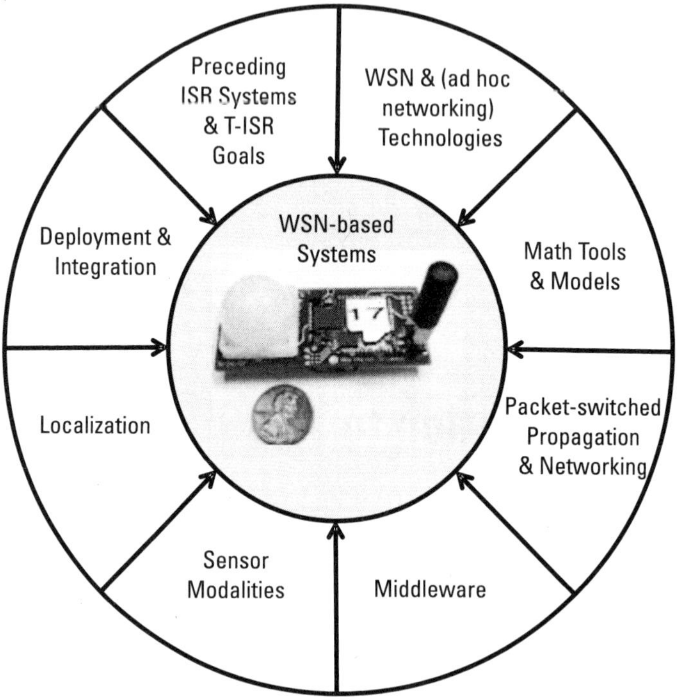

Figure 11.1 WSN topics: convergence toward actual T-ISR system designs.

the chapter discusses lessons learned from previous WSN T-ISR tests and demonstrations, along with established capabilities. The critical evaluation events presented in the chapter focus on aspects of WSN technologies illustrated in Figure 11.1.

11.1 Conceptualizing the Use of WSN for Military Applications

Research on distributed sensor networks began in earnest in 1978 with the Distributed Sensor Net (DSN) program sponsored by DARPA. Under DSN, the RAND Corporation researched core T-ISR functions with the objective of developing and evaluating effective target surveillance and tracking using spatially distributed but interconnected sensors and processing resources [1, 2]. DSN involved teams from the University of Michigan (UMI), University of California-Los Angeles (UCLA), and MIT/Lincoln Laboratories during the 1980s.

In 1992, a conceptual workshop convened at RAND [3], which resulted in the formalization of what was to become known as WSN. In concert with these efforts, several concepts were identified by the University of California-Berkeley (UCB) eventually funded by DARPA in a program known as Smart Dust, which delved into the adaptation of MEMS technology to provide sensor capabilities and communication solutions for extremely small sensors (dust-mote-sized) that could form an ad hoc network and operate as a collaborative entity. In parallel, the DARPA Wireless Integrated Network Sensors (WINS) program, initiated in 1993 at UCLA, produced the first generation of WSN devices and software to be field-tested in 1996. WINS proved the feasibility of multihop, self-assembled, wireless networks and was the first effort to demonstrate algorithms for operating wireless sensor nodes and networks at micropower levels. UCLA (with Rockwell Science Center of Thousand Oaks, California) developed a modular development platform to enable the evaluation of more sophisticated networking and signal processing algorithms required for various sensors. Results from this effort underscore the importance of separating low-power real-time functions from higher-level functions that require extensive software development [4].

Throughout the late 1990s, WSN research gained momentum at various laboratories, universities, and military organizations. In 1998, DARPA funded the SensIT program [5], which focused on middleware regarding network management functions and node-based processing of sensor data. In 1998, the U.S. Air Force research Laboratory (AFRL, Kirkland Florida) funded a program with UCB called PicoRadio [6, 7] to address the SoC implementation [3] of a sensor node capable of ad hoc communication networking, onboard computation, and geolocation. The PicoRadio program embodied key principles of DARPA initiatives for advanced computing, including adaptive computing and power-awareness, leading to the design, fabrication, and testing of low-power, low-cost radio transceiver chipsets [8].

In 1999, UCB responded to a DARPA business announcement (BAA 99-07) with an initiative, "The Endeavour Expedition: Charting the Fluid Information Utility." This activity addressed the development of RTOSs for sensor nodes and by 2000, resulted in a RTOS solution, TinyOS. TinyOS minimized the use of processing resources (i.e., machine cycles, memory, and processor activity) and strived to preclude the introduction of coding errors [9]. Built using a newly developed C-language, denoted as the nesC language, produced TinyOS version 1.0, which was refined throughout the next few years with versions 1.X (2002–2005) and 2.X (2005 to present). TinyOS was the go-to RTOS for motes during the WSN developmental decade 2000–2010 along alternate RTOSs, Contiki and Arduino.

11.2 I&T of WSN Systems

With requisite technologies, middleware development, and network processing for sensor node design achieving reasonable maturity, evaluation and verification of WSN capabilities became the focus of DARPA network sensor research. With results from WINS, Smart Dust, PicoRadio, and SensIT efforts, the question became, what could be accomplished using commercial-off-the-shelf (COTS) components, real-time algorithms, and middleware implementation to answer T-ISR problems?

11.2.1 DARPA Smart Dust: The 29-Palms Demonstrations

In March 2000, a proof-of-concept experiment was conducted by UCB, with the support of the USMC. A mote-based T-ISR system was designed to detect vehicle activity at an isolated desert intersection near Palm Springs (California). This demonstration employed the Rene mote platform with a two-axis Honeywell (HMC1002) magnetometer to detect vehicular activity on a roadway. Six motes were airdropped to a location approximately 20m from the road on 5-m centers. The motes were deployed aerially, using a UAV flying a GPS route [11]. The 5-ft-wingspan UAV, equipped with cameras, was preconfigured with a multipoint flight plan that included a low altitude to deploy motes along the road [12]. Once deployed, the motes formed a multihop network using 916.5-MHz ISM-band transceivers with simple on-off keying (OOK) modulation. Internal clocks were synchronized to allow correlation of discrete sensor observations obtained by each node. As large test vehicles (e.g., HMMWVs and Dragon Wagons) drove past sensor nodes, onboard magnetometers detected deviations in the magnetic field caused by each vehicle and determined the closest point of approach for each vehicle. Detection events were time-tagged as a vehicle event and communicated to neighboring nodes. The UAV returned to the motefield and queried the motefield. Data was transmitted to the UAV, which acted as an exfiltration relay. Upon landing at the operations center, data was downloaded off the UAV for subsequent processing and display.

The UCB 29-Palms experiment revealed the advantage of ad hoc network autonomy and data volume handling by providing in-network processing to support network capacity and system power management. Each individual sensor node processed data, which resulted with reduced network congestion and transmission times by individual nodes. The distributed processing balanced data loading across the network, which also conserved motefield power. Of note is that the RTOS (TinyOS) would be continually refined to lower power consumption during the test period. With these modifications, the revised TinyOS successfully halved total power consumption by the conclusion of the demonstration [11].

11.2.2 DARPA: A Line in the Sand Demonstrations

Having proven the potential for WSN technology to implement T-ISR functions [13, 14], DARPA initiated the NEST project in 2001 to coalesce the lessons learned and to extend research into the aspects that showed potential. In 2003, under the NEST program, OSU conducted a relevant demonstration using 90 motes (UCB MICA2 platform) at MacDill Air Force Base (Tampa, Florida) that successfully conducted T-ISR functions of detection, tracking, and classification of various target types of interest; namely, unarmed personnel, armed personnel (defined as, steel being present), and vehicles. For this demonstration, the motes were hand-emplaced and employed two sensor modalities, an onboard radar (Advantaca TWR-ISM-002 pulse-Doppler) and magnetometer.

This test, Line in the Sand, emphasized the versatility of WSN motes to successfully set up a tripwire power-saving configuration capable of target classification [15]. Furthermore, this demonstration addressed a number of concerns, such as real-time data compression, information exfiltration, and network management. Nodes preprocessed sensor signals with final target classification processing achieved at the MOC. A successful classification process was achieved through the implementation of a concept based on the definition of an influence field as a spatial statistic. Through the use of the OSU influence construct, the deployed motefield successfully realized target discrimination, even in the presence of node failures and network unreliability [16].

11.2.2.1 ExSCAL Demonstrations

With previous demonstrations, motefields were scaled for testing hardware, not sized to reflect what was required to support actual T-ISR missions. By 2004, DARPA now turned toward proving a capability required by military users: scalability. DARPA initiated Project Extreme Scale (ExSCAL) under NEST, with an objective to configure a 10,000-mote system capable of operating as an autonomous network to detect, track, and classify multiple intruders of different types within a 10 km x 1 km monitored area (AOI) [17]. In December 2004, OSU began by fielding a multinetwork (tiered) consisting of approximately 1,000 specially designed motes covering an area of 1.3 km x 300m [Avon Park Air Force Range (APAFR)]. This motefield was constructed by interconnecting multiple 200-node, peer-to-peer, ad hoc networks.

The mote design employed in this ExSCAL test was specifically designed with a focus on long-lived persistent vigilance through the use of passive sensor modalities. At this time, no widely available experimental mote platform supported the performance required. Additionally, a sensor node tailored to ExSCAL had to be imbued with characteristics that readily addressed ExSCAL objectives. This included sensor and RF coverage, ability to provide identification of various targets, and power-management capabilities such as software-controlled ADC reference voltages. This specialized design also addressed the use of an isotropic antenna design, over-the-air (multihop) programmability, insertion of a watchdog (grenade) timer, and physical packaging designed for easy setup. This OSU design, manufactured by Crossbow beginning in 2004, became known as the eXtreme Scale Mote (XSM100CB). The XSM mote (later, XSM2) was designed to be compatible with the MICA2 by employing the same Atmega 128L processor.

Figure 11.2 presents the XSM mote (top) and XSM circuit board (bottom) with T-ISR modalities indicated: acoustic, magnetometer, and four PIR detectors. The XSM mote used a transceiver tuned to 433±25 MHz, and was powered by 2AA batteries (approximately 6,000-mW-hr). The XSM design incorporated hardware/software supportive of code-error recovery through the implementation of a grenade timer circuit. As discussed previously, a grenade timer guarantees that if a processor loses control, the bootloader would eventually be provided the means to regain control when a predetermined timeout occurs. By properly setting the processor fusemap, the bootloader, interrupt vectors, and handlers would be protected from application code; a situation not afforded by TinyOS. How the bootloader is able to recognize obsolete or erroneous boot images and perform recovery is detailed in XSM design literature [18].

To support large-scale ExSCAL motefields, the WSN architecture operated as a three-tiered system. The fundamental tier was composed of XSM nodes. A second tier consisted of a Linux-based gateway and GPS receiver [denoted the eXtreme Scale Stargate (XSS)], also designed by OSU and operated with 20–50 XSM nodes each. A third and final tier was the master base station (i.e., exfiltration point or AP). The ExSCAL motefield was managed using two distinct application management approaches. At the relay and gateway (XSS) tiers, a multitier command-and-control framework enabled an operator to perform management operations from the base station (relay). Commands were disseminated using a service over the tier-2 network to a

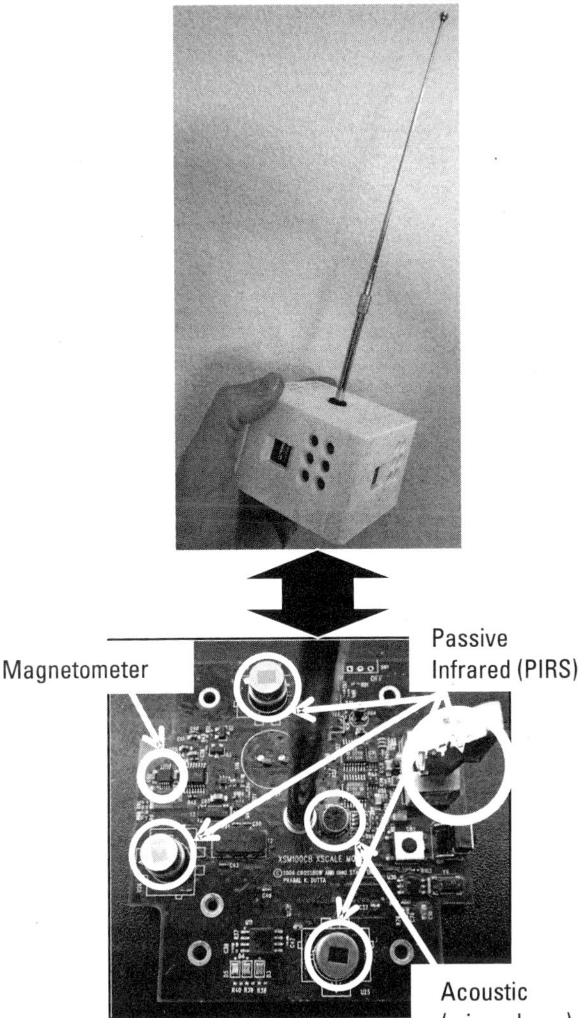

Figure 11.2 The XSM, with the main sensor modalities indicated.

management daemon that ran on each XSS gateway. Based on the command type and scoping rules, XSS gateways would either locally execute the command or invoke a Tier-1 management process. At the fundamental XSM mote level, autonomous management existed with individual nodes allowing access to local middleware services to detect and correct low-level faults and to maintain node operability. From multiple ExSCAL demonstrations, OSU was able to estimate ExSCAL success and reliability performance. The end-to-end routing yield at the tier 1 level was estimated as 86.72% of the field, while for tier 2, the yield was 98.32%. Deployment faults were responsible for a 5.37% failure rate. Localization faults were observed (11.4%) as well as reprogramming faults (5.5%), with both uniformly distributed across the XSM networks. Protocol design was capable of tolerating faults, and with policy-based managers identifying and resolving nodes that experienced contract-violation faults, it produced an overall tier 1-to- tier 2 reliability estimate of approximately 73% [17].

Under the auspices of the NEST program, DARPA tasked the University of Virginia (UVA) to build and test an integrated sensor network system to perform a surveillance mission emphasizing energy-efficiency and stealth. The result was the VigilNet system, an event-driven system designed initially to operate using MICA2 motes [19]. The premise of the VigilNet design was to characterize the tactical surveillance of target events as rare. This translates into the expectation of long periods with a reduced need for active sensing, processing, or messaging to occur. However, during the rare event of targets being present, intense data collection, processing, and messaging are required and take precedence over energy conservation. To embrace such an energy profile associated with surveillance, the VigilNet design incorporated the concept of a tripwire and sentry configuration to extend the system lifetime.

The partitioning of a motefield into tripwire partitions was defined from the mission-driven layout of the motefield and the average number of hops for each node-to-relay. These partitions would be divided into active and dormant sections, with dormant sections motes sequenced to a power-saving mode (sleep) whereas active sections had all motes actively operating. Multiple partitioning approaches exist within VigilNet based on estimating high-quality links by considering minimal hop counts or distance to the connected relay. The sentry configuration used a two-phase process, beginning with the determination of single-hop nearest neighbors followed by a second phase where each node sets a timer value based on a weighing of node energy reading and coverage rank, which is the number of neighboring nodes within a node's sensing range. Using the sentry configuration meant that a percentage of the motes were reserved to act as sentries (bellringers). While the preponderance of motes in a segment would be placed into low-power (sleep) mode, sentries would remain vigilant and ready to trigger a wake-up message to all motes upon sensing a target event. VigilNet could be configured to operate with tripwires, sentries, or a combination. To test VigilNet, UVA ported VigilNet to XSM motes and conducted persistent surveillance tests using 200 motes at Avon Park area. Results indicated that VigilNet, employing the energy conservation approach of tripwire and sentries, provided a motefield with significant energy savings when employing all power saving techniques; extending an initial duration from four days to 200 days [20].

During December 2004, a pivotal demonstration defined as a Clean Point event was conducted to demarcate the end of the NEST program's research-intensive objectives to shift the transfer of NEST technology to address issues associated with military users. With the emphasis on research ending and the verification and validation (V&V) process beginning, DARPA contracted the NGIT TASC division to conduct the V&V task to support the technology transfer to USSOCOM and the DIA under the Extreme Scale Adaptive Network Technologies (ExANT) project. In parallel, USSOCOM and DIA jointly contracted NGIT to conduct the Adaptive Netted Sensor Collection & Dissemination (ANSCD) Program Project 2888 to investigate how NEST technology could be operated within existing T-ISR systems. ANSCD was to consider how best to integrate a WSN-based system with existing SSUs and other T-ISR sensors, such as a laser vibrometer sensor system (LVSS) designed to uniquely identify targets through their vibration signature [21].

Figure 11.3 summarizes the ExSCAL/ExANT/ANSCD timeline associated with the NEST program, showing WSN development periods, test events, and major demonstrations. Figure 11.3 illustrates the overall progression of the DARPA NEST as it evolved from basic R&D programs, through the development of ExSCAL OSU and UVA contributions, to follow-up efforts that collectively address the transfer of the NEST technology to military end users. Abbreviations used in Figure 11.3 are defined as follows:

- FA: Field assessment (five in total across multiple DARPA programs);
- C2PC: NGIT development of NEST injector capability to connect ExANT and ANSCD motefield data to the USMC C2PC [22] ISR display capability;
- ExA: The NGIT ExANT program event;
- I&T: The initial integration and testing of ExSCAL technology for use by the ExANT program.

11.2.2.2 EXANT and ANSCD Field Testing

Throughout 2005–2007, operational evaluations and assessments were conducted by NGIT in the field. Over this two-year period, ExANT used the Green Swamp (Florida) area for large-scale (up to 4,000-mote) testing, and an area near the Moody USAF base area (Avon Park, FL) for smaller

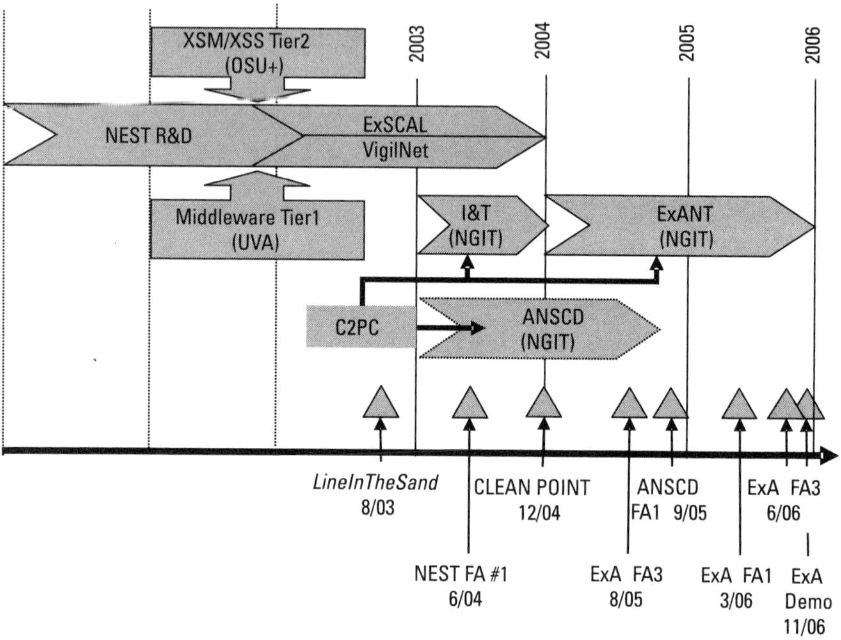

Figure 11.3 Timeline of NEST WSN programs: R&D to ExSCAL to ExANT.

(100–200-mote) tactical tests. Both test areas were considered representative of the environments end users of T-ISR mote systems would have to contend with.

At the Avon Park location, unexpected RF characteristics due to local soil composition conspired to limit near-ground RF propagation. Even commercial VHF communications (49–108 MHz) connectivity at this test site was intermittent; and in particular areas, GPS L-band operation; was sporadic even though the particular locations had access to open skies. Heavy rains, thick vegetation, and abundant animal life continually disrupted testing at both test sites. Although these test areas represented the intended use case, XSM motes were designed for research, not for harsh environments, where heavy rain, humidity, and extreme thermal variations occur. The attrition rate of motes increased, and a requirement to set up and tear down the motefield became an excessively time-consuming activity. To reduce manpower and time spent on large motefield set-up, maintenance, and shutdown, the ExANT large field-testing was relocated to a large (> 18,000-sq.-ft.) warehouse building by late 2006. Despite the field test setbacks, verification of both middleware

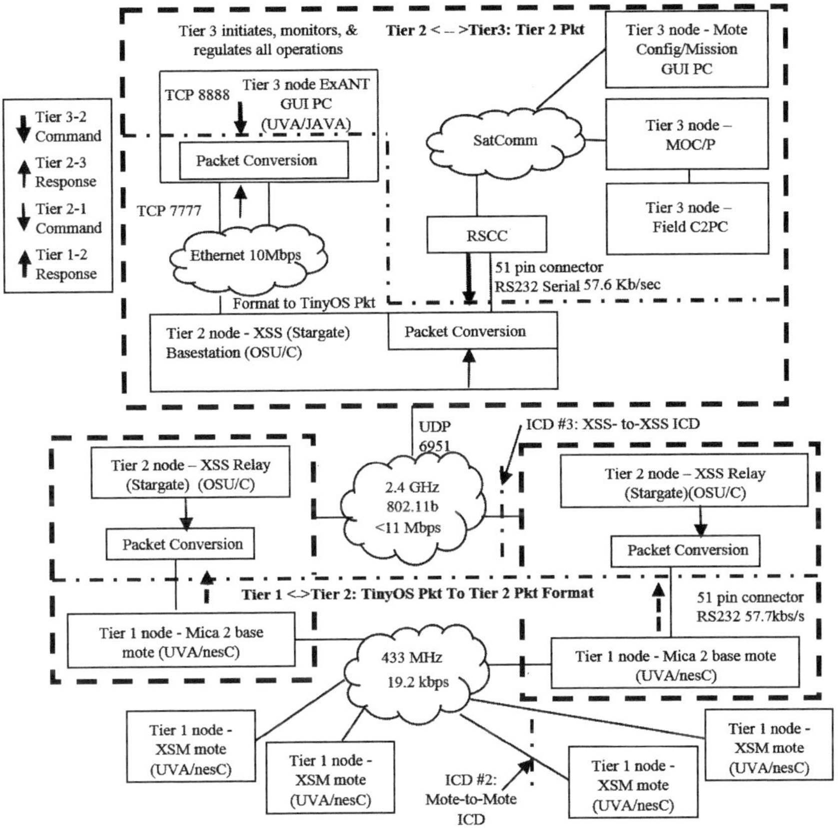

Figure 11.4 ExANT System block diagram (and ICD definitions).

and node capability were conducted, and measurements of reliability and discrepancies logged.

Figure 11.4 presents the block diagram employed for the large-scale (> 1,000 motes) EXANT test events. Shown are the three tier levels used for these large motefields with the top tier being the base-station (relay), the middle gateways (XSS), and the bottom tier (tier 1) using XSM motes (XSM). Also shown in Figure 11.4 (upper right) is the network hardware used to support WSN-based T-ISR through the USSOCOM architecture. MOC/P represents an impromptu operations center and processing and RSCC refers to a Remote Sensor Command & Control, which is a ground-to-satellite duplex link connecting USSOCOM field equipment to the worldwide network (e.g.,

DODIN). In addition to testing, ExANT required NGIT to modularize middleware code for submission into a DARPA-controlled code repository for availability to other approved WSN efforts, which was derived from both OSU and UVA ExSCAL efforts.

For verification testing, the relay (referred to as the base station or RSCC) transferred data from and commands to the sensor field directly from a remote operations control computer (laptop) using the DoDIN net (which employed an ISR satellite communications, via the Iridium SATCOM system). Target data extracted from a test motefield would be presented to T-ISR mission operators using a standard Intelligence Operations Workstation (IOW) that employed the USMC Command and Control Personal Computer (C2PC) display, complete with MIL-STD-2525D symbology. Control of the motefield

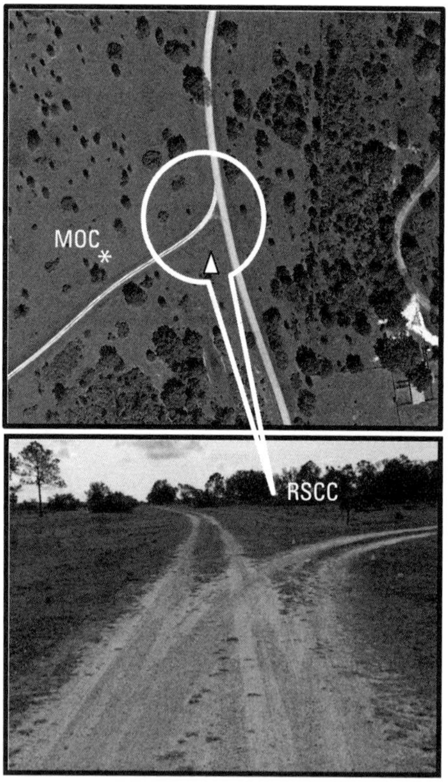

Figure 11.5 Avon Park test area. The asterisk marks denote the operations center and the triangle marks the motefield relay location.

would be handled using this same connectivity. At Avon Park, ANSCD test configurations stretched kilometers along a remote roadway and included a Y-intersection. Figure 11.5 presents aerial and ground views of the Y-intersection test area (Austin Hammock location, within APAFR). Beyond having terrain and conditions expected for T-ISR system operations, this area was selected because it provided a reasonably large AOI, challenging terrain, and characteristics considered typical of areas that a T-ISR capable of tracking targets might be required to operate within. With the Y-intersection, evaluation of tracking latency could be performed with vehicle targets continuing using two different directions (the bottom photo in Figure 11.5).

During the ANSCD project, NGIT modified the XSM mote design to: (1) decrease visual presence in alignment with ISR sensors used by USSOCOM, (2) increase RF range through redesign of transceiver subsystem, and (3) increase the capability of motes to operate in harsh (wet) environments. The result was the Tactical-XSM (TXSM), which eliminated the reflective 15.5-inch XSM antenna, reduced visibility associated with the XSM ($3.5 \times 3.5 \times 3$ inch^3) white cube via camouflage skin covering, and modified the RF circuitry by replacing the dipole with a more compact PCB antenna and stage. Figure 11.6 compares the physical appearance of the XSM and TXSM (upper photo). The use of TXSM for smaller (<200 motes) ANSCD tests adapted extremely well with the test area (lower photo of Figure 11.6).

Several scenarios were evaluated with the ANSCD field (Figure 11.5). A combination of targets (unarmed personnel, armed personnel, and vehicles) was used with each test sequence. During initialization phases, middleware would automatically determine those motes to be placed into a low-energy mode, powered down, or operated in sentry mode. In addition to testing motefields, ANSCD field tests were performed to demonstrate interoperability between the motefield and USSOCOM SSU (sensors). TXSM motes were configured to perform as bellringers to alert and activate the more energy-demanding SSU camera systems. Tests were conducted where vehicles would enter the monitored roadway, and through the motefield, would be detected, tracked, and classified.

At detection, the motefield would also alert SSU cameras (or seismic/acoustic sensors) to power up and acquire visible and IR video of the targets. The RSCC system segment would route mote and SSU information (including imagery data) to the MOC segment using an ISR SATCOM. Figure 11.7 presents the end-to-end architecture used in this SSU-cueing test. Within Figure 11.7 are photos of two SSU cameras employed during this testing (visible and infrared). The architecture was segmented as the fielded sensors (motefield

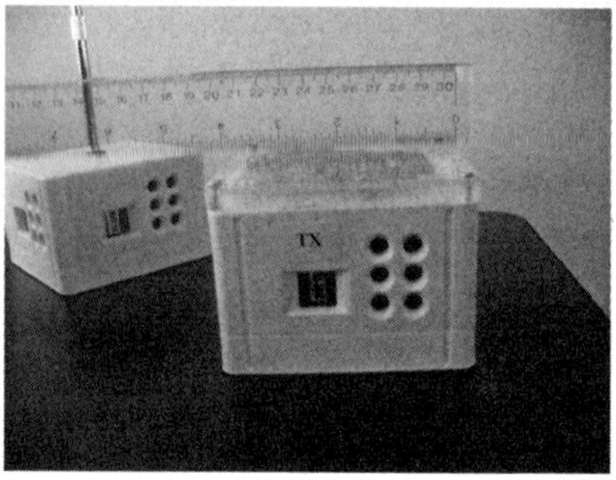

(a)
XSM background, and TXSM foreground;
with PA and antenna mods (within Lucite area)

(b)
TXSM, deployed in test area

Figure 11.6 TXSM, lower visibility, compact antenna significantly increase RF range (link reliability) (a) The original XSM2 (background) with a redesigned T-XSM (foreground), and (b) the T-XSM in the field.

plus SSUs), the global RSCC (relay), the ground data routing segment, and the mission operations segment (MOC).

Not unexpectedly, deployed motes would fail in significantly different ways. Lessons learned in performing ANSCD and ExANT field-testing developed practical debugging procedures to uncover failed motes and/or links. In addition, diagnostic tools were devised to rapidly identify mote failures. Figure 11.8 presents two monitoring tools successfully used during WSN testing. A tower-mounted receiver would intercept RF messages, identify the sending mote (Mote ID), and generate a histogram of messages sent per mote per fixed period of time. Figure 11.8(a) shows the resulting graphic of

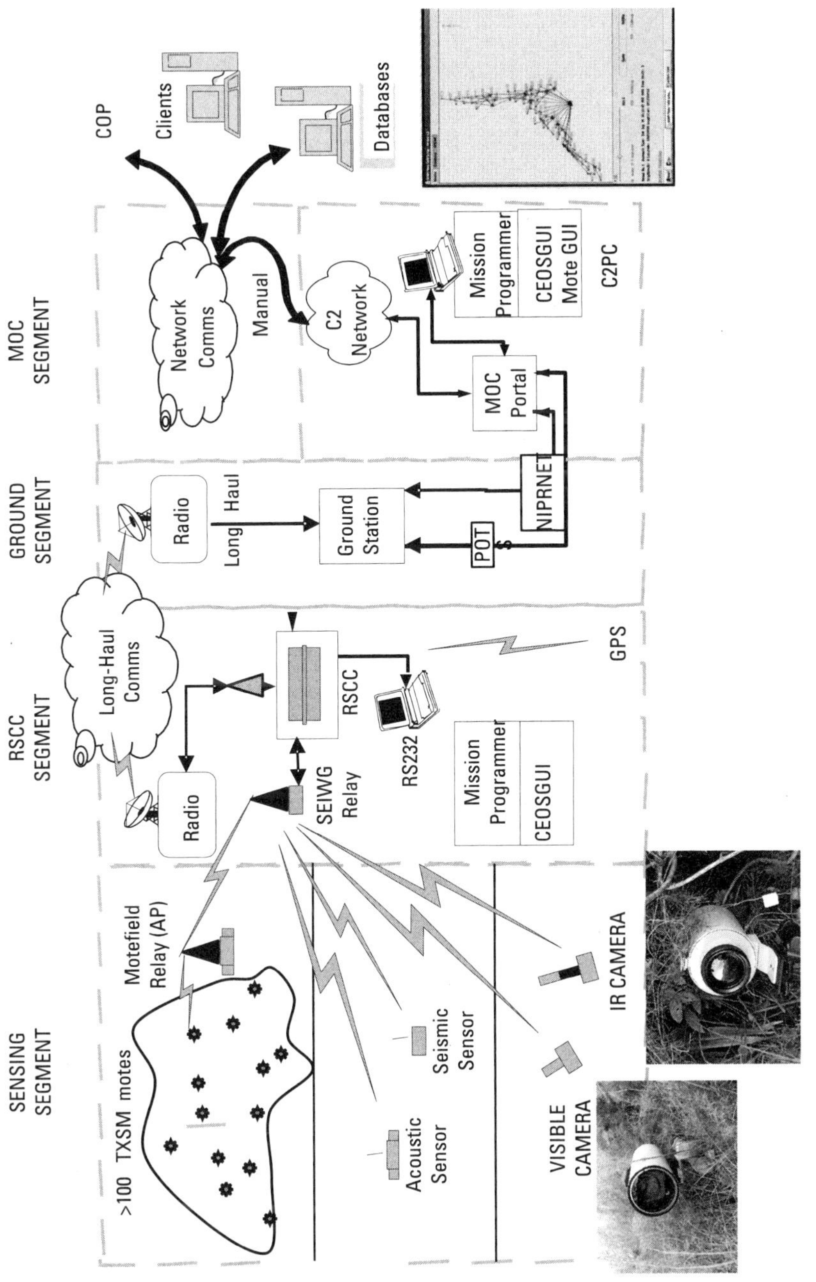

Figure 11.7 End-to-end ExANT/ANSCD test architecture, cueing SSU devices

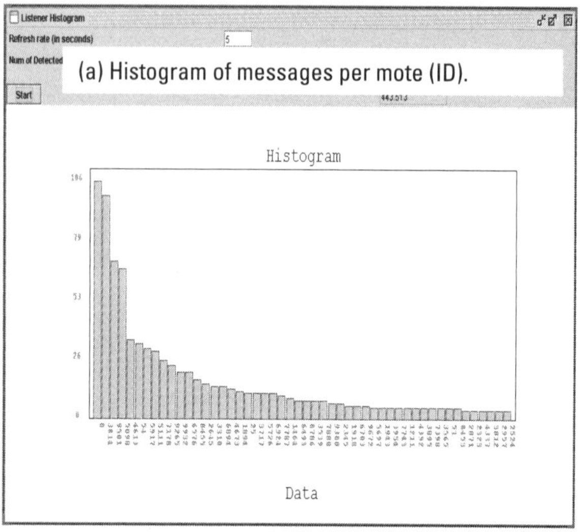

(b) TinyOS tool, network initialization phase.

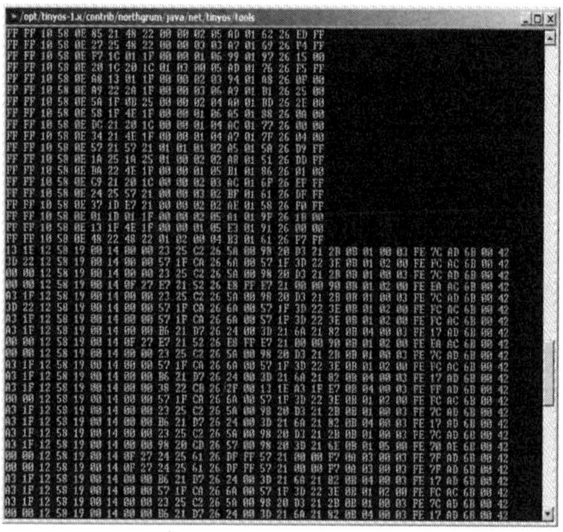

Figure 11.8 Monitoring stations during testing: (a) message/node histogram and (b) initialization progress (TinyOS Tool).

messages binned by mote. Using this tool, if a mote would successfully join the network but begin generating superfluous message traffic, this graphic display would indicate so. Self-healing would be allowed to try to resolve the issue. If unsuccessful, a command would be sent to power the mote off, and if that action failed, the mote would be located by its ID and manually turned off. These talker motes would disrupt the formation of the network by monopolizing the links with useless message traffic, creating congestion such as that seen during DoS attacks.

Motes that failed to power up were easiest to handle as they did not disrupt or jam the network; they simply failed to join the network (or provide any useful sensor information). Figure 11.8(b) shows a TinyOS tool that depicts the initialization progress by middleware phase for each mote. A third failure mode observed during these tests consisted of motes that tuned on, began to transmit, but did not join the network. These motes would transmit on the 433-MHz ISM channel, jamming the nearest neighbors and disrupting the RF links. These jammer motes were located with a portable RF spectral analyzer used to range-triangulate on the offending mote. Figure 11.9 depicts the receiver tower used to intercept network messages. The receiver was a mote with diagnostic software loaded to capture net traffic and produce a variety of displays useful in finding and resolving mote failures.

11.2.2.3 ExANT Large-Scale Demonstrations

With large-scale testing relocated to an 18,000-ft enclosed area, modifications and development to ExANT middleware development accelerated. In June 2007, the ExANT project was ending, and a final series of tests and demonstrations were conducted. Within the enclosure, a large (>2,400 motes) motefield was established using 1x1-m spacing on floors and walls to maximize mote count. Figure 11.10 presents the large test area used. The photo in Figure 11.10(a) shows one of the large areas where XSM motes are aligned in arrays on the floor and the walls. Figure 11.10(b) is a graphic that shows the two main areas, an external smaller linear motefield consisting of 69 motes, and a SOF camera system (SSU) used to demonstrate an IED scenario. In

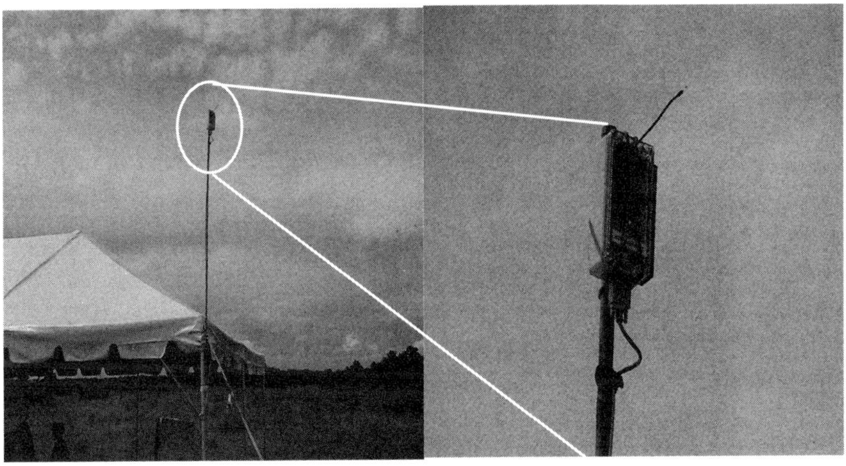

Figure 11.9 Monitoring stations during testing; field listening mote (tower).

(a)
Photo of enclosed test area

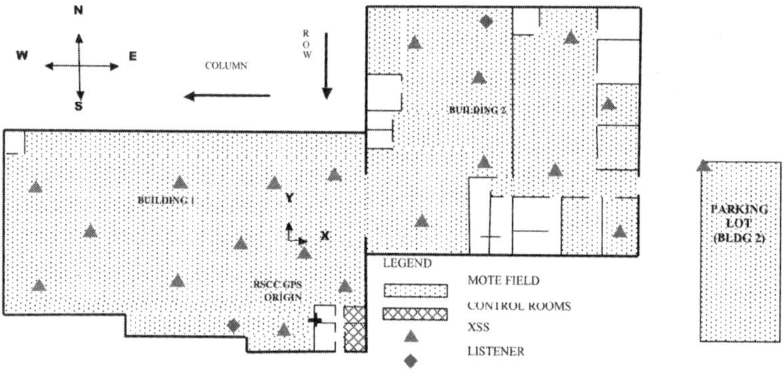

(b)
Graphic of test setup, including the two mote fields, and test/system components.

Figure 11.10 ExANT final 2400-mote demonstration layout: (a) photo of test area and (b) graphic of test areas 1 & 2 (parking lot IED simulation).

Figure 11.10, the triangular icons indicate the tier 2 gateways used to form the ExANT motefield and the singular relay used to extract data from the exterior motefield.

Both systems were operated and controlled using the same MOC, which was colocated at the test area. To accommodate the significant XSM mote

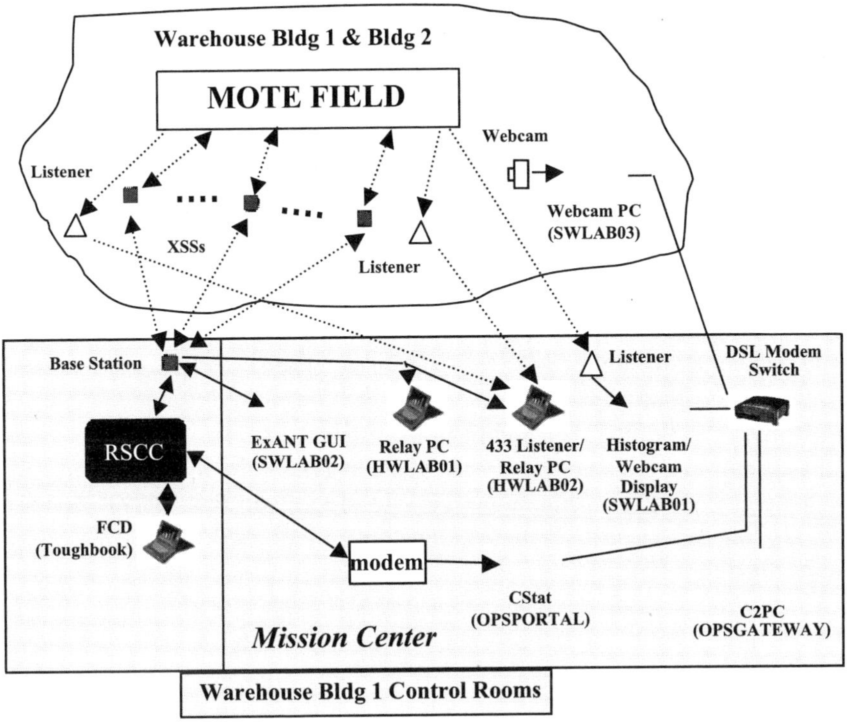

Figure 11.11 Communication and control architecture for the ExANT program final test.

density and to avoid saturation of RF signals in the large field, all mote antennas were not extended as seen in the test setup photo, Figure 11.10(a). For the large field, the three tiers were employed (XSM, XSS, and relay), as with VigilNet deployments. The connectivity of the test MOC and the main motefield is depicted in Figure 11.11. For the 2400-mote motefield, a USSOCOM RSCC relay was used to connect the motefield to the MOC via a SATCOM network. Using the labels in Figure 11.11, the status of each tier during the test was monitored using the ExANT GUI display (SWLAB02) and Relay processor (HWLABB01). Receiving sensor data and performing the formatting and overlay onto a common operating picture was performed by a C2PC laptop (OPSGATEWAY). Status of the RSCC and global network was provided through a standard C2PC CStat PC (OPSPORTAL). Test diagnostics, as previously described, were used continuously. The 433-MHz Listener/Relay PC (HWLAB02) and the Histogram/Webcam Display PC (SWLAB01) intercepted network traffic and provided alerts, graphics, and logs based on operator command.

The external demonstration field (IED scenario) consists of a UGS camera connected via RSCC to the linear XSM motefield. Figure 11.12(a) depicts the deployed motefield and SSU camera. The XSM motes lined a roadway with a visible camera positioned at the far end of the field. Cuing of the visible camera was demonstrated based on detection of a vehicle, or personnel, moving into the monitored area, then stopping for a period of time. If this loitering period exceeded a preset (commanded) time, the motes would cue the visible camera to power up, obtain a sequence of images, and forward images and alert messages to the MOC. The image shown in Figure 11.12(b) was taken using the SSU visible camera that was cued when the vehicle stopped and personnel dismounted to place an object (black box, left of the vehicle) at the side of the roadway.

From these final tests and demonstrations, several aspects of motefield capability were evaluated using personnel and a robotic vehicle (infrared and/or magnetic target) within the large field. To ensure that NGIT ExANT testing focused on military applications, individuals with significant special operations force (SOF) experience were directly involved with all test phases and activities, including tests conducted at Avon Park and Green Swamp and the in Final Demonstration events. With various targets, target numbers,

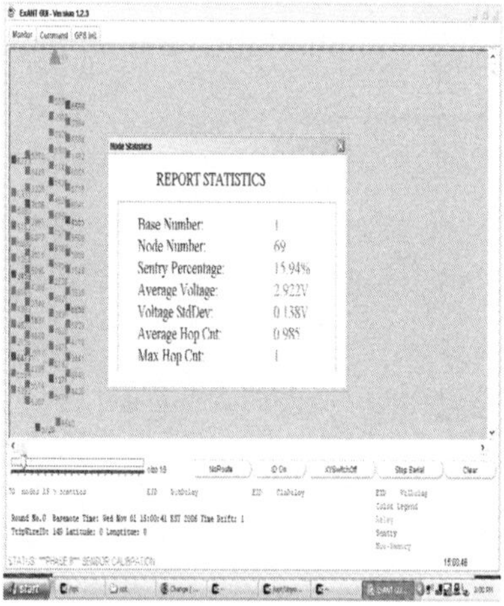

(a)

IED scenario motefield screen shot (69 motes).

(b)

Image captured by visible-cam (UGS) upon cue by motes, with X-second time delay.

Figure 11.12 ExANT final demonstration, IED test area (100-mote motefield acting as a delayed cue to the UGS visible camera). (a) The mote filed setup, and (b) image acquired through activation of the motefield of the test target.

and target spacing, the large motefield successfully detected, classified, and tracked and reported events in real-time using a C2PC display. In addition, during one phase of the test, two groups of personnel entered and triggered the system at opposite ends of the warehouse. Each group was aware of the other group's location through a link displayed in real time using a C2PC mobile display on Sprint BlackJack phones (Samsung SGH-i607). This information was pulled from the C2PC processor and forwarded to the handheld devices via Wi-Fi links. The target tracks on these devices were developed using mote data with screen icons generated using a KML overlay [23].

At this time, this ExANT motefield was the largest successfully connected and operating WSN system, with over 2,200 motes networked across 20 XSS devices. Figure 11.13(a, b) present displays of networked and operating motefields of 2,226 motes (June 26) and 2,195 motes (June 27). The Final Demonstration two-day event successfully showed that WSN can support T-ISR functions in real time and process and disseminate results and anywhere in the world within minutes. The tests also revealed the seamless integration of WSN for two remote independent areas operating with a single MOC system.

Several middleware applications were successfully evaluated, including the following:

- NMS, including self-organizing and self-healing network operation;
- Individual node localization;
- Target georeferencing, through reference of the track data from the motefield;
- Node and network power management;
- Over-the-air (OTA) on-demand and self-initiated reprogramming;
- Autonomous target detection, tracking, and classification;
- Secure communication measures evaluated (TinyPK, discussed in Section 11.2.2.4);
- Hand-over sensor queuing to legacy (in operation) sophisticated sensors (UGS);
- Direct injection of WSN sensor field data to various, operational ISR displays, including C2PC, COT, and Falcon View.

11.2.2.4 WSN Mote Security Research

As discussed in Chapter 8, WSN is quite vulnerable to security breaches through its fundamental nature: to operate effectively, it consists of a large number of spatially distributed, wirelessly connected, nodes. Numerous attacks are possible with WSN through physical means, disruption of the RF links, introduction of malicious messages or code, or simply through eavesdropping.

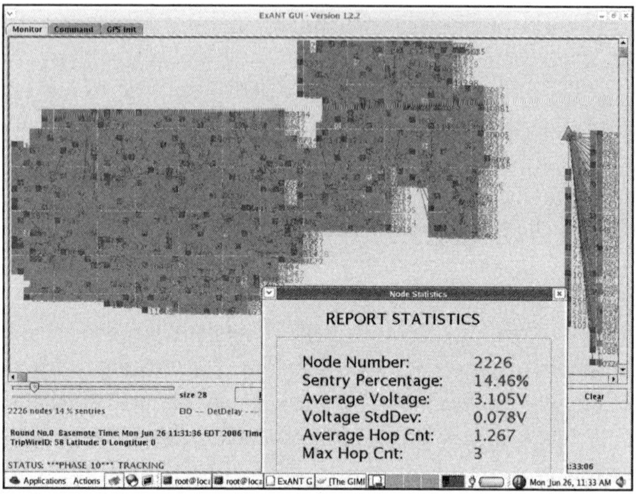

(a) 6/26/07—2226 XSM motes networked.

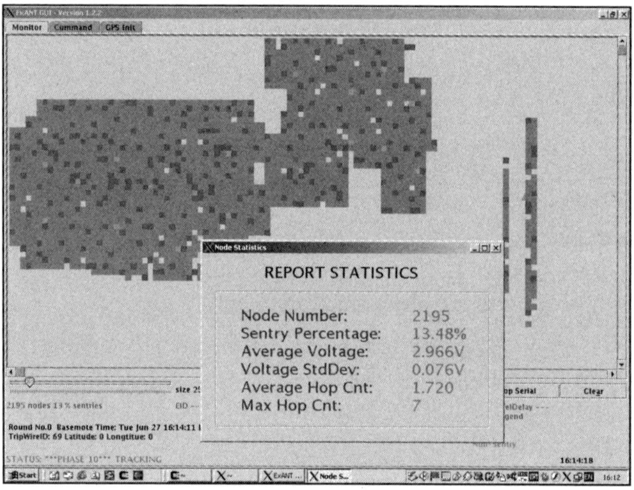

(b) 6/27/07—2195 XSM motes.

Figure 11.13 ExANT final demonstration layout: (a) 2,226 motes, June 26, 2007, and (b) 2,195 motes, June 27, 2007.

With poorly protected WSN elements, large amounts of confidential information may be accessed, destroyed, or tampered with. A major barrier in combating these attacks by deploying security elements on sensor nodes is their limited computation and communication capabilities. Cryptographic algorithms are not simple and pose nontrivial processing challenges for low-power microcontroller systems.

During development of NEST technologies, such security gaps did not go unnoticed by DARPA and associated research teams. These WSN teams addressing security issues turned to packet-switched networks for answers. With conventional networks, the dominant traffic pattern is end-to-end communication with intermediate routers requiring reading of message headers only. Message authenticity, integrity, and confidentiality are achieved by end-to-end security mechanisms such as secure shell (SSH), secure socket layer(SSL), or Internet Protocol security (IPSec) [24]. For WSN, the dominant traffic patterns are many-to-one, with multiple nodes communicating over a multihop topology to access an exfiltration relay.

Neighboring nodes in sensor networks usually incur the same or correlated events, and if each node sends a packet to the relay in response, energy and bandwidth are wasted. To reduce redundant messages, WSN systems perform aggregation and duplicate elimination, which requires intermediate nodes to access, modify, and suppress message content. As a result, end-to-end security mechanisms from sensor nodes to relay are not feasible as with wired digital networks. End-to-end security mechanisms are also vulnerable to various denial of service attacks. If message integrity is checked only at the final destination, the network may route packets injected by an adversary several hops before being detected, consuming energy and bandwidth [25].

Early in WSN development, UCB developed TinySec, the first fully implemented link layer security architecture for WSN systems. TinySec was designed using lessons learned from design vulnerabilities within security protocols associated with other wireless networks (e.g., 802.11b and GSM). With WSN, TinySec was constrained by limited mote resources that required trade-offs; however, such network limitations also presented an inherent advantage to be used in the design of packet overhead and resource requirements. Link-layer security architecture can detect unauthorized packets when they are first injected into the network. Link-layer security mechanisms guarantee authenticity, integrity, and confidentiality of messages between neighboring nodes, while permitting in-network processing.

Using the link layer, TinySec was designed to provide two different security options: encryption with identity authentication and authentication only. With identity authentication encryption, data is encrypted and an identity authentication code (using the MAC) is added to the packet. However, in the authentication-only option, data is not encrypted. According to UCB, energy consumption in the most resource-intensive and secure mode accounted for an additional power draw of 10%. Similarly, TinySec presented a low impact on bandwidth and latency, proving to be a viable solution for extreme resource limitations [25]. Used pervasively in WSN, TinySec does not solve all security issues, such as prevention of message retransmission attacks [26]. Subsequent versions of TinySec incorporated NSA's Skipjack, which uses simple and fast block cipher algorithms and furthers the protection provided.

With NEST funding, BBN Technologies conducted research on how to secure WSN systems with the development of a conventional modular arithmetic cryptosystem for motes, TinyPK. Public key (PK) technology is widely used to support symmetric key management for internet hosts and interconnections. Although thought to be too complex for motes, the venerable and widely used Rivest–Shamir–Adleman (RSA) PK cryptosystem was considered by BBN, along with implementation of Diffie-Hellman key agreement techniques. The result was that BBN demonstrated that a PK-based protocol was feasible for resource-constrained sensor networks. Incorporating the use of TinySec symmetric encryption service for mote networks, the TinyPK system provided functionality required by a mote to mutually authenticate and communicate securely. With TinyPK, a single stolen and reverse-engineered mote cannot be used to impersonate other motes with different credentials—a level of protection achieved with very little overhead. Unfortunately, TinyPK does not handle the problem of revocation of compromised private keys. Also, TinyPK was designed with limited protection against DoS attacks; however, improved radio hardware including extension of IEEE 802.15.4 spread spectrum capability with low probability of detection and interception may eventually resolve the DoS problems. As mote networks scale to larger sizes, the use of multiple session keys will be inevitable, and the WSN systems themselves will need to internally generate and deploy the new keys [27].

11.3 Integration of WSN with Sensor Web Services

Over the past decade, research, development, and refinement of WSN capabilities and technologies has continued. Associated technologies have matured rapidly and reshaped the battlespace. Perhaps the most notable of these are the appearance of low-cost unmanned vehicles (e.g., UAVs) and soldier-wearable networked sensors. With an increasing influence of unattended autonomous platforms and vehicles (aerial, maritime, ground-based), new and even more intriguing capabilities are expected to appear on the horizon for T-ISR. Sensor networks have increased by orders of magnitude beyond 10,000-node motefields, and as a result, managing networks and data operations, such as searching and querying, have become extremely difficult tasks. To address and alleviate this complexity, semantics in declarative descriptions of sensors, nodes, domains, and networks have been introduced and used to further organize and control gargantuan networks.

The superb effort and successful achievements resulting from the DARPA NEST programs continue to influence research and analyses. Numerous papers have been authored and published that address findings resulting from DARPA-sponsored WSN research. Of particular note is that with the advancement in RTOS for sensor nodes, current use of TinyOS remains significant [28].

11.3.1 Semantic Sensor Web

As noted in Chapter 1, data volume continues to be an issue for sensor networks. Excessive data volume without a comprehensive understanding of how best to process it leads to inefficiencies and perhaps to a complete loss of key correlations. An approach to alleviating excessive volumes of data involves annotation of sensor data with semantic metadata to increase interoperability among heterogeneous sensor networks and to provide the contextual information essential for SA. Capturing this design strategy has been the driving force behind Semantic Sensor Web (SSW) techniques designed to aid data integration and discovery [29]. The Sensor Web is a special type of web-centric information infrastructure for collecting, modeling, storing, retrieving, sharing, manipulating, analyzing, and visualizing information about sensors and sensor observations of phenomena. Sensor Web benefitted from the 1994 formation of the OGC, an international consortium of industry, academic, and government organizations tasked with developing open geospatial standards [30].

With standardized declarative descriptions for sensors, nodes, domains, and networks, complexity in the management of networks and data operations is significantly reduced. Furthermore, sensors operating within a semantic sensor web (SSW) can discover new sensors and autonomously share sensor data (time stamp and spatial coordinates) associated with the newly discovered sensors. The encoding of sensor descriptions and sensor observation data with semantic web languages provides expressive representation, advanced access, and formal analysis of sensor resources. With SSW, the merging of sensor and semantic web technologies has occurred [29], and future T-ISR systems will benefit tremendously from the inclusion of this harmonization tool.

11.3.2 DHS (Customs and Border Patrol) Cueing Demonstration

In addition to new mobile platforms for WSNs, such as UAVs and personnel, intelligent video has emerged amid growing concern about border and area (e.g., airport) security. Unfortunately, numerous barriers have limited the effectiveness of surveillance video in large-area surveillance, including the number of cameras needed to provide adequate coverage, that produce large data sets, thereby overwhelming system users. Using funding from the Department of Homeland Security (DHS) Science & Technology Directorate, a prototype surveillance system was codeveloped by intuVision Inc. and NGIT to leverage smart sensor motes, intelligent-based video cameras, and sensor web technologies to aid in large-area monitoring operations to enhance the security of borders and critical infrastructures. The prototype system concept addressed these surveillance barriers through the use of a smart video node (SVN) in a sensor web framework.

SVN is an IP video camera with automated event detection capability that can be cued by external sensors. Using motefield observations, alerts are formed and intelligently routed using NGIT PULSENet services [31] to SVN cameras to provide real-time directional information and directional guidance to the SVN scanning mechanism. PULSENet provides a standards-based framework for discovery, access, use, and control of heterogeneous sensors and their metadata and observation data. The prototype border security and surveillance system was implemented and tested by leveraging intuVision's intelligent video, NGIT's PIR-based motes (PIRmotes), and PULSENet technologies [32]. Figure 11.14 depicts the readiness testing of the PIRmote used by the DHS tests. Figure 11.14(a) depicts the author field-testing PIRmotes for a comprehensive WSN/SVN/SWE system demonstration. Figure 11.14(b) presents a screenshot of the PULSENet analyst screen indicating camera slew angles (black lines) to a location provided via a deployed PIRmote motefield. The image at the top left of the laptop screen, Figure 11.14(b), is a live smart video (SVN) feed.

11.4 WSN as IoBT

WSN-based systems were designed to leverage capabilities through access of worldwide communication architectures and internet-based services, and because of this, WSN is considered a subset of the IoBT. This presents the prominent concern for DoD regarding IoBT (WSN): vulnerability to adversarial threats. Ongoing efforts exist, and solutions continue to be evaluated, including NSA's Comply to Connect (C2C) [33], a network security platform that autonomously monitors device discovery and access control to keep pace with the exponentially growing network of entities.

Security of past T-ISR (ISR) systems was reliant on physical access, protection, and anti-spoofing of the incoming target signals. The design of

(a) Pre-test checkout of PIRmotes

(b) PULSENet display, of camera slewed via PIRmote cues

Figure 11.14 (a) Preparation and end-to-end testing and (b) in situ real-time cues to perform slewing to threat activity.

advanced T-ISR systems requires consideration of how best to infuse sophisticated situation awareness to provide a WSN system the logical complexity to autonomously operate successfully, with limited information. T-ISR systems must become capable of inferring a comprehensive overview of observed events, including security attacks, based on a degraded or incomplete set of inputs.

11.5 Examples of Ongoing DoD Activities

Research in WSN technology within the military has regained a considerable amount of support from the armed forces [34]. Of particular note is protection of U. S. Army large-scale forward operating bases (FOBs) and tactical operations centers (TOCs). WSN has become an integrated capability within perimeter defense systems, where smart technologies such as WSN can improve persistent monitoring, including the identification of potential threats. A problem that persists is the lack of transparent interoperability, which challenges seamless integration of IoBT capabilities. In response, the U.S. Army Combat Capabilities Development Command's ARL is working on long-range WAN (LoRaWAN) coverage to include urban areas, associate with smart installations, and interconnect with existing communications infrastructures [35]. To test these concepts, the Army evaluated LoRaWAN as a way to enhance the ability to transmit and receive data in urban environments, which would result in major consequences for Military Operation on Urbanized Terrain (MOUT) warfare in urban environments [36]. ARL addressed other WSN-related problems, including research into improvements to sensor modalities, such as ultra-sensitive vibration (seismic) sensors based on the Pacinian corpuscle [37].

Wearable connectivity and context-aware sensor awareness for direct use by individual soldiers continues. Interconnectivity among various manned and unmanned platforms has been and continues to expand the definition of MANET systems, while interoperability is being improved through intelligent filtering of data based on priority settings and relevance. Concepts, such as DARPA's MOSAIC Warfare [38], provide a systems of systems approach to military warfare with the strategy to confuse and overwhelm adversary forces by deploying low-cost, adaptable, and expendable systems that can operate in multiple roles and coordinate actions with one another. Promising to be a disruptive technology, MOSAIC seeks to complicate the decision-making process for the enemy.

References

[1] "Proceedings of a Workshop on Distributed Sensor Nets," Information Processing Techniques Office, *DARPA Teport AD-A143 691*, hosted by Carnegie-Mellon University, December 1978.

[2] MIT/Lincoln Laboratories, "Distributed Sensor Network," *Semi-Annual Technical Summary Report AD-A182 216*, 1986.

[3] Cook, B. W., et al., "SoC Issues for RF Smart Dust," *Proceedings of the IEEE*, 2006.

[4] Pottie, G., and W. J. Kaiser, "Wireless Integrated Network Sensors," *Communications of the ACM*, 2000.

[5] Kumar, S., "SensIT: Sensor Information Technology for the Warfighter," in *Proc. 4th Int. Conf. on Information Fusion*, 2001, pp. 1–7.

[6] Merrill W., K. Sohrabi, and G. J. Pottie, "Pico Wireless Integrated Network Sensors (PicoWINS): Investigating the feasibility of Wireless Tactical Tags," *U.S. Army Soldier and Biological Chemical Command, Final Report*, 2002.

[7] Rabaey, J., et al., "Picoradio Communication/Computation Piconodes for Sensor Networks," *ARFL Report VS-TR-2003-1013*, 2003.

[8] Hill, J., "System Architecture for Wireless Sensor Networks," dissertation submitted in partial satisfaction of the requirements for the degree of Doctor of Philosophy in Computer Science, University of California-Berkeley, 2003.

[9] Levis, P., "Experiences from a Decade of TinyOS Development," *OSDI'12: Proceedings of the 10th USENIX conference on Operating Systems Design and Implementation*, 2012.

[10] Tan, S., and B.A. Nguyen, "Survey and Performance Evaluation of Real-Time Operating Systems (RTOS) for Small Microcontrollers," *IEEE Micro*, 2009.

[11] "29 Palms Fixed/Mobile Experiment," University of California—Berkeley and MLB Company, 2004.

[12] Weng L., et al., "Neural-Memory Based Control of Micro Air Vehicles (MAVs) with Flapping Wings," in *Advances in Neural Networks,* D. Liu, et al. (eds), *Lecture Notes in Computer Science,* Vol. 4491, Berlin: Springer, 2007.

[13] Alexander, J., statement to the Subcommittee on Emerging Threats and Capabilities, Armed Services Committee, U.S. Senate, 2001.

[14] Tether, A., "Multidisciplinary Research," submitted to the Committee on Science U.S. House of Representatives, 2005.

[15] Arora, A., et al., "A Line in the Sand: A Wireless Sensor Network for Target Detection, Classification, and Tracking," *Computer Networks: The International Journal of Computer and Telecommunications Networking*, 2004.

[16] Madhuri, V., S. Umar, and P. Veeraveni "A Study on Smart Dust (MOTE) Technology," *IJCSET*, 2013.

[17] Arora, A., et al., "ExScal: Elements of an Extreme Scale Wireless Sensor Network," *11th IEEE International Conference on Embedded and Real-Time Computing Systems and Applications (RTCSA'05)*, 2005.

[18] "Chapter 3: The Extreme Scale Mote (XSM)," http://www2.ece.ohio-state.edu/~bibyk/ee582/XscaleMote.pdf.

[19] He, T., et al., "VigilNet: An Integrated Sensor Network System for Energy-Efficient Surveillance," *ACM Transactions on Sensor Networks*, 2006.

[20] Vicaire, P., et al., "Achieving Long-Term Surveillance in VigilNet," *ACM Transactions on Sensor Networks*, 2009.

[21] Cole, T. D., and A. S. El-Dinary, "Estimation of Target Vibration Spectra from Laser Radar Backscatter Using Time-Frequency Distributions," *SPIE Applied Laser Radar Technology*, 1993.

[22] Parks, E., "Integrating the Target Workflow System (TWS) with the Command and Control Personal Computer (C2PC) System: Proof of Concept," MITRE white paper, 1999.

[23] "KML 2.1 Reference—An OGC Best Practice," *Open Geospatial Consortium Inc., Report #OGC-07-039r1*, Google, 2007.

[24] "SSL, SSH and IPSec," Swarthmore Briefing, https://www.cs.swarthmore.edu/~mgagnel /teaching/2016_17/cs91/SSL_IPsec.pdf, Accessed 5/27/2020.

[25] Karlof, K., N. Sastry, and D. Wagner, "TinySec: A Link Layer Security Architecture for Wireless Sensor Networks," *ACM SenSys'04*, 2004.

[26] Dener, M., Security Analysis in Wireless Sensor Networks," *International Journal of Distributed Sensor Networks*, 2014.

[27] Watro, R., et al., "TinyPK: Securing Sensor Networks with Public Key Technology," *Proceedings of the 2nd ACM Workshop on Security of Ad Hoc and Sensor Networks SASN '04*, 2004.

[28] Xie, X., "Developing a Wireless Sensor Network Programming Language Application Guide Using Memsic Devices and LabVIEW," *Master of Technology Management Plan II Graduate Projects*, College of Technology, Architecture and Applied Engineering, Bowling Green State University, 2014.

[29] Broring, A., et al., "New Generation Sensor Web Enablement," *Sensors*, 2011.

[30] Sheth, A., C. Henson, and S. S. Sahoo, "Semantic Sensor Web," *IEEE Internet Computing*, 2008.

[31] Fairgrieve, S., J. A. Makuch, and S. R. Falke, "PULSENet™: An Implementation of Sensor Web Standards," *International Symposium on Collaborative Technologies and Systems*, 2009.

[32] Guler, S., et al., "Border Security and Surveillance System with Smart Cameras and Motes in a Sensor Web," *Proceeding of SPIE Independent Component Analyses, Wavelets, Neural networks, Biosystems, and Nanoengineering VIII*, 2010.

[33] National Security Agency (NSA), "Comply-to-Connect," Information *Assurance Symposium (IAS)*, 2016.

[34] Castiglione, A., et al., "Context Aware Ubiquitous Biometrics in Edge of Military Things," *IEEE Cloud Computing*, 2017.

[35] Kanowitz, K., "Army Tests Smart-City Communications Tool, *GCN*, 2019.

[36] Saccone, L., "Army Studies Smart Cities for New Communication Methods," *In Compliance*, 2019.

[37] U.S. Army CCDC Army Research Laboratory Public Affairs, "Army Researchers Develop Innovative Sensor Inspired by Elephant," 2020.

[38] Grayson, T., "Mosaic Warfare," DARPA/STO Brief, 2018.

About the Author

Timothy D. Cole is a leading authority on active and passive sensor systems with 40 years of experience during which he successfully designed, developed, and deployed sensing systems for military, space-based, and biomedical applications. At Teledyne-Brown, Cole developed sensor systems to perform exoatmospheric target tracking and identification using long-wave infrared and laser radar instrumentation. While at The Johns Hopkins University/Applied Physics Laboratory, Cole successfully designed and delivered the ground control and processing facility for the U.S. Navy's GEOSAT-1 Ku-band altimeter and was the designer/developer of NASA's Near-Earth Rendezvous laser radar. At Northrop Grumman, Cole initiated, developed, and delivered WSN-based sensors, including a microlaser mote and passive infrared mote. He designed and conducted demonstrations using WSN mote fields, UGSs, and laser radars to solve issues associated with border monitoring, secure facility protection, and high-value target (HVT) targeting, tracking, and locating, (TTL). Most recently, Cole was the lead calibration scientist for the NASA/GSFC (Goddard) photon-counting laser altimeter for NASA's ICESat-2 mission (launched September 2018). Cole was twice awarded the NASA Achievement Award, and while at Northrop Grumman, he was recognized as a Technical Fellow.

Index

The Artech House Intelligence and Information Operations Series

Ed Waltz, Series Editor

The Art and Science of Military Deception, Hy Rothstein and Barton Whaley

Aviation Security Engineering: A Holistic Approach, Rainer Kölle, Garik Markarian, and Alex Tarter

Concepts, Models, and Tools for Information Fusion, Éloi Bossé, Jean Roy, and Steve Wark

Cyberwarfare: An Introduction to Information-Age Conflict, Isaac R. Porche III

Data Fusion Support to Activity-Based Intelligence, Richard T. Antony

Designing Wireless Sensor Network Solutions for Tactical ISR, Timothy D. Cole

Electronic Intelligence: The Analysis of Radar Signals, Second Edition, Richard G. Wiley

Electronic Warfare for the Digitized Battlefield, Michael R. Frater and Michael Ryan

Electronic Warfare in the Information Age, D. Curtis Schleher

Electronic Warfare Target Location Methods, Second Edition, Richard A. Poisel

EW 101: A First Course in Electronic Warfare, David Adamy

High-Level Information Fusion Management and Systems Design, Erik Blasch, Éloi Bossé, and Dale A. Lambert, editors

Homeland Security Technology Challenges: From Sensing and Encrypting to Mining and Modeling, Giorgio Franceschetti and Marina Grossi, editors

Information Fusion and Analytics for Big Data and IoT, Éloi Bossé and Basel Solaiman

Information Warfare and Organizational Decision-Making, Alexander Kott, editor

Information Warfare Principles and Operations, Edward Waltz

Introduction to Communication Electronic Warfare Systems, Second Edition, Richard A. Poisel

Knowledge Management in the Intelligence Enterprise, Edward Waltz

Mathematical Techniques in Multisensor Data Fusion, Second Edition, David L. Hall and Sonya A. H. McMullen

Modern Communications Jamming Principles and Techniques, Second Edition, Richard A. Poisel

Modern Communications Receiver Design and Technology, Cornell Drentea

Principles of Data Fusion Automation, Richard T. Antony

Sensor Management in ISR, Kenneth J. Hintz

Strategem: Deception and Surprise in War, Barton Whaley

Statistical Multisource-Multitarget Information Fusion, Ronald P. S. Mahler

Tactical Communications for the Digitized Battlefield, Michael Ryan and Michael R. Frater

Target Acquisition in Communication Electronic Warfare Systems, Richard A. Poisel

For further information on these and other Artech House titles, including previously considered out-of-print books now available through our In-Print-Forever® (IPF®) program, contact:

Artech House
685 Canton Street
Norwood, MA 02062
Phone: 781-769-9750
Fax: 781-769-6334
e-mail: artech@artechhouse.com

Artech House
16 Sussex Street
London SW1V 14RW UK
Phone: +44 (0)20-7596-8750
Fax: +44 (0)20-7630-0166
e-mail: artech-uk@artechhouse.com

Find us on the World Wide Web at: www.artechhouse.com